Parameter Redundancy and Identifiability

Chapman & Hall/CRC Interdisciplinary Statistics Series

Series editors: B.J.T. Morgan, C.K. Wikle, P.G.M. van der Heijden

Recently Published Titles

Correspondence Analysis in Practice, Third Edition
M. Greenacre

Capture-Recapture Methods for the Social and Medical Sciences
D. Böhning, P. G. M. van der Heijden, and J. Bunge (editors)

The Data Book: Collection and Management of Research Data
M. Zozus

Modern Directional Statistics
C. Ley and T. Verdebout

Survival Analysis with Interval-Censored Data: A Practical Approach with Examples in R, Sas, and Bugs
K. Bogaerts, A. Komarek, and E. Lesaffre

Statistical Methods in Psychiatry and Related Field: Longitudinal, Clustered and Other Repeat Measures Data
R. Gueorguieva

Bayesian Disease Mapping: Hierarchical Modeling in Spatial Epidemiology, Third Edition
A.B. Lawson

Flexbile Imputation of Missing Data, Second Edition
S. van Buuren

Compositional Data Analysis in Practice
M. Greenacre

Applied Directional Statistics: Modern Methods and Case Studies
C. Ley and T. Verdebout

Design of Experiments for Generalized Linear Models
K.G. Russell

Model-Based Geostatistics for Global Public Health: Methods and Applications
P.J. Diggle and E. Giorgi

Statistical and Econometric Methods for Transportation Data Analysis, Third Edition
S. Washington, M.G. Karlaftis, F. Mannering and P. Anastasopoulos

Parameter Redundancy and Identifiability
Diana J. Cole

For more information about this series, please visit: https://www.crcpress.com/Chapman--HallCRC-Interdisciplinary-Statistics/book-series/CHINTSTASER

Parameter Redundancy and Identifiability

Diana J. Cole

CRC Press
Taylor & Francis Group
Boca Raton London New York

CRC Press is an imprint of the
Taylor & Francis Group, an **informa** business

A CHAPMAN & HALL BOOK

CRC Press
Taylor & Francis Group
6000 Broken Sound Parkway NW, Suite 300
Boca Raton, FL 33487-2742

International Standard Book Number-13: 978-1-498-72087-8 (Hardback)
978-0-367-49321-9 (Paperback)

Library of Congress Cataloging-in-Publication Data

Names: Cole, Diana, 1978- author.
Title: Parameter redundancy and identifiability / Diana Cole.
Description: Boca Raton : CRC Press, [2020] | Includes bibliographical
references and index. | Summary: "Statistical and mathematical models are
defined by parameters that describe different characteristics of those models.
Ideally it would be possible to find parameter estimates for every parameter
in that model, but in some cases this is not possible. For example, two
parameters that only ever appear in the model as a product could not be
estimated individually; only the product can be estimated. Such a model
is said to be parameter redundant, or the parameters are described as
non-identifiable. This book explains why parameter redundancy and
non-identifiability is a problem and the different methods that can be used
for detection, including in a Bayesian context"-- Provided by publisher.
Identifiers: LCCN 2020001141 (print) | LCCN 2020001142 (ebook) |
ISBN 9781498720878 (hardback) | ISBN 9781315120003 (ebook)
Subjects: LCSH: Parameter estimation. | Mathematical models. |
Ecology--Statistical methods. | Ecology--Mathematical models.
Classification: LCC QA276.8 .C54 2020 (print) | LCC QA276.8 (ebook) |
DDC 519.5/44--dc23
LC record available at https://lccn.loc.gov/2020001141
LC ebook record available at https://lccn.loc.gov/2020001142

Visit the Taylor & Francis Web site at
http://www.taylorandfrancis.com

and the CRC Press Web site at
http://www.crcpress.com

Contents

Preface   xi

Acknowledgements   xiii

Author   xv

Symbols   xvii

Index of Examples   xix

1 Introduction   1
  1.1 An Introduction to Mathematical and Statistical Models   1
  1.2 An Introduction to Parameter Redundancy   4
    1.2.1 The Basic Occupancy Model   5
    1.2.2 The Population Growth Model   5
    1.2.3 The Linear Regression Model   6
    1.2.4 The Mark-Recovery Model   6
  1.3 An Introduction to Identifiability   8
    1.3.1 Linear Regression Model   8
    1.3.2 Three Compartmental Model   9
    1.3.3 Multiple-Input Multiple-Output (MIMO) Model   10
  1.4 Detecting Parameter Redundancy and Non-Identifiability   11

2 Problems With Parameter Redundancy   13
  2.1 Likelihood Inference with Parameter Redundant Models   13
    2.1.1 Basic Occupancy Model   16
    2.1.2 Mark-Recovery Model   16
    2.1.3 Mixture Model   18
    2.1.4 Three Compartment Model   19
    2.1.5 Binary Logistic Regression Model   21
  2.2 Fitting Parameter Redundant Models   22
    2.2.1 Binary Latent Class Model   23
    2.2.2 Three Compartment Model   25
    2.2.3 Mark-Recovery Model   25
    2.2.4 N-Mixture Model   26
  2.3 Standard Errors in Parameter Redundant Models   28
    2.3.1 Binary Latent Class Model   29

2.3.2 Mark-Recovery Model 30
2.4 Model Selection 31
    2.4.1 Information Criteria 31
    2.4.2 The Score Test 32
        2.4.2.1 Mark-Recovery Model 32
2.5 Conclusion 34

3 **Parameter Redundancy and Identifiability Definitions and Theory** **35**
3.1 Formal Definitions of Parameter Redundancy and Identifiability 35
    3.1.1 Identifiability 35
    3.1.2 Parameter Redundancy 36
    3.1.3 Near-Redundancy, Sloppiness and Practical Identifiability 37
    3.1.4 Exhaustive Summaries 38
        3.1.4.1 Basic Occupancy Model 38
        3.1.4.2 Binary Latent Class Model 39
        3.1.4.3 Three Compartment Model 40
        3.1.4.4 Mark-Recovery Model 40
    3.1.5 The Use of Exhaustive Summaries 41
3.2 Detecting Parameter Redundancy or Non-Identifiability using Symbolic Differentiation Method 42
    3.2.1 Basic Occupancy Model 44
    3.2.2 Binary Latent Class Model 45
    3.2.3 Three Compartment Model 46
    3.2.4 Checking for Extrinsic Parameter Redundancy 46
    3.2.5 Finding Estimable Parameter Combinations 48
        3.2.5.1 Basic Occupancy Model 50
        3.2.5.2 Binary Latent Class Model 51
    3.2.6 Generalising Parameter Redundancy Results to Any Dimension 51
        3.2.6.1 The Extension Theorem 52
        3.2.6.2 Mark-Recovery Model 53
    3.2.7 Essentially and Conditionally Full Rank 56
        3.2.7.1 Mark-Recovery Model 58
        3.2.7.2 Checking the Rank using PLUR Decomposition 60
3.3 Distinguishing Local and Global Identifiability 61
    3.3.1 Global Identifiability Results for Exponential Family Models 62
        3.3.1.1 Linear Regression Model 62
    3.3.2 General Methods for Distinguishing Local and Global Identifiability 63
        3.3.2.1 Three Compartment Model 63

3.4 Proving Identifiability Directly 64
   3.4.1 The Population Growth Model 64
   3.4.2 Closed Population Capture-Recapture Model 65
3.5 Near-Redundancy, Sloppiness and Practical Identifiability . . 67
   3.5.1 Mark-recovery Model 68
3.6 Discussion 70

4 Practical General Methods for Detecting Parameter Redundancy and Identifiability   **71**
4.1 Numerical Methods 71
   4.1.1 Hessian Method 71
      4.1.1.1 Basic Occupancy Model 73
      4.1.1.2 Mark-Recovery Model 74
      4.1.1.3 Linear Regression Model 76
   4.1.2 Simulation Method 77
      4.1.2.1 Basic Occupancy Model 78
   4.1.3 Data Cloning 79
      4.1.3.1 Basic Occupancy Model 81
      4.1.3.2 Population Growth Model 82
   4.1.4 Profile Log-Likelihood Method 83
      4.1.4.1 Mark-Recovery Model 84
      4.1.4.2 Profiles to check Locally Identifiability . . . 85
4.2 Symbolic Method 86
4.3 Hybrid Symbolic-Numerical Method 90
   4.3.1 Basic Occupancy Model 92
   4.3.2 Complex Compartmental Model 92
4.4 Near-Redundancy and Practical Identifiability 93
   4.4.1 Detecting Near-Redundancy using the Hessian Matrix 94
      4.4.1.1 Mark-Recovery Model 95
   4.4.2 Practical Identifiability using Log-Likelihood Profiles . 96
      4.4.2.1 Mark-Recovery Model 96
4.5 Discussion and Comparison of Methods 97

5 Detecting Parameter Redundancy and Identifiability in Complex Models   **101**
5.1 Covariates 101
   5.1.1 Conditional Mark-Recovery Model 102
   5.1.2 Latent Class Model with Covariates 104
5.2 Extended Symbolic Method 105
   5.2.1 Simpler Exhaustive Summaries 105
   5.2.2 Reduced-Form Exhaustive Summaries 107
      5.2.2.1 Capture-Recapture Model 109
      5.2.2.2 Complex Compartmental Model 110
      5.2.2.3 Age-Dependent Fisheries Model 112

5.2.3 General Results using Extension and
Reparameterisation Theorems 115
5.2.3.1 CJS Model 116
5.3 Extended Hybrid Symbolic-Numerical Method 117
5.3.1 Subset Profiling 117
5.3.1.1 Mark-Recovery Model 120
5.3.2 General Results 122
5.3.2.1 Mark-Recovery Model 123
5.4 Conclusion 123

6 Bayesian Identifiability   **125**
6.1 Introduction to Bayesian Statistics 125
6.2 Definitions 126
6.3 Problems with Identifiability in Posterior Distribution 128
6.3.1 Basic Occupancy Model 129
6.3.2 Mark-Recovery Model 129
6.3.3 Logistic Growth Curve Model 132
6.4 Detecting Weak Identifiability using Prior and Posterior
Overlap 135
6.4.1 Basic Occupancy Model 137
6.4.2 Mark-Recovery Model 138
6.4.3 Logistic Growth Curve Model 142
6.5 Detecting Bayesian Identifiability using Data Cloning 143
6.5.1 Basic Occupancy Model 143
6.5.2 Logistic Growth Curve Model 146
6.6 Symbolic Method for Detecting Bayesian Identifiability . . . 146
6.6.1 Basic Occupancy Model 148
6.6.2 Logistic Growth Curve Model 150
6.6.3 Mark-Recovery Model 151
6.7 Conclusion 153

7 Identifiability in Continuous State-Space Models   **155**
7.1 Continuous State-Space Models 155
7.1.1 Two-Compartment Linear Model 156
7.1.2 Three-Compartment Linear Model 157
7.1.3 Michaelis Menten Kinetics Model 158
7.2 Methods for Investigating Identifiability in Continuous
State-Space Models 158
7.2.1 Transfer Function Approach 158
7.2.1.1 Two-Compartment Linear Model 159
7.2.1.2 Three-Compartment Linear Model 160
7.2.2 Similarity of Transform or Exhaustive Modelling
Approach 161
7.2.2.1 Two-Compartment Linear Model 162
7.2.2.2 Three-Compartment Linear Model 162

      7.2.3   Taylor Series Expansion Approach 163
            7.2.3.1   Two-Compartment Linear Model 164
            7.2.3.2   Three-Compartment Linear Model 164
            7.2.3.3   Michaelis Menten Kinetics Model 165
      7.2.4   Other Methods 166
  7.3   Software 167
  7.4   Discussion 167

8 Identifiability in Discrete State-Space Models   169
  8.1   Introduction to Discrete State-Space Models 169
      8.1.1   Abundance State-Space Model 170
      8.1.2   Multi-Site Abundance State-Space Model 171
      8.1.3   SIR State-Space Model 171
      8.1.4   Stochastic Volatility State-Space Model 172
      8.1.5   Stochastic Ricker Stock Recruitment SSM 173
  8.2   Hessian and Profile Log-Likelihood Methods 174
      8.2.1   Stochastic Ricker Stock Recruitment SSM 174
  8.3   Prior and Posterior Overlap 175
      8.3.1   Abundance State-Space Model 175
  8.4   Data Cloning 176
      8.4.1   Abundance State-Space Model 176
  8.5   Symbolic Method with Discrete State-Space Models 177
      8.5.1   Z-Transform Exhaustive Summary 178
      8.5.2   Expansion Exhaustive Summary 179
      8.5.3   Stochastic Ricker Stock Recruitment SSM 180
      8.5.4   Stochastic Volatility Model State-Space Model 181
      8.5.5   Multi-Site Abundance State-Space Model 182
  8.6   Prediction 183
      8.6.1   SIR State-Space Model 184
  8.7   Links to Continuous State-Space Models 185
  8.8   Hidden Markov Models 185
      8.8.1   Poisson HMM 186
  8.9   Conclusion 187

9 Detecting Parameter Redundancy in Ecological Models   189
  9.1   Marked Animal Populations 189
      9.1.1   Mark-Recovery Model 189
      9.1.2   Capture-Recapture Model 191
      9.1.3   Capture-Recapture-Recovery Model 194
      9.1.4   Multi-State Model 196
      9.1.5   Memory Model 199
  9.2   Integrated Population Models 201
      9.2.1   Four Data Set Integrated Model 205
      9.2.2   Posterior and Prior Overlap in Integrated Population
            Model 209

9.3   Experimental Design   210
      9.3.1   Robust Design   211
              9.3.1.1   Robust Design Occupancy Model   211
              9.3.1.2   Robust Design Removal Model   211
      9.3.2   Four Data Set Integrated Model   212
9.4   Constraints   212
      9.4.1   Multi-State Model   213
      9.4.2   Memory Model   213
9.5   Discussion   215

10 Concluding Remarks   **217**
10.1 Which Method to Use and When   217
10.2 Inference with a Non-Identifiable Model   218

A Maple Code   **221**
A.1   Checking Parameter Redundancy / Non-Identifiability   221
      A.1.1   Basic Occupancy Model   221
A.2   Finding Estimable Parameter Combinations   222
      A.2.1   Basic Occupancy Model   223
A.3   Distinguishing Essentially and Conditionally Full Rank Models   224
A.4   Hybrid Symbolic-Numerical Method   224
      A.4.1   Basic Occupancy Model   224

B Winbugs and R Code   **227**
B.1   Hessian Method   227
B.2   Data Cloning   230

Bibliography   **233**

Index   **249**

Preface

This book discusses the problems associated with identifiability, focusing on identifiability of parameters in mathematical and statistical models and parameter redundancy.

I first encountered the problems associated with parameter redundancy in the context of statistical models used in ecology. (This is why many examples in this book are from ecology.) Ecological models are increasingly more realistic whilst simultaneously data collection methods are becoming more sophisticated. The result is models that are more complex, and it is no longer obvious whether a model is non-identifiable and which parameters can be estimated. However this problem is not unique to ecological models; identifiability could potentially be an issue in any model.

This book is designed to be applicable to a wide audience from MSc students to researchers, from mathematicians and statisticians to practitioners using mathematical or statistical models. Whilst some preliminary background material is included in most chapters, other books are recommended if more detail is wanted on some mathematical or statistical concepts.

If you are a practitioner, such as an ecologist, with little mathematical or statistical background, it is recommended that you read Chapter 2, to understand the problems that parameter redundancy can cause, then read Chapter 4 to get a practical overview of all the different methods that can be used to detect parameter redundancy.

If you have a mathematical or statistical background, then Chapters 3, 5 and 6, which cover more theoretical aspects of parameter redundancy and identifiability, will be more accessible. Although it is also recommended that you read Chapter 2 and understand the problems that parameter redundancy can cause, and Chapter 4 if you are interested in numerical methods.

Chapters 7, 8 and 9 deal with specific examples in the areas of continuous and discrete state-space models and statistical ecology. They are both of interest as applications to those specific area, but also offer further in-depth examples of the different techniques discussed in earlier chapters.

This book makes use of various computer packages, including Maple, R and Winbugs. Code for these programs is available on the publisher's website: https://www.crcpress.com/Parameter-Redundancy-and-Identifiability/Cole/p/book/9781498720878 and on the author's website: https://www.kent.ac.uk/smsas/personal/djc24/parameterredundancy.html.

Acknowledgements

Thank you to every colleague who has discussed parameter redundancy and identifiability with me. In particular my interest in parameter redundancy and identifiability started with a post doc position with Byron Morgan. I'm very grateful to Bryon's support. Thank you to Remi Choquet for introducing and working with me on his hybrid symbolic-numerical method. The chapter on discrete state-space models came out of an identifiability workshop given at Casa Matematica, Oaxaca, as part of the New Perspectives on State Space Models workshop. Thank you to the organisers, Marie Auger-Methe, David Campbell and Len Thomas and to all the participants who all helped to shape the material included in this chapter. My knowledge of identifiability continuous state-state space models came about from two workshops organised by Michael Chappell and Neil Evans. The chapter on ecological models includes work shaped by PhD students and colleagues. In particular thank you to Ben Hubbard, Chen Yu, Ming Zhou, Rachel McCrea, Eleni Matechou, Marina Jimenez-Munoz and Chloe Nater.

Thank you to the publishers and reviewers with their help in shaping this book.

Finally, a huge thank you to Peter Cole, who proofread many iterations of this book.

Author

Diana J. Cole is a Senior Lecturer in Statistics at the University of Kent. She has written and co-authored 15 papers on parameter redundancy and identifiability, including general theory and ecological applications.

Symbols

Below is a list of notations and symbols used in this book. We follow the standard convention that a matrix of vector is represented in bold.

Symbol Description

$\log(x)$	The natural logarithm of x.	$\mathbf{H}$	hessian matrix
$\Gamma(x)$	The gamma function of x, $\Gamma(x) = \int_0^\infty t^{x-1} \exp(-t)dt$.	$M(\boldsymbol{\theta})$	representation of a model with parameters θ
$\mathbf{x}^T$	The transpose of matrix $\mathbf{x}$.	$\kappa(\boldsymbol{\theta})$	exhaustive summary of a model with parameters θ
$\det(\mathbf{x})$	The determinant of matrix $\mathbf{x}$.	$\mathbf{D}$	derivative matrix
$\boldsymbol{\theta}$	vector of parameters	r	rank of a derivative matrix
q	number of parameter	$\dim(\mathbf{x})$	The length of vector $\mathbf{x}$
$f(y; \boldsymbol{\theta})$	probability density function (pdf), with parameters θ	$f(\boldsymbol{\theta}|\mathbf{y})$	posterior distribution, with parameters θ and data $\mathbf{y}$
$L(\boldsymbol{\theta}|\mathbf{y})$	likelihood function, with parameters θ and data $\mathbf{y}$	$f(\mathbf{y}|\boldsymbol{\theta})$	the sampling model for data $\mathbf{y}$, given parameters $\boldsymbol{\theta}$, also referred to as the likelihood
$l(\boldsymbol{\theta}|\mathbf{y})$	log-likelihood function, with parameters θ and data $\mathbf{y}$	$f(\boldsymbol{\theta})$	the prior distribution for parameters $\boldsymbol{\theta}$
$\hat{\theta}$	maximum likelihood estimate of parameter θ	$\text{logit}(x)$	the logit function with $\text{logit}(x) = \log\{x/(1-x)\}$
$\mathbf{J}$	Fisher information matrix		

Index of Examples

The following examples are used in this book. Where applicable the names of computer code files are included.

Example	Section(s)	Computer Code
Basic occupancy model	1.2.1, 2.1.1, 3.1.4.1, 3.2.1, 3.2.5.1, 4.1.1.1, 4.1.2.1, 4.1.3.1, 4.3.1, 6.3.1, 6.4.1, 6.5.1, 6.6.1, A.1.1, A.2.1, A.4.1, B.1	basicoccupancy.mw bassicocc.R
Population growth model	1.1, 1.2.2, 3.4.1, 4.1.3.2	popgrowth.mw pop.R
Linear regression model	1.2.3, 1.3.1, 4.1.1.3	linear.R
Mark-recovery model	1.2.4, 2.1.2, 2.2.3, 2.3.2, 2.4.2.1, 3.1.4.4, 3.2.4, 3.2.6.2, 3.5.1, 4.1.1.2, 4.1.4.1, 4.2, 4.4.1.1, 4.4.2.1, 5.2.1, 5.3.1.1, 5.3.2.1, 6.3.2, 6.4.2, 6.6.3, 9.1.1	markrecovery.R scoretesteg.mw, markrecovery.mw mrmsteps.mw markrecbay.R
Three compartment model	1.3.2, 2.1.4, 2.2.2, 3.1.4.3, 3.2.3, 3.3.2.1	compartmental.R compartmental.mw
Multiple-input multiple-output model	1.3.3, 3.2.5	MIMO.mw
Mixture model	2.1.3	
Binary regression model	2.1.5	
Binary latent class model	2.2.1, 2.3.1, 3.1.4.2 3.2.2, 3.2.5.2	latentvariable.mw
N-mixture model	2.2.4	
Capture-recapture with trap dependence	3.2.7.2	capturerecapture.mw
Closed population capture-recapture Model	3.4.2	betabin.mw
Complex compartmental model	4.3.2, 5.2.2.2	complexcomp.mw

Example	Section(s)	Computer Code
Conditional mark-recovery model	5.1.1	conditional.mw
Latent class model with covariates	5.1.2	latentcovariates.mw
Capture-recapture	5.2.2.1, 5.2.3.1, 9.1.2	CJS.mw
Age-dependent fisheries model	5.2.2.3	fisheries.mw
Logistic growth curve model	6.3.3, 6.4.3, 6.5.2, 6.6.2	logisticgrowth.mw, logisticgrowth.R
Two-compartment linear model	7.1.1, 7.2.1.1, 7.2.2.1, 7.2.3.1	twocompartmentmodel.mw
Three-compartment linear model	7.1.2, 7.2.1.2, 7.2.2.2, 7.2.3.2	threecompartmentmodel.mw
Michaelis Menten kinetics model	7.1.3, 7.2.3.3	nonlinearmodel.mw
Abundance state-space model	8.1.1, 8.3.1, 8.4.1, 8.5.1, 8.5.2	abundance.R, abundance.mw
Multi-site abundance state-space model	8.1.2, 8.5.5	multisiteabundance.mw
SIR state-space model	8.1.3, 8.6.1, 8.7	SIRmodel.mw
Stochastic volatility state-space model	8.1.4, 8.5.4	volatility.mw
Stochastic Ricker stock recruitment state-space model	8.1.5, 8.2.1, 8.5.3	ricker.R, ricker.mw
Poisson HMM	8.8.1	poissonHMM.mw
Capture-recapture-recovery model	9.1.3	CRR.mw
Multi-state model	9.1.4, 9.4.1	4stateexample.mw
Memory model	9.1.5, 9.4.2	memory.mw
Joint census and mark-recovery	9.2	basicIPM.mw
Four data set integrated model	9.2.1, 9.3.2	
Immigration integrated model	9.2.2	immIPM.R immIPM.mw
Robust design occupancy model	9.3.1.1	
Robust design removal model	9.3.1.2	

1

Introduction

Mathematical and statistical models are used to describe different systems or provide a representation of a process. For example, a dynamic system could be used to describe a pharmacokinetic system or a statistical model could be used to describe an ecological population. These models are defined by parameters that describe different characteristics that underlie the model. In a pharmacokinetic system, a parameter could represent the infusion rate of a drug, or in an ecological model a parameter could describe the survival probability of an animal. Determining or estimating these parameters will provide key information about the underlying process. In some models, however, it is not possible to estimate every parameter. For example, two parameters that only ever appear in the model as a product can not be estimated individually; only the product can be estimated. Such a model is said to be parameter redundant, or the parameters are described as non-identifiable.

This Chapter starts with an introduction to mathematical and statistical models, in Section 1.1, and then introduces the concepts of parameter redundancy and identifiability through a series of examples. Section 1.2 provides an introduction to parameter redundancy. Section 1.3 explains identifiability. Lastly Section 1.4 outlines the different methods that can be used to detect parameter redundancy and investigate identifiability.

1.1 An Introduction to Mathematical and Statistical Models

In a mathematical or statistical model the representation of a system or a process is expressed in terms of an algebraic expression. The form of this expression depends on the type of model. One common characteristic of mathematical and statistical models is that they depend on parameters which describe characteristics of the process. How the model is expressed and the parameters that appear in the model is key to understanding identifiability and parameter redundancy, so below we explain how parameters are expressed in these models.

In statistics we are interested in understanding a population through data collected about that population. Typically we do not measure every

1

member of a population, so there will be variability in the data that is collected. In a statistical model the uncertainty associated with the data is represented by a probability distribution. A probability distribution, represented as $f(y; \boldsymbol{\theta})$, depends on a variable y, which represents a single observation within a set of data. The probability distribution also depends on one or more parameters, $\boldsymbol{\theta}$.

The form of the probability distribution depends on the type of data collected. Suppose that an experiment has two possible outcomes: success (represented by a 1) or failure (represented by a 0). A possible probability distribution is the binary distribution, which is expressed as

$$f(y; \theta) = \theta^y (1 - \theta)^{1-y}, \ y = 0, 1, \ 0 < \theta < 1.$$

In this example there are only two options, $y = 1$ or $y = 0$. The probability of success $(y = 1)$ is $f(1) = \theta^1 (1-\theta)^{1-1} = \theta$ and the probability of failure $(y = 0)$ is $f(0) = \theta^0 (1 - \theta)^{1-0} = 1 - \theta$. The parameter θ represents the probability of success.

Alternatively suppose that data could take on any value (in a given range), for example the height of an individual. A commonly used distribution is the normal distribution which has probability distribution

$$f(y; \mu, \sigma) = \frac{1}{\sigma\sqrt{2\pi}} \exp\left\{ -\frac{1}{2} \left(\frac{y - \mu}{\sigma} \right)^2 \right\}, \ -\infty < y < \infty.$$

The parameter μ represents the mean of the distribution. Two example plots of normal probability distributions are given in Figure 1.1. The plots show the normal distribution is symmetric about the mean, $\mu = 5$. How variable the distribution is about the mean in controlled by the second parameter, σ, the standard deviation. A larger σ results in more variability about the mean.

There is usually more than one way of expressing parameters in a model. These are known as different parameterisations. For example, there are two commonly used parameterisations of the gamma distribution. A gamma distribution may be a more applicable model for height data, as it is only defined for positive values, and does not require the shape of the distribution to be symmetrical, unlike the normal distribution. The first parameterisation of the gamma distribution has a probability distribution

$$f(y) = \frac{\beta^\alpha}{\Gamma(\alpha)} y^{\alpha-1} \exp(-\beta y), \ y > 0, \ \alpha > 0, \ \beta > 0,$$

where $\Gamma(z)$ is the gamma function with $\Gamma(z) = \int_0^\infty x^{z-1} \exp(-x) dx$. The parameters here are α and β. The distribution has mean α/β. The second parameterisation of the gamma distribution has a probability density function

$$f(y) = \frac{1}{\Gamma(a)b^a} y^{a-1} \exp\left(-\frac{y}{b}\right), \ y > 0, \ a > 0, \ b > 0,$$

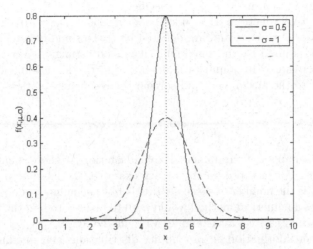

FIGURE 1.1
Plots of normal probability distributions. Both probability distributions have $\mu = 5$, which is the mean of the distribution, represented by the dotted line. The solid line represents a normal distribution with $\sigma = 0.5$, which is less variable than the normal distribution represented by dashed line where $\sigma = 1$.

with parameters a and b and mean ab. The second parameterisation could be described as a reparameterisation of the first option, with $a = \alpha$ and $b = 1/\beta$. These are only two different parameterisations; an alternative parameterisation could have one parameter representing the mean of the distribution, $\mu = \alpha/\beta$. Different parameterisations can be used if there is an interest in a particular parameter, or they can be used to simplify parameter estimation (see, for example, chapter 5 of Morgan, 2009). The parameterisation is also important in parameter redundancy, as we explain in Section 1.2 below.

Typically we collect more than one observation to form a set of data, $\mathbf{y}$. In statistical inference a likelihood function, $L(\boldsymbol{\theta}|\mathbf{y})$, is formed for parameters $\boldsymbol{\theta}$ based on data $\mathbf{y}$. If we had n independent observations y_i, each with a probability distribution $f(y_i; \boldsymbol{\theta})$, then the likelihood is

$$L(\boldsymbol{\theta}|\mathbf{y}) = \prod_{i=1}^{n} f(y_i; \boldsymbol{\theta}).$$

To simplify calculations, it is usually simpler to work with the log-likelihood function, which is the natural logarithm of the likelihood:

$$l(\boldsymbol{\theta}|\mathbf{y}) = \log\{L(\boldsymbol{\theta}|\mathbf{y})\} = \sum_{i=1}^{n} \log\{f(y_i; \boldsymbol{\theta})\}.$$

Estimates for the parameters, for a specific data set, are found by maximising the likelihood function, or log-likelihood function. These are called maximum likelihood estimates and are denoted by $\hat{\boldsymbol{\theta}}$. To simplify calculation, the log-likelihood is used to find maximum likelihood estimates. Maximum likelihood estimates can be found by differentiating the log-likelihood function with respect to the parameters and solving the resulting equations

$$\frac{\partial l(\hat{\theta}_j)}{\partial \theta_j} = 0, j = 1, \ldots, q,$$

where q is the number of parameters. Explicit solutions to these equations exist in some cases. For instance, in the binary example, the maximum likelihood estimate of θ is the number of successes divided by the number of observations. In other cases a numerical method would need to be used to find the maximum likelihood estimates (see, for example, Morgan, 2009).

For further information on probability distributions and maximum likelihood see, for example, Davison (2003), Hogg et al. (2014) or Morgan (2009).

Alternatively mathematical models may have no uncertainty, unlike a statistical model. A mathematical model describes the relationship between one or more variables to represent a specific problem (see Berry and Houston, 1995). For example, a population size, $P(t)$, at time t, could be represented by the differential equation model

$$\frac{dP(t)}{dt} = (\theta_b - \theta_d)P(t),$$

where θ_b is the birth rate and θ_d is the death rate. The differential equation can be solved to give that the population at time t is

$$P(t) = P_0 \exp\{(\theta_b - \theta_d)t\},$$

where P_0 is the population at time 0. In this model the parameters are θ_b and θ_d.

1.2   An Introduction to Parameter Redundancy

The parameterisation of a model is important in considering whether a model is parameter redundant. Parameter redundancy occurs when a model can be reparameterised in terms of a smaller number of parameters. This could be due to the inherent structure of the model, or could be caused by the data, or both. In Sections 1.2.1 to 1.2.4 below, the concept of parameter redundancy is demonstrated using a series of simple examples which are obviously parameter redundant. A more formal definition of parameter redundancy is given later in Section 3.1.2.

1.2.1 The Basic Occupancy Model

Occupancy experiments examine whether or not a species is present or absent from a site (MacKenzie et al., 2005). In the simplest experiment, sites are visited once with two possible outcomes: either the species is observed at a site, or the species is not observed at a site. The latter case could occur because either the species was actually absent or because the species is present but not observed. As well as considering whether the site is occupied or not, the basic occupancy model also needs to consider the detectability of the species. Suppose the probability that the species is present (or a site is occupied) is ψ and the probability that the species is detected is p, then the probability of observing the species is

$$\Pr(\text{species observed}) = \psi p.$$

There are two cases which result in the species not being observed: either the species is present but not observed, with probability $\psi(1-p)$, or the species is absent with probability $1-\psi$. By combining these two probabilities, the probability of the species not being observed is

$$\Pr(\text{species not observed}) = \psi(1-p) + 1 - \psi = 1 - \psi p.$$

Note that in the two possible outcomes of the experiment the two parameters only ever appear as the product ψp. This means that, in terms of what is measured, it is only ever possible to estimate that product and never the two parameters ψ and p separately. Therefore the model can be described as parameter redundant. This model is actually the binary distribution, which was discussed in Section 1.1, with the probability of success θ, replaced by ψp.

To be able to separate the probability that a species is present, ψ, from the probability a species is detected, p, a specific experimental design was developed, where sites are surveyed more than once within a short space of time, assuming that ψ does not change between these surveys (MacKenzie et al. 2002).

1.2.2 The Population Growth Model

As discussed in Section 1.1, a population size, $P(t)$, at time t, can be modelled by the differential equation

$$\frac{dP(t)}{dt} = (\theta_b - \theta_d)P(t), \ P(0) = P_0,$$

where θ_b is the birth rate and θ_d is the death rate (see, for example, Chapter 2, Berry and Houston, 1995). It is possible to reparameterise this model in terms of the parameter $\beta = \theta_b - \theta_d$. It would only ever be possible to uniquely determine β rather than θ_b and θ_d individually. For example a birth rate of $\theta_b = 1.5$ and a death rate of $\theta_d = 0.5$ will give the same population size, as a birth rate of $\theta_b = 2$ and a death rate of $\theta_d = 1$ (for a given time and given initial population size, P_0). In fact this would be true for any birth and death rate that satisfied $\beta = \theta_b - \theta_d = 1$.

1.2.3   The Linear Regression Model

Linear regression models are used to describe how an observed variable, y_i, depends on one or more covariates, $x_{i,j}$, where i is the observation number and j is the covariate number. We could, for example, have a very small survey with two observations: $y_1 = 8.5$ and $y_2 = 12.5$, and with two covariates: $x_{1,1} = 2$, $x_{1,2} = 3$, $x_{2,1} = 7$ and $x_{2,2} = 2$. The linear regression model, which includes both covariates, has expected value

$$E[y_i] = \beta_0 + \beta_1 x_{i,1} + \beta_2 x_{i,2}.$$

Of course it would not be sensible to try and fit any linear regression model with only two observations, however it does provide a simple example where parameter redundancy is caused by having more parameters than there are observations. This example has three parameters and only two observations. The parameter values $\beta_0 = 6.9$, $\beta_1 = 0.8$ and $\beta_2 = 0$ result in $E[y_i] = y_i$ for $i = 1, 2$. However this is not the only set of parameters that gives this result, the parameter values $\beta_0 = 5.2$, $\beta_1 = 0.9$ and $\beta_2 = 0.5$ also do, as does any parameters that satisfy the equations $\beta_0 = 6.9 - 3.4\beta_2$ and $\beta_1 = 0.8 + 0.2\beta_2$. We can set $\beta_2 = 0$ (or $\beta_1 = 0$) and still get $E[y_i] = y_i$ for $i = 1, 2$. As a result, this model is parameter redundant. In more realistic data sets, parameter estimates are found using maximum likelihood, and interest would not normally lie in the model for which $E[y_i] = y_i$, which is known as the saturated model. See for example, Chapter 8, Davison (2003).

1.2.4   The Mark-Recovery Model

Marking of animal populations is used in ecology for estimating various aspects of population dynamics including the survival probabilities of wild animals. Data collection involves marking animals, then subsequently when the animal dies, some of the marked animals are recovered (see for example, Freeman and Morgan, 1990 and Freeman and Morgan, 1992). As an illustration, consider a small subset of mark-recovery data on the Northern Lapwing, *Vanellus vanellus*, from Catchpole et al. (1999). In 1963, 1147 new born lapwings, were marked with rings on their feet, 14 of which were recovered dead that year. A further 4 were recovered dead in 1964. That year 1285 new born lapwings were marked, 16 of which were recovered dead that year. This process continues into subsequent years. The first three years of data are summarised below, where **F** is a vector where each entry is the number of animals marked each year and **N** is a matrix of the number recovered dead each year. The rows correspond to each successive year of marking and the columns correspond to each recovery year.

$$\mathbf{F} = \begin{bmatrix} 1147 \\ 1285 \\ 1106 \end{bmatrix}, \mathbf{N} = \begin{bmatrix} 14 & 4 & 1 \\ 0 & 16 & 4 \\ 0 & 0 & 11 \end{bmatrix}.$$

Mark-recovery models can be used to estimate the survival probability of the animals, but we also need to consider the fact that not every animal that dies will be recovered. Suppose that survival of the first year of life is different to subsequent years and let ϕ_1 be the probability of surviving the first year and ϕ_a be the probability of surviving subsequent years. Age-dependence is also assumed for the probability of recovery with λ_1 and λ_a representing the probability that a dead bird's mark is recovered in the first and subsequent years respectively. The probability that a bird was marked and recovered dead in 1963 is $(1 - \phi_1)\lambda_1$, as it did not survive their first year of life and they were recovered dead in their first year. The probability that a bird was marked in 1963 and then recovered dead in 1964 is $\phi_1(1 - \phi_a)\lambda_a$, as it survived its first year of life but did not survive its next year of life as an older bird, and was then recovered dead as an older bird. Similar probabilities can be found to correspond to each of the entries in the matrix $\mathbf{N}$, which are summarised in the matrix $\mathbf{P}$ below.

$$\mathbf{P} = \begin{bmatrix} (1 - \phi_1)\lambda_1 & \phi_1(1 - \phi_a)\lambda_a & \phi_1\phi_a(1 - \phi_a)\lambda_a \\ 0 & (1 - \phi_1)\lambda_1 & \phi_1(1 - \phi_a)\lambda_a \\ 0 & 0 & (1 - \phi_1)\lambda_1 \end{bmatrix}.$$

The entries in $\mathbf{P}$, $P_{i,j}$, correspond to the probability that an animal marked in year i is recovered dead in year j. As well as the probabilities of being marked in a given year and then recovered dead in the same or subsequent year, a mark-recovery model also needs to include the birds that were marked and then not recovered dead for each marking year. This could occur either because the animal is still alive or it died and was not recovered. The probability of a marked bird never being recovered is $1 - \sum_{j=i}^{3} P_{i,j}$. This means that $\mathbf{P}$ provides all the information on the parameters for the mark-recovery models.

Now consider the reparameterisation with $\beta_1 = (1 - \phi_1)\lambda_1$ and $\beta_2 = \phi_1\lambda_a$. We can rewrite the probabilities of $\mathbf{P}$ in terms of β_1, β_2 and ϕ_a, giving

$$\mathbf{P} = \begin{bmatrix} \beta_1 & \beta_2(1 - \phi_a) & \beta_2\phi_a(1 - \phi_a) \\ 0 & \beta_1 & \beta_2(1 - \phi_a) \\ 0 & 0 & \beta_1 \end{bmatrix}.$$

The original parameterisation of $\mathbf{P}$ was in terms of four parameters, ϕ_1, ϕ_a, λ_1 and λ_a, however we can rewrite $\mathbf{P}$ in terms of the three parameters β_1, β_2 and ϕ_a. This model is therefore parameter redundant because the model can be reparameterised in terms of a smaller number of parameters. Parameter redundancy in mark-recovery models is discussed in Cole et al. (2012).

1.3   An Introduction to Identifiability

Identifiability of models refers to whether there is a unique representation of a model or whether there are multiple representations of a model. Models can either be classified as globally identifiable, locally identifiable or non-identifiable.

- A model is said to be globally identifiable if there is only one set of parameter values that result in a single model output.

- A model is said to be locally identifiable if there are a countable set of parameter values that results in a single model output. If the parameter space is restricted in a locally identifiable model there will be only one set of parameter values that result in a single model output.

- Non-identifiability caused by parameters occurs in models where there are infinite parameter values that result in a single model output, which occurs when a model is parameter redundant.

More formal definitions of global, local and non-identifiability are given in Section 3.1.1. To introduce these concepts we consider three examples in Sections 1.3.1 to 1.3.3. The first is globally identifiable, the second is locally identifiable and the third is non-identifiable.

1.3.1   Linear Regression Model

Identifiability of linear regression models is a well known field (see, for example, Fisher, 1966). Continuing the example in Section 1.2.3, the basic linear regression model is used to illustrate the concept of global identifiability.

Consider the model with only one covariate. This has expectation $E[y_i] = \beta_0 + \beta_1 x_{i,1}$. Using the same data as in Section 1.2.3 the values $\beta_0 = -7.93$, $\beta_1 = 1.65$ provide a perfect fit to the data, that is $E(y_i) = y_i$, as illustrated in Figure 1.2a. There is only one line that goes between the two data points, therefore there is only one set of parameters that specify this linear regression model. As there is unique set of parameters, that specify this condition, this model is globally identifiable.

Typically a linear regression model would be fitted to a larger data set. In Figure 1.2b the sample size is extended to ten. The linear regression model with $\beta_0 = -1.34$ and $\beta_1 = 1.10$ is no longer a perfect fit to the set of data, but can be considered as the line of best fit. This line of best fit, or least squares regression line, is calculated to minimise the squared vertical distances between the observations and the line. The values of the parameters also correspond to the maximum likelihood estimates. There is only one least squares regression line (or only one set of parameters) satisfying these criteria,

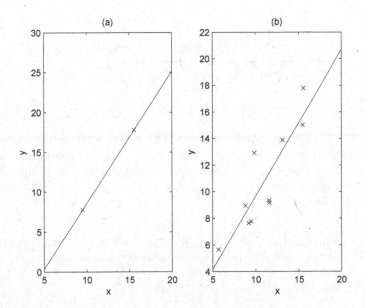

FIGURE 1.2
Linear regression example with observation, y and covariate $x = x_1$. Figure
(a) shows an example with only two observations and the line that results in
perfect fit. Figure (b) shows an example with ten observations and the line of
best fit.

therefore this model is also globally identifiable. In general, a linear regression model with one covariate and at least two (different) observations will be
globally identifiable. This result stems from the identifiability condition for
linear regression models, which is discussed in Section 3.3.1.1.

1.3.2 Three Compartmental Model

Compartmental models are deterministic models used to represent a system where materials or energies are transmitted between multiple compartments. They are used in areas such as pharmacokinetics, epidemiology and
biomedicine (see, for example, Godfrey, 1983). One type of compartmental
models, linear compartmental models, have the output function:

$$\mathbf{y}(t, \boldsymbol{\theta}) = \mathbf{C}(\boldsymbol{\theta})\mathbf{x}(t, \boldsymbol{\theta}), \text{ with } \frac{d}{dt}\mathbf{x}(t, \boldsymbol{\theta}) = \mathbf{A}(\boldsymbol{\theta})\mathbf{x}(t, \boldsymbol{\theta}) + \mathbf{B}(\boldsymbol{\theta})\mathbf{u}(t),$$

where $\mathbf{x}$ is the state-variable function, $\boldsymbol{\theta}$ is a vector of unknown parameters,
$\mathbf{u}$ is the input function and t is the time recorded on a continuous time scale.
The matrices $\mathbf{A}, \mathbf{B}, \mathbf{C}$ are the compartmental, input and output matrices
respectively.

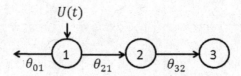

FIGURE 1.3
A representation of a compartmental model. The circles represent the compartments and the arrows represent the movement between compartments or the environment, with rate θ_{ij}. This example, from Godfrey and Chapman (1990), has three compartments with input into compartment 1, $U(t)$. Only compartment 3 is observed.

An example from Godfrey and Chapman (1990) with three compartments is summarised in Figure 1.3. This linear compartmental model has

$$
\mathbf{A} = \begin{bmatrix} -\theta_{01} - \theta_{21} & 0 & 0 \\ \theta_{21} & -\theta_{32} & 0 \\ 0 & \theta_{32} & 0 \end{bmatrix}, \mathbf{B} = \begin{bmatrix} 1 \\ 0 \\ 0 \end{bmatrix}, \mathbf{C} = \begin{bmatrix} 0 & 0 & 1 \end{bmatrix},
$$

with parameters $\boldsymbol{\theta} = [\theta_{01}, \theta_{21}, \theta_{32}]$. The parameter θ_{ij} represents the rate of transfer to compartment i from compartment j, with 0 representing the environment outside the compartments.

The solution of this set of differential equations is

$$
y(t) = \frac{\theta_{21} \left\{ \theta_{32} e^{-(\theta_{01} + \theta_{21})t} - (\theta_{01} + \theta_{21}) e^{-\theta_{32}t} + \theta_{01} + \theta_{21} - \theta_{32} \right\}}{(\theta_{01} + \theta_{21} - \theta_{32})(\theta_{01} + \theta_{21})},
$$

(see Maple file `compartmental.mw`). The parameters $\theta_{01} = 0.5$, $\theta_{21} = 2$ and $\theta_{32} = 0.5$ give

$$
y(t) = \frac{1}{5} e^{-5t/2} - e^{-t/2} + \frac{4}{5}.
$$

This value of $y(t)$ can also be obtained from the parameters $\theta_{01} = 0.1$, $\theta_{21} = 0.4$ and $\theta_{32} = 2.5$. There are similar pairs of parameter sets that result in the same value of $y(t)$. As there are two possible sets of parameter values that give the same value of $y(t)$, the model is locally identifiable, rather than globally identifiable. A formal proof of this generic result is given in Section 3.3.2.1 or in Godfrey and Chapman (1990).

1.3.3 Multiple-Input Multiple-Output (MIMO) Model

Multiple-Input Multiple-Output (MIMO) models are deterministic models that have been used in machine learning and elsewhere in numerous and widespread applications. They are defined by the equations

$$
y_i = g_i(\mathbf{u}, \boldsymbol{\theta}),
$$

where $\mathbf{u} = [u_1, \ldots, u_n]$ is the input vector of length n, $\mathbf{y} = [y_1, \ldots, y_m]$ is the output vector of length m, g_i for $i = 1, \ldots, m$ are known functions, and $\boldsymbol{\theta} = [\theta_1, \ldots, \theta_q]$ is a vector of q parameters on which the model depends (see, for example, Ran and Hu, 2014).

Consider a specific example from Ran and Hu (2014) with

$$\begin{cases} y_1(\mathbf{u}, \boldsymbol{\theta}) = \exp(-\theta_2\theta_3 u_1) + \dfrac{\theta_1}{\theta_2}\left\{1 - \exp(-\theta_2\theta_3 u_1)\right\}u_2 \\ y_2(\mathbf{u}, \boldsymbol{\theta}) = \theta_1\theta_3 u_1 + (1 - \theta_2\theta_3)u_2 \end{cases},$$

where $\mathbf{u} = [u_1, u_2]$ is the input vector of length $n = 2$, $\mathbf{y} = [y_1, y_2]$ is the output vector of length $m = 2$ and $\boldsymbol{\theta} = [\theta_1, \theta_2, \theta_3]$ is a vector of $q = 3$ parameters.

Consider the output

$$\begin{cases} y_1 = \exp(-2u_1) + \frac{1}{2}\left\{1 - \exp(-2u_1)\right\}u_2 \\ y_2 = u_1 - u_2 \end{cases}.$$

This could have occurred if $\theta_1 = 0.5$, $\theta_2 = 1$ and $\theta_3 = 2$ or if $\theta_1 = 4$, $\theta_2 = 8$ and $\theta_3 = 0.25$. There are in fact infinite other ways of creating this output, that satisfy the equations $2\theta_1 = \theta_2$ and $1 = \theta_1\theta_3$. As there are infinite parameter values that result in one output this model is non-identifiable.

Alternatively we can show the model is parameter redundant, so therefore it must be non-identifiable, by reparameterising the model with just two parameters $\beta_1 = \theta_2/\theta_1$ and $\beta_2 = \theta_1\theta_3$, giving

$$\begin{cases} y_1 = \exp(-\beta_1\beta_2 u_1) + \dfrac{1}{\beta_1}\left\{1 - \exp(-\beta_1\beta_2 u_1)\right\}u_2 \\ y_2 = \beta_2 u_1 + (1 - \beta_1\beta_2)u_2 \end{cases}.$$

1.4 Detecting Parameter Redundancy and Non-Identifiability

Determining whether or not a model is parameter redundant is crucially important, as information and conclusions drawn from a parameter redundant model may be biased and lead to incorrect results. This is discussed in detail in Chapter 2.

In some models, parameter redundancy is obvious, such as two parameters only ever appearing as a product. In other cases where it is not so obvious, various methods exist for detecting parameter redundancy. Parameter redundancy can be detected using symbolic algebra, numerical methods or a combination of the two. These methods, which are discussed in detail in Chapters 3 to 5, are generic and apply to any model which has an explicit form. As non-identifiability of parameters occurs when a model is parameter redundant, the same methods can be used to check for non-identifiability. Chapter 3 also explains how to distinguish between local and global identifiability.

One method for detecting parameter redundancy or non-identifiability, involves calculating a derivative matrix by differentiating some representation of the model with respect to the parameters. The rank of this derivative matrix is then calculated and if this is less than the number of parameters the model is parameter redundant (see, for example, Cole et al., 2010). The theory for this method is discussed in Section 3.2. A practical guide for using this method is given in Section 4.2.

This symbolic method was described as the gold standard for detecting parameter redundancy in Bailey et al. (2010). As well as detecting parameter redundancy this method can also be used to:

- find combinations of parameters that can be estimated in parameter redundant models (Section 3.2.5);

- obtain general results for any dimension of a model (Section 3.2.6);

- find parameter redundant nested models in models that are not parameter redundant (Section 3.2.7).

Whilst these methods can detect parameter redundancy, and therefore non-identifiability of parameters in a model, they cannot distinguish whether a model is locally or globally identifiable. The further steps needed to achieve this are discussed in Section 3.3.

To detect parameter redundancy or non-identifiability there are alternative numerical methods, which are discussed in Section 4.1. There is also a hybrid symbolic-numerical method, which combines the symbolic method with a numeric method (Choquet and Cole, 2012), and this is explained in Section 4.3. Section 4.4 discusses issues of near redundancy, where models behave as parameter redundant models, because they are in some way close to a parameter redundant model. The various different methods for detecting parameter redundancy are compared in Section 4.5.

In more complex models the symbolic method can fail because of a lack of computer memory calculating the rank of the derivative matrix (see, for example, Jiang et al., 2007, Forcina, 2008, Hunter and Caswell, 2009). In such cases there is an extended symbolic method (Cole et al., 2010), which is discussed in Chapter 5. The hybrid symbolic-numeric method is also extended in Chapter 5, and all the methods are compared.

Chapter 6 explains the issues of identifiability in Bayesian statistics. Then the book finishes with a series of chapters on different applications. Chapters 7 to 9 consider identifiability in continuous state-space models, discrete state-space models and ecological models respectively. Finally Chapter 10 concludes with recommendations of which method to use and when, as well as discussing how to proceed if a model is non-identifiable.

2

Problems With Parameter Redundancy

This chapter illustrates why parameter redundancy is a problem in modelling and inference. Section 2.1 explores the issues caused by parameter redundancy in maximum likelihood parameter estimation, then in Section 2.2 we show the practical problems with fitting parameter redundant models. Standard errors do not exist for parameter redundant models; this is discussed in Section 2.3. Finally, in Section 2.4, we show how model selection is effected by parameter redundancy.

2.1 Likelihood Inference with Parameter Redundant Models

The purpose of maximum likelihood estimation is to find the most probable parameter estimates for parameters in a model, given a set of data. For example, in an ecological population, maximum likelihood estimation can be used to find an estimate of the probability of survival. However in a parameter redundant model there is not a single parameter estimate for every parameter. Certain parameters will have multiple maximum likelihood estimates, with no global maximum, severely reducing the usefulness of the model.

Consider a parameter redundant model with q_θ parameters, $\boldsymbol{\theta}$, and log-likelihood $l(\boldsymbol{\theta}|\mathbf{y}) = \log\{L(\boldsymbol{\theta}|\mathbf{y})\}$. As the particular parameterisation of the model is parameter redundant, it can be reparameterised in terms of a smaller number of parameters, $\boldsymbol{\beta}$. The new set of parameters, $\boldsymbol{\beta}$, consists of q_β parameters where $q_\beta < q_\theta$. We can write $\boldsymbol{\beta} = g(\boldsymbol{\theta})$ for some appropriate function g. Now assume that the model parameterised in terms of $\boldsymbol{\beta}$ is not parameter redundant and has log-likelihood $l(\boldsymbol{\beta}|\mathbf{y}) = \log\{L(\boldsymbol{\beta}|\mathbf{y})\}$. The two likelihood and two log-likelihood functions are equal, that is $L(\boldsymbol{\theta}|\mathbf{y}) = L(\boldsymbol{\beta}|\mathbf{y})$ and $l(\boldsymbol{\theta}|\mathbf{y}) = l(\boldsymbol{\beta}|\mathbf{y})$. The second model is not parameter redundant therefore we can find a maximum likelihood estimate using standard maximum likelihood theory. The likelihood $L(\boldsymbol{\beta}|\mathbf{y})$, is maximised at $\hat{\boldsymbol{\beta}}$, where

$$\left.\frac{\partial l(\boldsymbol{\beta}|\mathbf{y})}{\partial \boldsymbol{\beta}}\right|_{\boldsymbol{\beta} = \hat{\boldsymbol{\beta}}} = 0.$$

13

Then for the parameter redundant parameterisation the likelihood, $L(\boldsymbol{\theta}|\mathbf{y})$, is maximised at all values of the parameters $\boldsymbol{\theta}$ that satisfy the equation

$$\hat{\boldsymbol{\beta}} = g(\boldsymbol{\theta})$$

(Dasgupta et al., 2007).

A consequence of this is that at least some parameters in a parameter redundant or non-identifiable parameterisation of a model will have multiple maximum likelihood estimates. This will result in a flat ridge in the likelihood surface (Catchpole and Morgan, 1997). In parameter redundant or non-identifiable models there will be infinite sets of maximum likelihood parameter estimates. If the maximum likelihood estimates exist, a locally identifiable model will have a countable number of maximum likelihood parameter estimates and a globally identifiable model will have a single set of maximum likelihood parameter estimates.

The differences in the log-likelihood surface for non-identifiable, locally identifiable and globally identifiable models are illustrated in Figure 2.1. Model (a) is parameter redundant or non-identifiable; model (b) is locally identifiable; and model (c) is globally identifiable. All three models have two parameters, θ_1 and θ_2. The first row in the figure shows three-dimensional plots of the log-likelihood surface for the two parameters. The flat ridge in the likelihood surface for the parameter redundant model is indicated by the dotted line. In the locally identifiable model there are two maximum likelihood estimates. Note that if the parameter space was restricted to a region around one maximum likelihood estimate, so that it excluded the other maximum likelihood estimate, the model would be globally identifiable. The globally identifiable model has a single maximum likelihood estimate.

An alternative to a three-dimensional plot is a contour plot of the log-likelihood surface. These are given in the second row of Figure 2.1. The x and y axes correspond to the two parameters, and the contours correspond to the log-likelihood values. The flat ridge in the likelihood surface for the parameter redundant model is indicated by the dotted line and the crosses indicate the maximum likelihood estimates in the locally and globally identifiable models.

If there are more than two parameters it is no longer possible to plot the log-likelihood surface directly. Instead the log-likelihood surface can be visualised by examining a log-likelihood profile for each parameter. A log-likelihood profile for a particular parameter is obtained by fixing the particular parameter across a range of values, and then maximising the log-likelihood over all the other parameters (see, for example, Section 4.3 of Morgan, 2009). This is illustrated in the third row of Figure 2.1. In the parameter redundant model the parameters have a flat log-likelihood profile, whereas there will typically be multiple peaks for a locally identifiable model (two in this illustrative example) and a single peak for a globally identifiable model.

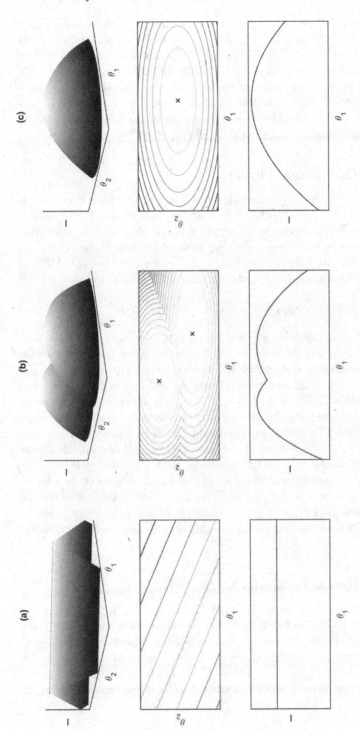

FIGURE 2.1
Log-Likelihood and profile log-likelihood plots for: (a) a parameter redundant, or non-identifiable model; (b) a locally identifiable model; (c) a globally identifiable model. All three models have θ_1 and θ_2. The rows show different representations of the log-likelihood surface, l. The first corresponds to a three-dimensional plot of the log-likelihood surface. The second corresponds to a contour plot of the log-likelihood surface. The third corresponds to a log-likelihood profile for θ_1.

We demonstrate the use of log-likelihood plots and profile log-likelihoods in a series of examples in Section 2.1.1 to 2.1.5. Sections 2.1.1 and 2.1.2 illustrate the log-likelihood surface and log-likelihood profiles respectively in parameter redundant models. Sections 2.1.3 and 2.1.4 illustrate profiles in locally identifiable models. The latter illustrates that the same issues arise using least squares rather than maximum likelihood. Section 2.1.5 illustrates log-likelihood profiles in a globally identifiable model, demonstrating a case where the data can influence whether or not the maximum likelihood estimate exists.

2.1.1   Basic Occupancy Model

Consider the basic occupancy model described in Section 1.2.1. This model has two parameters: ψ, the probability that the species is present and p, the probability that the species is detected. Suppose that $n_1 + n_2$ sites were visited and the species was observed at n_1 sites and was not observed at n_2 sites. As the probability of observing the species is ψp and the probability of not observing the species is $1 - \psi p$, the likelihood for this model is

$$L(\boldsymbol{\theta}|\mathbf{y}) \propto (\psi p)^{n_1}(1 - \psi p)^{n_2} = \beta^{n_1}(1 - \beta)^{n_2},$$

where $\beta = \psi p$. The parameters are $\boldsymbol{\theta} = [\psi, p]$, and the data, $\mathbf{y}$, can be summarised by the number of sites where the species was observed, n_1, and the number of sites where the species was not observed, n_2. This model is in fact a binary distribution, with n_1 successes and n_2 failures, with a maximum likelihood estimate of $\hat{\beta} = n_1/(n_1 + n_2)$. The parameterisation in terms of β has a single maximum likelihood estimate. However in terms of the original parameterisation the maximum likelihood estimates satisfy the equation $\hat{p} = n_1/\{\hat{\psi}(n_1 + n_2)\}$. Regardless of the data collected there will always be a ridge in the likelihood surface of the model with parameters $\boldsymbol{\theta} = [\psi, p]$. Figure 2.2 shows a three-dimensional plot of this log-likelihood surface for an arbitrary example with $n_1 = n_2 = 45$. The flat ridge of the log-likelihood surface is the white area at the top of the figure. This is more clearly seen in the contour plot, where the maximum likelihood estimates correspond to the dotted line.

2.1.2   Mark-Recovery Model

Consider the mark-recovery model discussed in Section 1.2.4. This model has two age classes, with survival and recovery both dependent on the age class. The parameters of this model are: ϕ_1, the probability of survival for juvenile animals in their first year of life; ϕ_a, the probability of survival for adult animals older than a year; λ_1, the probability of recovery for juvenile animals; and λ_a, the probability of recovery for adult animals. The R code `markrecovery.R`

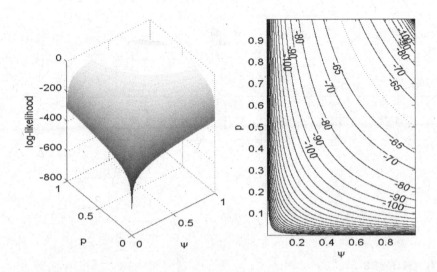

FIGURE 2.2
The first figure corresponds to the log-likelihood surface for the basic occupancy model with parameter p and ψ and with $n_1 = n_2 = 45$. The second figure corresponds to a contour plot for the log-likelihood surface, where the dotted line indicates the maximum likelihood estimates along the ridge.

finds log-likelihood profiles, given in Figures 2.3a to 2.3d. The flat profile log-likelihoods for the parameters ϕ_1, λ_1 and λ_a indicate that these parameters are not identifiable. The parameter ϕ_a does not have a flat profile so is individually identifiable.

Figures 2.3e and 2.3f are alternative plots of the results found when Figure 2.3a was created. Figure 2.3e corresponds to fixing ϕ_1 and plotting it against the maximum likelihood estimates of λ_1. Similarly, Figure 2.3f corresponds to fixing ϕ_1 and plotting it against the maximum likelihood estimates of λ_a. These maximum likelihood estimates show the relationship $\lambda_1 = 0.16/(1 - \phi_1)$ and $\lambda_a = 0.18/\phi_1$. Rearranging these equations gives $\lambda_1(1 - \phi_1) = 0.16 = \hat{\beta}_1$ and $\lambda_1\phi_1 = 0.18 = \hat{\beta}_2$, which corresponds to a reparameterisation of the model, in terms of ϕ_a, $\beta_1 = (1 - \phi_1)\lambda_1$ and $\beta_2 = \phi_1\lambda_a$ the model. This reparameterised model is no longer parameter redundant. For the example presented here, the maximum likelihood estimates are $\hat{\phi}_a = 0.65$, $\hat{\beta}_1 = (1 - \hat{\phi}_1)\hat{\lambda}_1 = 0.16$ and $\hat{\beta}_2 = \hat{\phi}_1\hat{\lambda}_a = 0.18$.

These graphs show the flat ridge in the likelihood surface that will occur in a parameter redundant model. This is discussed as a method for detecting

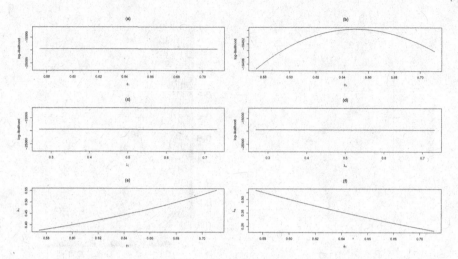

FIGURE 2.3
Figures (a)-(d) show log-likelihoods profiles for the parameters ϕ_1, ϕ_a, λ_1, λ_a, for the mark-recovery model with two age classes, with survival and recovery both dependent on the age class. Figures (e) and (f) show the maximum likelihood for the parameters λ_1 and λ_a which correspond to fixing ϕ_1.

parameter redundancy in Section 4.1.4. Figures 2.3e and 2.3f also give a method for finding a reparameterisation that results in a model that is no longer parameter redundant. This method is used in Eisenberg and Hayashi (2014) and is explained in more detail in Section 4.3.

The log-likelihood profiles for the reparameterised model with parameters ϕ_a, $\beta_1 = (1 - \phi_1)\lambda_1$ and $\beta_2 = \phi_1\lambda_a$ are given in Figure 2.4. It can be seen from the plots that in this reparameterised model the profile log-likelihoods are not flat, and all the parameters are identifiable.

2.1.3   Mixture Model

A mixture distribution can be used to model data where the underlying population consists of two or more sub-populations, but the data does not contain information on which sub-population the data point belongs to. A mixture of two distributions may be locally identifiable. For example, consider a mixture of two gamma distributions. The probability density function is given by

$$f(x) = \pi \frac{1}{\Gamma(a_1)b_1^{a_1}} x^{a_1 - 1} \exp\left(-\frac{x}{b_1}\right) + (1 - \pi) \frac{1}{\Gamma(a_2)b_2^{a_2}} x^{a_2 - 1} \exp\left(-\frac{x}{b_2}\right).$$

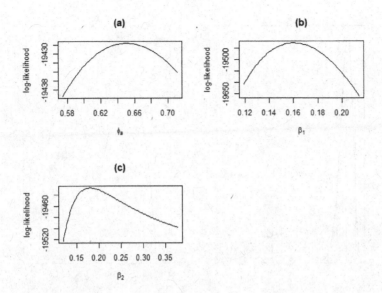

FIGURE 2.4
Profile likelihoods for the mark-recovery model, with two age classes, with survival and recovery both dependent on the age class, reparameterised so that $\beta_1 = (1 - \phi_1)\lambda_1$ and $\beta_2 = \phi_1\lambda_a$. The other parameter, ϕ_a, is unchanged.

Figure 2.5a shows a histogram of a set of (simulated) data. This example has two sets of maximum likelihood estimates. The first has parameter estimates $\pi = 0.71$, $a_1 = 8.67$, $a_2 = 20.72$, $b_1 = 0.12$ and $b_2 = 0.23$. The second has parameter estimates $\pi = 0.29$, $a_1 = 20.72$, $a_2 = 8.67$, $b_1 = 0.23$ and $b_2 = 0.12$. The log-likelihood for both sets of parameter estimates is $l = -125.47$. The profile log-likelihood for the mixture parameter π is given in Figure 2.5b. This is an example of label switching, where the parameters can be permuted and still give the same log-likelihood value (see, for example, Redner and Walker, 1984). This model is therefore locally identifiable when the parameter space allows $0 < \pi < 1$. If it is known that $\pi > 0.5$, then we could restrict the parameter space to this region making the model globally identifiable.

Local identifiability in mixture models is discussed in Kim and Lindsay (2015).

2.1.4 Three Compartment Model

Parameters do not have to be estimated using maximum likelihood. One alternative method of determining parameters involves using least squares. Suppose that there are n observations of a process, z_i at time t_i, and the output of a

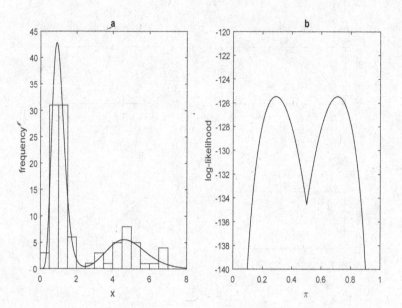

FIGURE 2.5
a. Histogram of a mixture model with the probability density function of
the maximum likelihood estimate superimposed. b. Profile log-likelihood for
mixture parameter π.

TABLE 2.1
Example Three Compartment Model Data. i is the
observation number, t_i is the time and z_i is the observation.

i	1	2	3	4	5	6	7	8	9
t_i	0.5	1	1.5	2	2.5	3	3.5	4	4.5
z_i	0.07	0.21	0.33	0.43	0.51	0.58	0.63	0.66	0.69

model at time t_i is $y(t_i)$, then least squares involves minimising the function

$$R^2 = \sum_{i=1}^{n} \{z_i - y(t_i)\}^2.$$

Similar to a log-likelihood profile plot, a profile plot of R^2 would be flat for non-
identifiable parameters, typically have one minimum for a globally identifiable
parameter and a countable number of minimums for a locally identifiable
parameter.

For example, consider the compartmental model described in Section 1.3.2.
Suppose the model represented a process that could be observed, with data
given in Table 2.1. Profile plots for R^2 are shown in Figure 2.6, which were

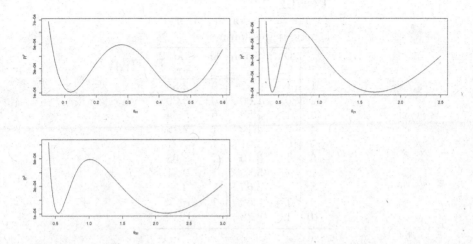

FIGURE 2.6
Three compartment model profile plots of R^2 for the parameters θ_{01}, θ_{21} and θ_{32}.

produced using the R code `compartmental.R`. As this example is locally identifiable with two possible solutions, there are two minimums in the profile plots with identical values of R^2.

2.1.5 Binary Logistic Regression Model

Whilst most globally identifiable models will have one maximum likelihood estimate (and locally identifiable models will have one maximum likelihood estimate within a specific region), in practice some data sets will have an undefined maximum likelihood estimate or maximum likelihood estimates might be outside the parameter range. We illustrate this in Figure 2.7, which shows profile log-likelihoods for a binary logistic regression model for two different data sets. The data sets are given in Table 2.2.

The model used for these data sets is

$$E(y_i) = \frac{1}{1 + \exp\{-(\beta_0 + \beta_1 x_i)\}},$$

where y_i is a binary observation and x_i is a covariate. The parameters are β_0 and β_1.

Whilst data set 1 has a single maximum likelihood estimate for both parameters, the estimates tend to $\pm\infty$ for the second data set. This is because

TABLE 2.2

Two data sets for the binary
logistic regression model.

i	Data Set 1 y_i	x_i	Data Set 2 y_i	x_i
1	1	0.27	0	2.85
2	0	1.00	0	3.29
3	1	2.53	0	5.34
4	0	4.10	1	7.19
5	0	5.07	1	7.27
6	1	5.10	1	8.68
7	1	5.85	1	9.31
8	1	6.05	1	9.54
9	1	6.52	1	10.58
10	1	8.38	1	12.75

there is perfect separation between the observation and covariate in the second data set. That is, when $y_i = 0$, $x_i \leq 5.34$ and when $y_i = 1$, $x_i > 5.34$. Albert and Anderson (1984) prove that maximum likelihood estimates do not exist if there is perfect separation.

2.2   Fitting Parameter Redundant Models

In Section 2.1 we demonstrated the problem with parameter redundant models having multiple maximum likelihood estimates. In more complex models, deterministic search method are used to estimate parameters. (The different search methods are discussed in Morgan, 2009.) Nearly all such methods require starting values for the parameter estimates, and depending on which starting value used, the algorithm will converge to one of the multiple maximum likelihood estimate (assuming that the algorithm does converge). If parameter redundancy is ignored and the likelihood surface is not explored this will result in incorrect and misleading parameter estimates. This is illustrated with three examples in Sections 2.2.1, 2.2.3 and 2.2.4 below.

Local identifiability can also be problematic in certain models. In a locally identifiable model we have to make an assumption that the locally identifiable parameter lies in a specific region. This could cause problems for constructing confidence intervals for parameters (Kim and Lindsay 2015). For example, for the mixture model in Section 2.1.3, if we assume $\pi < 0.5$, the maximum likelihood estimate is the left peak in Figure 2.5b. A confidence interval with an upper limit of 0.4 would be fine, but a confidence interval with an upper limit of 0.6 would be outside the region of $\pi < 0.5$ and would not be identifiable.

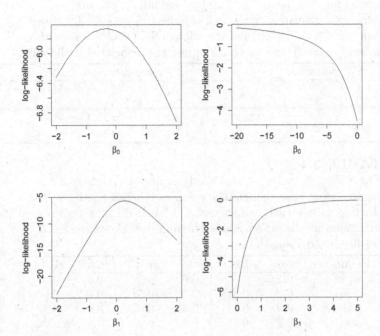

FIGURE 2.7
Profile log-likelihood plots for two data sets from the binary logistic regression
model $E(y_i) = \frac{1}{1+\exp\{-(\beta_0+\beta_1 x_i)\}}$.

2.2.1 Binary Latent Class Model

Latent class models allow for a structure to exist between associated variables using an unobserved or latent variable. Forcina (2008) considers a simple binary latent class model for two binary response variables, Y_1 and Y_2, which are conditionally independent given a binary latent variable Z. Let $\theta_{jk} = Pr(Y_j = 1|Z = k)$ and $p = Pr(Z = 1)$. As Y_j and Z are binary variables $Pr(Y_j = 0|Z = k) = 1 - \theta_{jk}$ and $Pr(Z = 0) = 1 - p$. An example set of observations along with the associated probabilities are given in Table 2.3.

Forcina (2008) states this model is well known to be parameter redundant. This is because there are five parameters, but only three independent combinations of probabilities. As the four probabilities sum to one, one of the four probabilities given in Table 2.3 could be rewritten as one minus the other

TABLE 2.3

Binary latent class model data and probabilities. The first two columns give the values of the two variables Y_1 and Y_2. The third column gives the number of observations (No. Obs.) for a set of (simulated) data. The final column gives the associated probability.

Y_1	Y_2	No. Obs.	Probability
0	0	397	$p(1 - \theta_{11})(1 - \theta_{21}) + (1 - p)(1 - \theta_{10})(1 - \theta_{20})$
0	1	301	$p(1 - \theta_{11})\theta_{21} + (1 - p)(1 - \theta_{10})\theta_{20}$
1	0	147	$p\theta_{11}(1 - \theta_{21}) + (1 - p)\theta_{10}(1 - \theta_{20})$
1	1	154	$p\theta_{11}\theta_{21} + (1 - p)\theta_{10}\theta_{20}$

TABLE 2.4

Three sets of maximum likelihood parameter estimates for a binary latent class model. The columns labelled Start give the starting values used, and the column labelled Est. gives the maximum likelihood estimate. The final row shows the log-likelihood value, l.

Parameter	Start	Est.	Start	Est.	Start	Est.
p	0.2	0.02	0.6	0.6	0.6	0.81
θ_{11}	0.2	0.31	0.5	0.51	0.6	0.64
θ_{10}	0.2	0.67	0.4	0.93	0.6	0.77
θ_{21}	0.2	0.56	0.3	0.25	0.6	0.26
θ_{20}	0.2	0.23	0.2	0.22	0.6	0.14
l		-1192.0		-1192.0		-1192.0

three probabilities. For example

$$
\begin{aligned}
1 - &\{p(1 - \theta_{11})\theta_{21} + (1 - p)(1 - \theta_{10})\theta_{20}\} \\
-&\{p\theta_{11}(1 - \theta_{21}) + (1 - p)\theta_{10}(1 - \theta_{20})\} \\
-&\{p\theta_{11}\theta_{21} + (1 - p)\theta_{10}\theta_{20}\} \\
= &\, p(1 - \theta_{11})(1 - \theta_{21}) + (1 - p)(1 - \theta_{10})(1 - \theta_{20})
\end{aligned}
$$

The maximum likelihood estimates are found using a numerical method, which requires starting values for the parameter estimates. In Table 2.4 we give the maximum likelihood estimates returned by the algorithm for different arbitrary starting values. In each case different parameter estimates are returned, but the log-likelihood, l, is identical.

Numerical methods that find maximum likelihood estimates will stop as soon as one set of maximum likelihood estimates has been identified. As there are infinite maximum likelihood estimates for a parameter redundant model, different starting values will give different maximum likelihood estimates. Without trying multiple starting values, there is no indication that the model is parameter redundant from finding maximum likelihood estimates numerically.

TABLE 2.5
Parameter estimates from fitting the
three compartment model using least
squares.

Parameter	Set 1	Set 2
θ_{01}	0.47	0.11
θ_{21}	1.68	0.43
θ_{32}	0.54	2.15
R^2	0.0001054	0.0001054

2.2.2 Three Compartment Model

Consider again the compartmental model described in Section 1.3.2 and determining parameter estimates using least squares in Section 2.1.4. In Table 2.5
we show the two sets of parameter estimates that minimise the function R^2. A
numerical algorithm is used to minimise R^2, in the R code `compartmental.R`.
Different starting values will converge to either parameter set 1 or parameter
set 2.

  If it is not known that the model is locally identifiable, it may be assumed
that there is just one set of parameters that are optimal, however this may
not be true.

2.2.3 Mark-Recovery Model

Consider the mark-recovery model introduced in Section 1.2.4. This mark-
recovery model has four parameters ϕ_1, ϕ_a, λ_1 and λ_a, and is parameter
redundant. Here we use simulation to explore fitting this parameter redundant
model, and the results are summarised in Table 2.6. In this study, 100 data sets
were simulated from this mark-recovery model, with parameter values given in
the true values column of Table 2.6. Each simulated data set represents 5 years
and marking and 5 years of recovery with 100 animals marked each year. For
the numerical method used to find the maximum likelihood estimate we need
starting values for the parameters. How these starting values are chosen would
depend on the example. In some cases there is no knowledge of the parameters,
so the starting values may be chosen at random. In other cases the sample
data or other knowledge can be used to inform the starting values. We use two
different starting values for each data set, to represent these different cases.
The first is a different random number between 0 and 1 for each parameter
and data set and the second is the true value.

  If everything is working correctly when you simulate from a model and
then fit that same model, the parameter estimates should be approximately
the same as the true values. However for this parameter redundant model
when random starting values are used, the estimates are biased for the non-
identifiable parameters, ϕ_1, λ_1 and λ_a. That is, the average estimate is not

TABLE 2.6
Simulation study with 100 simulations from the parameter
redundant mark-recovery model with parameters ϕ_1, ϕ_a, λ_1 and
λ_a. The true values are the parameter values used to simulate the
data. 'Average' is the average parameter estimate. 'StDev.' is the
standard deviation of the parameter estimate. The third and
fourth columns correspond to using random starting values. The
fifth and sixth columns correspond to using true values as
starting values.

Parameter	True Value	Random		True	
		Average	StDev.	Average	StDev.
ϕ_1	0.4	0.56	0.34	0.40	0.018
ϕ_a	0.7	0.71	0.049	0.70	0.058
λ_1	0.1	0.35	0.31	0.10	0.013
λ_2	0.2	0.30	0.30	0.20	0.031

the same as the true parameter value used in simulating the data. This bias
can disappear if the true values are used rather than random values.

This example demonstrates that fitting a parameter redundant model can
result in biased, unreliable parameter estimates.

2.2.4   N-Mixture Model

The N-mixture model is a widely used model for estimating the abundance
of animals from a set of count data (Royle, 2004). The animals are counted
at $t = 1, \ldots, T$ sampling visits across $i = 1, \ldots, R$ sites, and are assumed to
come from a population that is closed to births, deaths, immigration and emigration. The observed counts at time t, n_{it}, are assumed to follow a binomial
distribution $\text{Bin}(N_i, p)$, where p is the probability that an animal is detected
and N_i is the unknown population size at site i. If N_i follows a Poisson distribution with mean λ, the resulting model is a mixture of Poisson distributions
with likelihood

$$L = \prod_{i=1}^{R} \left[\sum_{N_i = \kappa_i}^{\infty} \left\{ \prod_{t=1}^{T} \binom{N_i}{n_{it}} p^{n_{it}} (1-p)^{N_i - n_{it}} \right\} \frac{\lambda^{N_i} \exp(-\lambda)}{N_i!} \right],$$

where $\kappa_i = \max_t(n_{it})$ (Royle, 2004). The likelihood contains an infinite sum.
To find the maximum likelihood estimator Royle (2004) use a large fixed value
of K to replace ∞ in the likelihood; this algorithm is executed in the computer
packages unmarked (Fiske and Chandler, 2011) and PRESENCE (Hines, 2011).
In unmarked it is possible to specify the value of K, however in PRESENCE K
is fixed. Dennis et al. (2015) have developed an alternative formation of the

TABLE 2.7

Two data sets simulated from a
Poisson N-Mixture Model. The
two data sets are labelled (a)
and (b).

(a)		(b)	
$t = 1$	$t = 2$	$t = 1$	$t = 2$
2	2	2	2
4	1	3	0
2	5	4	1
0	3	3	2
1	1	4	2
0	3	5	3
2	2	2	0
2	6	3	3
3	1	1	1
1	0	5	4
2	1	1	1
2	3	2	3
1	4	3	2
4	3	0	1
7	2	3	6
3	2	2	2
3	3	3	0
2	3	2	3
0	2	3	2
1	3	1	2

same model, which has likelihood

$$L = e^{-(2\theta_1+\theta_0)} \prod_{i=1}^{R} \left\{ \frac{\theta_1^{n_{i1}+n_{i2}}}{n_{i1}!n_{i2}!} \sum_{u=0}^{\min(n_{i1},n_{i2})} \binom{n_{i1}}{u} \binom{n_{i2}}{u} u! \left(\frac{\theta_0}{\theta_1^2}\right)^u \right\},$$

where $\theta_1 = \lambda p(1 - p)$ and $\theta_0 = \lambda p^2$. This form of the likelihood avoids the
need for finding an infinite sum. Table 2.7 shows two (simulated) data sets
with $T = 2$ visits across $R = 20$ sites. PRESENCE, unmarked and code from
Dennis et al. (2015) are then used to estimate the parameters λ and p. For
unmarked three different values of K of 100, 500 and 1000 have been chosen.
The results are displayed in Table 2.8.

The different programs give consistently similar estimates of λ and p for
data set (b). However for data set (a) the estimates of λ are very different,
varying as K is changed. In fact, for data set (a), the maximum likelihood
estimates are $\hat{p} = 0$ and $\hat{\lambda} = \infty$. Numerical maximisation algorithms will only
tend to ∞, so the very large estimate of $\lambda = 5.35 \times 10^{10}$ and the almost
zero estimate of $p = 4.30 \times 10^{-11}$ from using the Dennis et al. (2015) code is

TABLE 2.8

Parameter estimates from fitting the N-Mixture model to the two data sets (a) and (b), using different computer packages.

Method	(a)		(b)	
	λ	p	λ	p
PRESENCE	45.26	0.05	7.98	0.29
unmarked $K = 100$	31.19	0.07	8.00	0.29
unmarked $K = 500$	68.72	0.03	8.00	0.29
unmarked $K = 1000$	862.64	0.003	8.00	0.29
Dennis et al. (2015)	5.35×10^{10}	4.30×10^{-11}	7.98	0.29

essentially correct. The results from the programs unmarked and PRESENCE, however, are incorrect.

Dennis et al. (2015) show how in some cases N-mixture models can produce unrealistically large estimates of abundance. For larger values of λ and small detection probabilities the model will behave as a parameter redundant model, and any parameter estimates will be unreliable. This is known as near-redundancy and is discussed in Section 4.4. Identifiabilty in the N-mixture model is discussed in Kery (2018). Dennis et al. (2015) provide a covariance diagnostic statistic for when this occurs. In the case of $T = 2$ sampling visits to R sites the covariance diagnostic is

$$\text{cov}^* = \frac{\sum_{i=1}^{R} n_{i1}n_{i2}}{R} - \left\{ \frac{1}{2}\left(\frac{\sum_{i=1}^{R} n_{i1}}{R} + \frac{\sum_{i=1}^{R} n_{i2}}{R} \right) \right\}^2.$$

If $\text{cov}^* \leq 0$ then the maximum likelihood estimate occurs at the boundary $p = 0$ and $\lambda = \infty$. For data set (b) $\text{cov}^* = 0.51$ so finite parameter estimates exist. However for data set (a) $\text{cov}^* = -0.29$. As this diagnostic statistic is negative the maximum likelihood estimate occurs at the boundary $p = 0$ and $\lambda = \infty$.

This example demonstrates that if a model is fitted when it is parameter redundant or near-redundant, parameter estimates can be biased and unreliable. If PRESENCE was used for data set (a) the wrong population size would have been concluded. If unmarked was used with increasing values of K, following Royle (2004)'s recommendation, the issue may have been identified. However knowledge of parameter redundancy and the covariance diagnostic will prevent incorrect parameter estimates.

2.3   Standard Errors in Parameter Redundant Models

The precision of a parameter estimate can be measured by finding the standard error or variance. The variance can be found using the inverse of the Fisher information matrix, $\mathbf{J}$, with

$$\mathbf{J} = -E \begin{bmatrix} \frac{\partial^2 l}{\partial\theta_1\partial\theta_1} & \frac{\partial^2 l}{\partial\theta_1\partial\theta_2} & \cdots & \frac{\partial^2 l}{\partial\theta_1\partial\theta_p} \\ \frac{\partial^2 l}{\partial\theta_2\partial\theta_1} & \frac{\partial^2 l}{\partial\theta_2\partial\theta_2} & \cdots & \frac{\partial^2 l}{\partial\theta_2\partial\theta_p} \\ \vdots & & & \vdots \\ \frac{\partial^2 l}{\partial\theta_p\partial\theta_1} & \frac{\partial^2 l}{\partial\theta_p\partial\theta_2} & \cdots & \frac{\partial^2 l}{\partial\theta_p\partial\theta_p} \end{bmatrix},$$

where $l = l(\boldsymbol{\theta}|\mathbf{y})$ is the log-is $\mathbf{J}^{-1}$ evaluated at the maximum likelihood estimate, $\hat{\boldsymbol{\theta}}$. The standard error for a parameter is the square root of the appropriate variance term on the diagonal of the matrix. Frequently, for ease of calculation, the Fisher information matrix is replaced by the observed information matrix or minus the Hessian matrix, $\mathbf{H}$, with

$$\mathbf{H} = \begin{bmatrix} \frac{\partial^2 l}{\partial\theta_1\partial\theta_1} & \frac{\partial^2 l}{\partial\theta_1\partial\theta_2} & \cdots & \frac{\partial^2 l}{\partial\theta_1\partial\theta_p} \\ \frac{\partial^2 l}{\partial\theta_2\partial\theta_1} & \frac{\partial^2 l}{\partial\theta_2\partial\theta_2} & \cdots & \frac{\partial^2 l}{\partial\theta_2\partial\theta_p} \\ \vdots & & & \vdots \\ \frac{\partial^2 l}{\partial\theta_p\partial\theta_1} & \frac{\partial^2 l}{\partial\theta_p\partial\theta_2} & \cdots & \frac{\partial^2 l}{\partial\theta_p\partial\theta_p} \end{bmatrix}.$$

The variance-covariance matrix is then approximately, $-\mathbf{H}^{-1}$ evaluated at the maximum likelihood estimate, $\hat{\boldsymbol{\theta}}$.

The Fisher information matrix is singular in a non-identifiable or parameter redundant model (Rothenberg, 1971). The Hessian matrix evaluated at the maximum likelihood estimate is also singular in a non-identifiable model (Viallefont et al., 1998). As it is not possible to invert a singular matrix, standard errors do not exist in non-identifiable models.

In many cases the exact Fisher information or Hessian matrix are difficult or impossible to calculate, so these are approximated using numerical methods. Such numerical approximations may result in a matrix that is not singular, in which case either large standard errors or imaginary standard errors are returned. (Imaginary standard errors occur when the Fisher information or Hessian matrix is not singular, so an inverse can be found, but the variance terms on the diagonal are negative.) This is demonstrated below for a binary latent class model and a mark-recovery model in Sections 2.3.1 and 2.3.2.

2.3.1 Binary Latent Class Model

Continuing the binary latent class model introduced in Section 2.2.1, Table 2.9 shows the standard errors for one set of maximum likelihood estimates. Approximate standard errors are then found by using finite differencing to approximate the Hessian matrix (see, for example, Morgan, 2009). To ensure the parameters remain in the range 0 to 1 a logistic transformation is used, with $\text{logit}(\theta) = \log\{\theta/(1-\theta)\}$. Maximum likelihood estimates and standard errors are found on the logistic scale and then transformed back to the standard scale. The delta method is used to find the approximate standard errors

TABLE 2.9
Parameter estimates given on the logit scale
($\text{logit}(\hat{\theta})$) and standard scale ($\hat{\theta}$), along with
standard errors (SE) for a binary latent
class model.

Parameter	$\text{logit}(\hat{\theta})$	SE	$\hat{\theta}$	SE
p	0.20	40.71	0.55	10.07
$\theta_{1,1}$	-0.21	24.85	0.49	6.15
$\theta_{1,0}$	-1.99	38.90	0.12	4.12
$\theta_{2,1}$	0.20	8.60	0.55	2.13
$\theta_{2,0}$	-0.66	21.29	0.34	4.78

on the standard scale, by multiplying the standard error on the logit scale by
$\theta(1 - \theta)$ (see, for example, Morgan, 2009). The parameter estimates and standard errors on the logistic and standard scale are given in Table 2.9. In this
case, the numerical approximation has found apparent standard errors. In fact
as this model is parameter redundant, standard errors should not exist. The
true Hessian matrix would be singular at the maximum likelihood estimate.
In a singular matrix there will be at least one eigenvalue that is zero. This
numerical approximation of the Hessian matrix is not singular but does have
standardised eigenvalues of 0.000003, 0.000004, 0.1189, 0.8661 and 1. (Standardised eigenvalues are the eigenvalues divided by the largest eigenvalue).
The fact that two eigenvalues are close to zero suggests that this numerical
approximation is close to being singular.

2.3.2   Mark-Recovery Model

Continuing the mark-recovery simulation from Section 2.2.3. Table 2.10 shows
the parameter estimates and numerical standard errors for three different simulations. (As with the previous example the maximum likelihood estimates
are found numerically on a logit scale and transferred back to the standard
scale.) For the first simulated data set the standard error is ∞, as the Hessian is singular. This is as expected for a parameter redundant model. In the
second simulated data set, the standard errors for ϕ_1, λ_1 and λ_a are imaginary, because the inverse of the Hessian had negative values on the diagonal,
so the variance is negative. This indicates that there is a problem with this
model, which could be parameter redundancy. In the third data set the stan-
dard errors are defined, but are large for ϕ_1, λ_1 and λ_a. This could occur
when a model is not parameter redundant and therefore it is not possible to
tell whether this model is parameter redundant from examining the standard
errors alone.

TABLE 2.10
Parameter estimates (Par. Est.) and standard errors (SE) from three simulated data sets. All three data sets are simulated from the parameter redundant mark-recovery model with parameters ϕ_1, ϕ_a, λ_1 and λ_a. Imag indicates that the standard error was imaginary.

Parameters	Data Set 1 Par. Est.	SE	Data Set 2 Par. Est.	SE	Data Set 3 Par. Est.	SE
ϕ_1	0.94	∞	0.75	Imag	0.11	0.15
ϕ_a	0.67	∞	0.71	0.062	0.73	0.066
λ_1	1.00	∞	0.27	Imag	0.078	0.016
λ_2	0.11	∞	0.10	Imag	0.62	0.84

2.4 Model Selection

For a particular data set more than one statistical model may be applicable. Model selection is a collection of methods that can be used to identify which is the best model(s) for a given data set. Most model selection methods are based on the various candidate models being identifiable. We discuss two model selection methods in Sections 2.4.1 and 2.4.2 below.

2.4.1 Information Criteria

One of the most straightforward methods of model selection, information criteria, involves the log-likelihood and a penalty for the number of parameters. The two most common information criteria are Akaike Information Criterion (AIC),

$$AIC = -2l + 2q$$

(Akaike, 1992) and Bayesian Information Criterion (BIC)

$$BIC = -2l + q\log(n),$$

(Schwarz, 1978), where l is the log-likelihood, q is the number of parameters and n is the sample size. The best model is the one with the smallest information criterion value, (see, for example, Burnham and Anderson, 2004 or Claeskens and Hjort, 2008).

This result holds if the model is identifiable. A parameter redundant model with q parameters can be reparameterised in terms of a smaller number of parameters where $q_e < q$. (Assume that the reparameterisation is such that the model with q_e parameters is not parameter redundant.) The likelihood for both models is identical, but the penalty will be different. Therefore, in AIC or BIC, q should be replaced by q_e, which is the number of estimable parameters in a model.

As part of the process of detecting parameter redundancy most methods will also give the number of estimable parameters, which is explained further in Chapters 3 to 5.

2.4.2   The Score Test

A score test is a model selection method for two nested models (see, for example, Morgan, 2009). Suppose the simpler model has maximum likelihood estimate $\hat{\boldsymbol{\theta}}_0$ and the more complex model has maximum likelihood estimate $\hat{\boldsymbol{\theta}}$. The test statistic is
$$S = \mathbf{U}(\hat{\boldsymbol{\theta}}_0)^T \mathbf{J}^{-1}(\hat{\boldsymbol{\theta}}_0) \mathbf{U}(\hat{\boldsymbol{\theta}}_0),$$
where $\mathbf{U}(\hat{\boldsymbol{\theta}}_0)$ is the score vector of the more complex model evaluated at $\hat{\boldsymbol{\theta}}_0$, that is
$$\mathbf{U}(\hat{\boldsymbol{\theta}}_0) = \left[\begin{array}{c} \frac{\partial l}{\partial \theta_1} \\ \frac{\partial l}{\partial \theta_2} \\ \vdots \\ \frac{\partial l}{\partial \theta_p} \end{array} \right]\Bigg|_{\boldsymbol{\theta}=\hat{\boldsymbol{\theta}}_0},$$
and where $\mathbf{J}(\hat{\boldsymbol{\theta}}_0)$ is the Fisher information matrix of the more complex model evaluated at $\hat{\boldsymbol{\theta}}_0$, that is
$$\mathbf{J}(\hat{\boldsymbol{\theta}}_0) = -E \left[\begin{array}{cccc} \frac{\partial^2 l}{\partial \theta_1 \partial \theta_1} & \frac{\partial^2 l}{\partial \theta_1 \partial \theta_2} & \cdots & \frac{\partial^2 l}{\partial \theta_1 \partial \theta_p} \\ \frac{\partial^2 l}{\partial \theta_2 \partial \theta_1} & \frac{\partial^2 l}{\partial \theta_2 \partial \theta_2} & \cdots & \frac{\partial^2 l}{\partial \theta_2 \partial \theta_p} \\ \vdots & & & \vdots \\ \frac{\partial^2 l}{\partial \theta_p \partial \theta_1} & \frac{\partial^2 l}{\partial \theta_p \partial \theta_2} & \cdots & \frac{\partial^2 l}{\partial \theta_p \partial \theta_p} \end{array} \right]\Bigg|_{\boldsymbol{\theta}=\hat{\boldsymbol{\theta}}_0}.$$

The more complex model is preferred if the score test statistic, S, is larger than $\chi^2_{d;\alpha}$, the 100α percentage point from a chi-square distribution with d degrees of freedom, where d is the difference in the number of parameters between the simpler and more complex model.

The score test involves finding the inverse of the Fisher information matrix. In a non-identifiable or parameter redundant model the Fisher information matrix and Hessian matrix are both singular, which means that the inverse of the Fisher information matrix is undefined, so the test statistic cannot be evaluated.

To use a score test for a parameter redundant model, the model needs to first be reparameterised with a smaller number of parameters to give a model that is no longer parameter redundant. This is demonstrated in the example in Section 2.4.2.1.

2.4.2.1   Mark-Recovery Model

Score tests in mark-recovery models are considered in Catchpole and Morgan (1996), although the models compared are not parameter redundant. Using the Lapwing data given in Section 1.2.4 we demonstrate here how to use a score test to test between a mark-recovery model with constant survival and recovery parameters, ϕ and λ, and a parameter redundant mark-recovery model, with two age classes for survival and recovery. For the parameter redundant model the survival parameters are ϕ_1 and ϕ_a and the recovery parameters are λ_1 and λ_a. This is the same parameter redundant model also examined in Sections 2.1.2, 2.3.2 and 2.2.3.

The probabilities of recapture, $P_{i,j}$ are given in Section 1.2.4 for the more complex model (with parameters ϕ_1, ϕ_a, λ_1 and λ_a). The simpler model is a nested model with $\phi = \phi_1 = \phi_a$ and $\lambda = \lambda_1 = \lambda_a$. The likelihood is

$$L \propto \left(\prod_{i=1}^{3} \prod_{j=i}^{3} P_{i,j}^{N_{i,j}} \right) \left\{ \prod_{i=1}^{3} \left(1 - \sum_{j=i}^{3} P_{i,j} \right)^{F_i - \sum_{j=i}^{3} N_{i,j}} \right\},$$

where $N_{i,j}$ and F_i are the data given in Section 1.2.4. Maple code for calculating the score test results below is given scoretesteg.mw.

Firstly, the simple model with parameters ϕ and λ is fitted to the data. This model is not parameter redundant and has maximum likelihood estimates $\hat{\phi} = 0.2805$ and $\hat{\lambda} = 0.0161$. Ordinarily we would next find the score test statistic by evaluating the Fisher information matrix and gradient of the more complex model at $\phi_1 = \phi_a = 0.2805$ and $\lambda_1 = \lambda_a = 0.0161$. However this more complex model is parameter redundant, so the Fisher information matrix will be singular. Instead we need to first reparameterise this model to create a model that is not parameter redundant. The model with parameters ϕ_a, $\beta_1 = (1 - \phi_1)\lambda_1$ and $\beta_2 = \phi_1\lambda_a$ is not parameter redundant. Therefore we instead evaluate the Fisher information matrix and gradient of the reparameterised complex model at $\phi_a = 0.2805$, $\beta_1 = (1 - 0.2805) \times 0.0161 = 0.0116$ and $\beta_2 = 0.2805 \times 0.0161 = 0.0045$, giving

$$\mathbf{J}(\boldsymbol{\theta}_0) = \begin{bmatrix} 20.28 & -8.86 & -1935.41 \\ -8.86 & 308549.69 & 2012.21 \\ -1935.41 & 2012.21 & 439758.80 \end{bmatrix} \text{ and } \mathbf{U}(\boldsymbol{\theta}_0) = \begin{bmatrix} -0.23 \\ -3.97 \\ 10.18 \end{bmatrix}.$$

This results in the score test statistic $S = 0.003$. There is one parameter difference between the two parameters in the simpler model and the three parameters in the reparameterised version of the more complex model. From the chi-squared distribution with one degree of freedom the p-value is 0.96. This suggests that the more complex parameter redundant model is not suitable for this data set.

This method requires that there is at least one degree of freedom difference between the reparameterised complex model and simpler model (that is, a difference of a least one between the numbers of parameters). Therefore it will not always be possible to use this method, e.g. if there was only one degree of freedom difference between the original parameter redundant model and the simpler model. The method also requires a suitable reparameterisation; how to find such a parameterisation is discussed in Sections 3.2.5.

2.5   Conclusion

In this chapter we have demonstrated some of the problems of fitting parameter redundant models.

Parameter estimates are unreliable and standard errors are undefined in parameter redundant models, and model selection methods need to be adapted. It is therefore recommended that parameter redundancy in a model should be explored as a matter of course as part of the model fitting process. Ideally parameter redundancy caused by the model structure or model observations should be explored before model fitting. The different methods of detecting parameter redundancy are explored in Chapters 3 to 5.

3

Parameter Redundancy and Identifiability
Definitions and Theory

This chapter examines the theory and theoretical methods used in detecting parameter redundancy and identifiability. Although this chapter has several illustrative examples it also includes formal theorems, definitions and mathematical theory behind the methods. In this chapter, we do not discuss numerical methods for detecting parameter redundancy. Instead, Chapter 4 gives a practical approach to detecting parameter redundancy, including alternative numerical methods.

Section 3.1 gives formal definitions of parameter redundancy and identifiability as well as definitions used in this area. Section 3.2 provides general symbolic methods for detecting parameter redundancy and non-identifiability. Section 3.3 demonstrates how you can distinguish between local and global identifiability. Section 3.4 discusses proving identifiability results directly and compares to the general methods discussed in Section 3.2 and 3.3. Section 3.5 discusses near-redundancy, sloppiness and practical identifiability.

3.1 Formal Definitions of Parameter Redundancy and Identifiability

This section provides formal definitions of local and global identifiability and parameter redundancy as well as definitions used in detecting parameter redundancy.

3.1.1 Identifiability

Any model will have some function that defines that model, $M(\boldsymbol{\theta})$. For example, in statistics this could be a likelihood, or in compartmental modelling this could be a set of differential equations. (Here we are using model to mean a single mathematical model, rather than a statistical model, which means a family of probability distributions). The model $M(\boldsymbol{\theta})$ depends on one or more unknown parameters $\boldsymbol{\theta} \in \Omega$, where Ω is a $\dim(\boldsymbol{\theta})$-dimensional vector space. Here $\dim(\boldsymbol{\theta})$ denotes the dimension or number of terms in a general vector $\boldsymbol{\theta}$.

A model is non-identifiable if different sets of parameter values result in the same model (Silvey, 1970). That is, two different sets of parameter values, $\boldsymbol{\theta}_1$ and $\boldsymbol{\theta}_2$ give the same output, $M(\boldsymbol{\theta})$. A model is identifiable if two different sets of parameter values do not result in the same model. Identifiability can be split into two categories: global identifiability and local identifiability. The former occurs if a model is identifiable for the whole parameter space; the latter occurs if a model is identifiable for a region of the parameter space. A formal definition of identifiability is given in Definition 1 below.

Definition 1 *A model is globally identifiable if $M(\boldsymbol{\theta}_1) = M(\boldsymbol{\theta}_2)$ implies that $\boldsymbol{\theta}_1 = \boldsymbol{\theta}_2$. A model is locally identifiable if there exists an open neighbourhood in the parameter space of $\boldsymbol{\theta}$ such that this is true. Otherwise a model is non-identifiable (see, for example, Rothenberg, 1971, Walter, 1982, Bekker et al., 1994, Cole et al., 2010).*

3.1.2   Parameter Redundancy

An obvious cause of non-identifiability is having too many parameters in a model. When this happens, the model can be reparameterised in terms of a smaller set of parameters, which is termed parameter redundancy (Catchpole and Morgan, 1997). More formally, a definition of parameter redundancy is given in Definition 2 below.

Definition 2 *A model is parameter redundant if we can write $M(\boldsymbol{\theta})$ as a function just of $\boldsymbol{\beta}$, where $\boldsymbol{\beta} = g(\boldsymbol{\theta})$ and $dim(\boldsymbol{\beta}) < dim(\boldsymbol{\theta})$, where dim is the dimension or length of the vector (Catchpole and Morgan, 1997, Cole et al., 2010).*

In Section 3.2 below we explain how parameter redundancy can be detected by examining the rank of a particular matrix. A model that is not parameter redundant will have the maximum rank possible for that matrix. Therefore a model that is not parameter redundant is termed full rank.

Definition 3 *A model that is not parameter redundant is termed full rank (Catchpole and Morgan, 1997, Cole et al., 2010).*

A model that is full rank is not necessarily full rank for every possible parameter value. Catchpole and Morgan (1997) use 'essentially full rank' and 'conditionally full rank' to distinguish between cases where a model is full rank for all possible parameter values or when there are points in the parameter space of a full rank model that are parameter redundant.

Definition 4 *An essentially full rank model is full rank for all $\boldsymbol{\theta}$. A conditionally full rank model is full rank for some but not all $\boldsymbol{\theta}$. (Catchpole and Morgan, 1997, Cole et al., 2010)*

A model that is essentially full rank will be at least locally identifiable (Catchpole and Morgan, 1997). This refers to the fact that a model could be only locally identifiable or it could be globally identifiable. If we restrict the parameter space of a conditionally full rank to exclude the values of θ that are not full rank, the resulting model θ is at least locally identifiable.

The inherent structure of a model can cause parameter redundancy. This is known as intrinsic parameter redundancy and is the general case of parameter redundancy. When we say that a model is parameter redundant, we are actually referring to intrinsic parameter redundancy of a model with a specific parameterisation. In such a model it would not matter how much data are collected, it would never be possible to estimate all the parameters in the model without reparameterising the model, or adding a constraints to specific parameters.

Parameter redundancy can also be caused by a specific data set. This is known as extrinsic parameter redundancy. Extrinsic parameter redundancy occurs when a model is not intrinsically parameter redundant, but for a specific data set we can reparameterise in terms of a smaller set of parameters (Gimenez et al., 2004). The definitions of intrinsic and extrinsic parameter redundancy are given by Definition 5 below.

Definition 5 *Intrinsic parameter redundancy refers to a parameter redundant model that can be reparameterised in terms of a smaller set of parameters, without consideration of a specific data set. Extrinsic parameter redundancy refers to a model that is parameter redundant when the likelihood can be reparameterised in terms of a smaller number of parameters for a specific data set, but is not parameter redundant in general (Gimenez et al., 2004).*

3.1.3 Near-Redundancy, Sloppiness and Practical Identifiability

Whilst a model may be identifiable or full rank, in practice there may still be issues with determining or estimating parameters. This could be a problem of near-redundancy, sloppiness or practical non-identifiability. Definitions for these three terms are given below.

Definition 6 *A near-redundant model is one that is formally full rank, but might behave as parameter redundant model, because the model is very similar to a model that is parameter redundant for a particular data set. (Catchpole et al., 2001, Cole et al., 2010)*

Definition 7 *A model is sloppy if the condition number of the Hessian or Fisher information matrix is large, where the condition number is the largest eigenvalue of a matrix divided by the smallest eigenvalue (Chis et al., 2016, Dufresne et al., 2018).*

Definition 8 *A parameter is practically non-identifiable if the log-likelihood has a unique maximum, but the parameter's likelihood-based confidence region tends to infinity in either or both directions (Raue et al., 2009).*

3.1.4 Exhaustive Summaries

Identifiability and parameter redundancy can be determined directly from the model $M(\boldsymbol{\theta})$. (This is discussed further in Section 3.4.) However, it is often more convenient to use a vector of parameter combinations that represents the model $M(\boldsymbol{\theta})$. This is known as an exhaustive summary, and it consists of a vector of particular combinations of parameters that fully specify the model. The concept of an exhaustive summary was first used in compartmental modelling (Walter and Lecourtier 1982, Walter, 1982). This is generalised in Cole et al. (2010) to any model. A vector $\boldsymbol{\kappa}(\boldsymbol{\theta})$ is an exhaustive summary if knowledge of $\boldsymbol{\kappa}(\boldsymbol{\theta})$ uniquely determines the model. That is, there is a one-to-one transformation between the specification of the model and the exhaustive summary. Ran and Hu (2014) provide a mathematical formation of an exhaustive summary. A formal definition of an exhaustive summary is given in Definition 9 below.

Definition 9 *A parameter vector $\boldsymbol{\kappa}(\boldsymbol{\theta})$ is an exhaustive summary if knowledge of $\boldsymbol{\kappa}(\boldsymbol{\theta})$ uniquely determines $M(\boldsymbol{\theta})$, so that $\boldsymbol{\kappa}(\boldsymbol{\theta}_1) = \boldsymbol{\kappa}(\boldsymbol{\theta}_2) \Leftrightarrow M(\boldsymbol{\theta}_1) = M(\boldsymbol{\theta}_2) \,\forall\, \boldsymbol{\theta}_1, \boldsymbol{\theta}_2 \in \boldsymbol{\Theta}$ (Cole et al., 2010 and Ran and Hu, 2014).*

Essentially an exhaustive summary is a vector of parameter combinations that can be used to infer results about parameter redundancy or identifiability. For any model there will be many different options for exhaustive summaries. A general exhaustive summary, for statistical models, consists of components of the log-likelihood. For example, if the log-likelihood consists of n independent data points and can be written in the form $l = \sum_{i=1}^{n} l_i$ then an exhaustive summary is $\boldsymbol{\kappa}^T = [l_1, l_2, \ldots, l_n]$. However it does not necessarily follow that this is the simplest form for the likelihood, or the ideal exhaustive summary to use. In general, any model $M(\boldsymbol{\theta})$ will have many options for suitable exhaustive summaries. Illustrative examples of exhaustive summaries are given in Sections 3.1.4.1 to 3.1.4.4.

To investigate extrinsic parameter redundancy we need to include the data in the exhaustive summary; for example, the log-likelihood exhaustive summary described above includes the data. We would need to assume perfect data when investigating intrinsic identifiability using an exhaustive summary including data. The dependence on the data could be dropped, and simpler exhaustive summaries derived, as demonstrated in Section 3.1.4.4.

3.1.4.1 Basic Occupancy Model

The basic occupancy model discussed in Section 1.2.1 consists of two options: either a species is detected at a site, or it is not detected. The probabilities of these two events form the exhaustive summary,

$$\kappa_{occ,1} = \left[\begin{array}{c} \psi p \\ 1 - \psi p \end{array} \right]. \tag{3.1}$$

An alternative representation of the model is the log-likelihood given in Section 2.1, which is

$$l = c + n_1 \log(\psi p) + n_2 \log(1 - \psi p),$$

for some constant c, where n_1 and n_2 are the number of sites observed and unoccupied respectively. An alternative exhaustive summary consisting of the log-likelihood terms is

$$\kappa_{occ,2} = \left[\begin{array}{c} n_1 \log(\psi p) \\ n_2 \log(1 - \psi p) \end{array} \right]. \tag{3.2}$$

3.1.4.2 Binary Latent Class Model

An exhaustive summary for the binary latent class model, introduced in Section 2.3.1, consists of the probabilities given in Table 2.3. These are the probabilities that the binary response variables, Y_1 and Y_2, are equal to each combination of 0 or 1. The exhaustive summary is therefore

$$\kappa_{bin,1} = \left[\begin{array}{c} \kappa_{bin,1,1} \\ \kappa_{bin,1,2} \\ \kappa_{bin,1,3} \\ \kappa_{bin,1,4} \end{array} \right]$$

$$= \left[\begin{array}{c} p\,(1 - \theta_{11})\,(1 - \theta_{21}) + (1 - p)\,(1 - \theta_{10})\,(1 - \theta_{20}) \\ p\,(1 - \theta_{11})\,\theta_{21} + (1 - p)\,(1 - \theta_{10})\,\theta_{20} \\ p\theta_{11}\,(1 - \theta_{21}) + (1 - p)\,\theta_{10}\,(1 - \theta_{20}) \\ p\theta_{11}\theta_{21} + (1 - p)\,\theta_{10}\theta_{20} \end{array} \right],$$

where $\theta_{jk} = Pr(Y_j = 1 | Z = k)$ and $p = Pr(Z = 1)$, for latent variable Z. The exhaustive summary can be simplified, as the first term $\kappa_{bin,1,1}$ can be rewritten as $\kappa_{b,1,1} = 1 - \kappa_{b,1,2} - \kappa_{b,1,3} - \kappa_{b,1,4}$. If a term can be rewritten in terms of other exhaustive summary terms, it can be removed to create a simpler exhaustive summary. This leads to the exhaustive summary,

$$\kappa_{bin,2} = \left[\begin{array}{c} p\,(1 - \theta_{11})\,\theta_{21} + (1 - p)\,(1 - \theta_{10})\,\theta_{20} \\ p\theta_{11}\,(1 - \theta_{21}) + (1 - p)\,\theta_{10}\,(1 - \theta_{20}) \\ p\theta_{11}\theta_{21} + (1 - p)\,\theta_{10}\theta_{20} \end{array} \right]. \tag{3.3}$$

Note that we could have chosen to remove any of terms of $\kappa_{bin,1}$ for the same reason.

3.1.4.3   Three Compartment Model

In linear compartmental models the transfer function can be used to create an exhaustive summary. Suppose the linear compartment model has the output function

$$\mathbf{y}(t, \boldsymbol{\theta}) = \mathbf{C}\mathbf{x}(t, \boldsymbol{\theta})$$

with

$$\frac{d}{dt}\mathbf{x}(t, \boldsymbol{\theta}) = \mathbf{A}(\boldsymbol{\theta})\mathbf{x}(t, \boldsymbol{\theta}) + \mathbf{B}(\boldsymbol{\theta})\mathbf{u}(t),$$

where $\mathbf{x}$ is the state-variable function, $\boldsymbol{\theta}$ is a vector of unknown parameters, $\mathbf{u}$ is the input function and t is the time recorded on a continuous time scale. $\mathbf{A}$, $\mathbf{B}$, $\mathbf{C}$ are the compartmental, input and the output matrices respectively. By taking Laplace transforms of $\mathbf{y}$, we obtain the transform function, $Q(s) = \mathbf{C}(s\mathbf{I} - A)^{-1}\mathbf{B}$, which is the ratio of the Laplace transformations of $\mathbf{y}$ and $\mathbf{u}$ (Bellman and Astrom, 1970). The transfer function is described in more detail in Section 7.2.1. The transfer function will be of the form,

$$Q(s) = \frac{a_0 + a_1 s + a_2 s^2 + \dots}{b_0 + b_1 s + b_2 s^2 + b_3 s^3 \dots}.$$

The coefficients of polynomials in s in the numerator and denominator, $a_0, a_1, a_2, \dots$ and $b_0, b_1, b_2, b_3, \dots$ form an exhaustive summary. Parameter independent values of a_i and b_i, (e.g. constant and zero values), can be ignored in the exhaustive summary, as they contain no information on parameters, therefore will contain no information on the identifiability of the model.

As an example, consider the three compartment model from Godfrey and Chapman (1990) given in Section 1.3.2. The transfer function is

$$Q(s) = \frac{\theta_{21}\theta_{32}}{s^3 + s^2(\theta_{01} + \theta_{21} + \theta_{32}) + s\theta_{32}(\theta_{01} + \theta_{21})}.$$

An exhaustive summary is then

$$\boldsymbol{\kappa}_{comp} = \begin{bmatrix} \theta_{21}\theta_{32} \\ \theta_{01} + \theta_{21} + \theta_{32} \\ \theta_{32}(\theta_{01} + \theta_{21}) \end{bmatrix}. \tag{3.4}$$

The transfer function approach is discussed further in Section 7.2.1. Alternative exhaustive summaries for compartmental models are also discussed in Section 7.2.

3.1.4.4   Mark-Recovery Model

The Mark-Recovery model, introduced in Section 1.2.4, has log-likelihood

$$l = c + \sum_{i=1}^{3}\sum_{j=i}^{3} N_{i,j}\log(P_{i,j}) + \sum_{i=1}^{3}\left(F_i - \sum_{j=i}^{3} N_{i,j}\right)\log\left(1 - \sum_{j=i}^{3} P_{i,j}\right)$$

where

$$\mathbf{P} = \begin{bmatrix} (1-\phi_1)\lambda_1 & \phi_1(1-\phi_a)\lambda_a & \phi_1\phi_a(1-\phi_a)\lambda_a \\ 0 & (1-\phi_1)\lambda_1 & \phi_1(1-\phi_a)\lambda_a \\ 0 & 0 & (1-\phi_1)\lambda_1 \end{bmatrix},$$

for parameters $\boldsymbol{\theta} = [\phi_1, \phi_a, \lambda_1, \lambda_a]$, where $N_{i,j}$ and F_i represent the data. $N_{i,j}$ is the numbers of animals recovered on given occasions, and F_i is the number marked.

One possible exhaustive summary can be formed from the terms that make up the log-likelihood with

$$\kappa_1 = \begin{bmatrix} N_{1,1}\log\{(1-\phi_1)\lambda_1\} \\ N_{1,2}\log\{\phi_1(1-\phi_a)\lambda_a\} \\ N_{1,3}\log\{\phi_1\phi_a(1-\phi_a)\lambda_a\} \\ N_{2,2}\log\{(1-\phi_1)\lambda_1\} \\ N_{2,3}\log\{\phi_1(1-\phi_a)\lambda_a\} \\ N_{3,3}\log\{(1-\phi_1)\lambda_1\} \\ \left(F_1 - \sum_{j=1}^{3} N_{1,j}\right)\log\left(1 - \sum_{j=1}^{3} P_{1,j}\right) \\ \left(F_2 - \sum_{j=2}^{3} N_{2,j}\right)\log\left(1 - \sum_{j=2}^{3} P_{2,j}\right) \\ (F_3 - N_{3,3})\log\{1 - (1-\phi_1)\lambda_1\} \end{bmatrix}.$$

In this case perfect data would consist of $N_{ij} \neq 0$, $\forall i \geq j$. This exhaustive summary can be used to examine both intrinsic and extrinsic parameter redundancy. Parameter redundancy results may change if some $N_{ij} = 0$ (see Section 3.2.4, and Cole et al. 2012).

Catchpole and Morgan (1997) show that as long as there is perfect data, that is, as long as $N_{ij} \neq 0$, $\forall i \geq j$, we can simplify the exhaustive summary to just the non-zero entries of $\mathbf{P}$. Repeated terms can be deleted. This gives the exhaustive summary

$$\kappa_2 = \begin{bmatrix} (1-\phi_1)\lambda_1 \\ \phi_1(1-\phi_a)\lambda_a \\ \phi_1\phi_a(1-\phi_a)\lambda_a \end{bmatrix}.$$

Alternatively we can form an exhaustive summary from the natural logarithms of these entries, with

$$\kappa_3 = \begin{bmatrix} \log\{(1-\phi_1)\lambda_1\} \\ \log\{\phi_1(1-\phi_a)\lambda_a\} \\ \log\{\phi_1\phi_a(1-\phi_a)\lambda_a\} \end{bmatrix}.$$

The latter exhaustive summary can be useful as it simplifies calculations, which involve differentiating a product of two terms.

3.1.5   The Use of Exhaustive Summaries

Exhaustive summaries are useful as they specify the structure of a model and offer a convenient summary of the model which can be used to check for parameter redundancy or identifiability. This is explained formally in Theorems 1 and 2 below. Finding an exhaustive summary is the first step in investigating parameter redundancy; see Section 3.2 below.

Theorem 1 *A model is globally (locally) identifiable if (there exists an open neighbourhood of θ such that) $\kappa(\theta) = \kappa(\theta') \Rightarrow \theta = \theta'$ (Cole et al., 2010).*

Theorem 2 *A model is parameter redundant if we can reparameterise $\kappa(\theta)$ as $\kappa(\beta)$, where $\beta = f(\theta)$ and $dim(\beta) < dim(\theta)$ (Cole et al., 2010).*

3.2 Detecting Parameter Redundancy or Non-Identifiability using Symbolic Differentiation Method

The symbolic differentiation method is a general purpose method that can be used to determine whether or not a model is parameter redundant or non-identifiable. It can be used for any model where an exhaustive summary can be created. We refer to this method as the symbolic method, due to the symbolic algebra involved.

The method has been used in different forms for different classes of models for several decades. This includes work by Koopmans et al. (1950), Wald (1950), Fisher (1959) and Fisher (1961) on non-linear simultaneous equations, generalised by Rothenberg (1971). Goodman (1974) used this method for latent-class models and Shapiro (1986) for non-linear regression models. Within the area of compartmental modelling and dynamic systems, use of this method include Thowsen (1978), Pohjanpalo (1982), Delforge (1989) and Chappell and Gunn (1998). In ecological statistics, the first use of the method is Catchpole and Morgan (1997). More recently, Cole et al. (2010) provided a general framework for detecting parameter redundancy in any model which has an explicit exhaustive summary, which we describe below. A similar method is presented in Rothenberg (1971) using reduced-form parameters.

The basis of determining the parametric structure is to form an appropriate derivative matrix by differentiating each of the exhaustive summary terms, with respect to each of the parameters,

$$
\mathbf{D} = \frac{\partial \kappa}{\partial \theta} = \begin{bmatrix} \dfrac{\partial \kappa_1}{\partial \theta_1} & \dfrac{\partial \kappa_2}{\partial \theta_1} & \cdots & \dfrac{\partial \kappa_n}{\partial \theta_1} \\ \dfrac{\partial \kappa_1}{\partial \theta_2} & \dfrac{\partial \kappa_2}{\partial \theta_2} & \cdots & \dfrac{\partial \kappa_n}{\partial \theta_2} \\ \vdots & & & \vdots \\ \dfrac{\partial \kappa_1}{\partial \theta_q} & \dfrac{\partial \kappa_2}{\partial \theta_q} & \cdots & \dfrac{\partial \kappa_n}{\partial \theta_q} \end{bmatrix},
$$

where κ_j is the jth element of exhaustive summary κ, which has length n, and θ_i is the ith of q parameters, θ. The derivative matrix can be used to examine parameter redundancy or identifiability by determining the rank of the derivative matrix, r. If $r < q$, the model is parameter redundant and non-identifiable. The deficiency of a model, which is described in Definition 10, can be used to classify parameter redundancy and identifiability results as laid out in Theorem 3.

Definition 10 *The deficiency of a model is the difference between the number of parameters, q, and the rank of a model's derivative matrix, r (Catchpole et al., 1998a).*

Theorem 3 *If the rank of $\mathbf{D}$ is equal to r, the deficiency of a model with q parameters is $d = q - r$. If $d > 0$ then the model is parameter redundant and the model will be non-identifiable. If $d = 0$ then the model is full rank, and will be at least locally identifiable for part of the parameter space (Theorem 2a of Cole et al., 2010).*

Note here we use a derivative matrix with each column corresponding to individual exhaustive summary terms and each row corresponding to individual parameters. However, Theorem 3 and subsequent theorems are all applicable if the transpose of the derivative matrix is used. This is known as the Jacobian matrix (see, for example, Evans and Chappell, 2000).

Theorem 3 involves differentiating the exhaustive summary and calculating a rank of a symbolic matrix. This is only feasible by hand for the simplest models. Generally a symbolic algebra package, such as Maple, is used to find the derivative matrix and its rank. A guide of how to use Maple to execute this method is given in Appendix A.1.

Alternatively in statistical models, with log-likelihood $l = l(\boldsymbol{\theta}|\mathbf{y})$, it is possible check for parameter redundancy or non-identifiability by calculating the rank of the Fisher Information matrix,

$$
\mathbf{J} = -E \begin{bmatrix}
\frac{\partial^2 l}{\partial\theta_1\partial\theta_1} & \frac{\partial^2 l}{\partial\theta_1\partial\theta_2} & \cdots & \frac{\partial^2 l}{\partial\theta_1\partial\theta_q} \\
\frac{\partial^2 l}{\partial\theta_2\partial\theta_1} & \frac{\partial^2 l}{\partial\theta_2\partial\theta_2} & \cdots & \frac{\partial^2 l}{\partial\theta_2\partial\theta_q} \\
\vdots & & & \vdots \\
\frac{\partial^2 l}{\partial\theta_q\partial\theta_1} & \frac{\partial^2 l}{\partial\theta_q\partial\theta_2} & \cdots & \frac{\partial^2 l}{\partial\theta_q\partial\theta_q}
\end{bmatrix},
$$

or the Hessian matrix,

$$
\mathbf{H} = \begin{bmatrix}
\frac{\partial^2 l}{\partial\theta_1\partial\theta_1} & \frac{\partial^2 l}{\partial\theta_1\partial\theta_2} & \cdots & \frac{\partial^2 l}{\partial\theta_1\partial\theta_q} \\
\frac{\partial^2 l}{\partial\theta_2\partial\theta_1} & \frac{\partial^2 l}{\partial\theta_2\partial\theta_2} & \cdots & \frac{\partial^2 l}{\partial\theta_2\partial\theta_q} \\
\vdots & & & \vdots \\
\frac{\partial^2 l}{\partial\theta_q\partial\theta_1} & \frac{\partial^2 l}{\partial\theta_q\partial\theta_2} & \cdots & \frac{\partial^2 l}{\partial\theta_q\partial\theta_q}
\end{bmatrix},
$$

evaluated at a maximum likelihood estimate, which we write as $\mathbf{H}(\hat{\boldsymbol{\theta}})$.

Similar to Theorem 3, Theorem 4 explains how to detect parameter redundancy and non-identifiability using the Fisher Information, $\mathbf{J}$, or the Hessian matrix evaluated at a maximum likelihood estimate, $\mathbf{H}(\hat{\boldsymbol{\theta}})$.

Theorem 4 *a Suppose the rank of $\mathbf{J}$ is equal to r_J, and the deficiency is $d_J = q - r_J$, for a model with q parameters. If $d_J > 0$ then the model is parameter redundant and the model will be non-identifiable. If $d_J = 0$ then the model is full rank, and will be at least locally identifiable for part of the parameter space. (Theorems 1 and 2 of Rothenberg, 1971).*
b Suppose the rank of $\mathbf{H}(\hat{\boldsymbol{\theta}})$ is equal to r_H, and the deficiency is $d_H = q - r_H$, for a model with q parameters. If $d_H > 0$ then the model is parameter redundant and the model will be non-identifiable. If $d_H = 0$ then the model is full rank, and will be at least locally identifiable for part of the parameter space. (Viallefont et al., 1998, Little et al., 2009).

The ranks of the matrices $\mathbf{D}$, $\mathbf{J}$ and $\mathbf{H}(\hat{\boldsymbol{\theta}})$ should all be identical, that is $r = r_J = r_H$. This has been proved to be the case for exponential family models in Bekker et al. (1994) and Catchpole and Morgan (1997). (The exponential family is a family of probability distributions. See McCullagh and Nelder, 1989 and Section 3.3.1.)

As discussed in Section 2.3, to find the standard errors we need to invert the Jacobian matrix (or the Hessian matrix). However, to be able to invert a matrix it needs to be full rank. Therefore a consequence of Theorem 4 is that standard errors do not exist for parameter redundant models, as discussed in Section 2.3.

We illustrate the use of this symbolic method for detecting parameter redundancy using three examples in Sections 3.2.1 to 3.2.3. In Section 3.2.4 we show how to check for intrinsic parameter redundancy. Then Section 3.2.5 shows how to find a reparameterisation of a parameter redundant model, that is no longer parameter redundant. In Section 3.2.6 we demonstrate how to generalise results to any dimension of the model. Finally in Section 3.2.7 we explain how it is possible to distinguish between essentially and conditionally full rank models.

3.2.1   Basic Occupancy Model

For the basic occupancy model, using the exhaustive summary $\kappa_{occ,1}$, given by equation (3.1) in Section 3.1.4.1, and the parameters $\boldsymbol{\theta} = [\psi, p]$, the derivative matrix is

$$\mathbf{D}_{occ,1} = \begin{bmatrix} \dfrac{\partial}{\partial \psi}\psi p & \dfrac{\partial}{\partial \psi}(1 - \psi p) \\ \dfrac{\partial}{\partial p}\psi p & \dfrac{\partial}{\partial p}(1 - \psi p) \end{bmatrix} = \begin{bmatrix} p & -p \\ \psi & -\psi \end{bmatrix}. \tag{3.5}$$

This derivative matrix has rank 1, but there are 2 parameters, therefore by Theorem 3 this model is parameter redundant with deficiency 1.

Maple code for this example is given in Appendix A.1.1 and Maple file `basicoccupancy.mw`. This Maple code also shows how using the exhaustive summary $\kappa_{occ,2}$, given by equation (3.2), results in the same rank. Regardless of the exhaustive summary used, the rank and deficiency will always be the same.

If there were n sites observed the Fisher information matrix is

$$\mathbf{J} = \begin{bmatrix} \frac{np}{\psi(1-p\psi)} & \frac{n}{1-p\psi} \\ \frac{n}{1-p\psi} & \frac{n\psi}{p(1-p\psi)} \end{bmatrix}.$$

(Calculation of the Fisher information matrix is shown in the Maple code.) This matrix has rank 1, but there are 2 parameters, therefore by Theorem 4a we have again shown the model is parameter redundant.

If the species was observed at n_1 out of n sites, the maximum likelihood estimates are any solutions to the equation $\hat{p}\hat{\psi} = n_1/n$. For example the maximum likelihood estimates could be $\hat{\psi} = 0.75$ and $\hat{p} = 4n_1/(3n)$. The Hessian matrix, at these maximum likelihood estimates, is

$$\mathbf{H}(\hat{p}, \hat{\psi}) = \begin{bmatrix} -\frac{16nn_1}{9(n-n_1)} & -\frac{n^2}{n-n_1} \\ -\frac{n^2}{n-n_1} & -\frac{9n^3}{16n_1(n-n_1)} \end{bmatrix}.$$

(The calculation of the Hessian matrix is shown in the Maple code.) The Hessian matrix has rank 1, therefore by Theorem 4b we have again shown the model is parameter redundant. Note that the Hessian would have rank 1 regardless of which set of maximum likelihood estimates was chosen.

3.2.2 Binary Latent Class Model

For the binary latent class model, introduced in Section 2.3.1, we use the exhaustive summary, $\kappa_{bin,2}$, given by equation (3.3) in Section 3.1.4.2. The parameters for this model are

$$\begin{bmatrix} p & \theta_{11} & \theta_{21} & \theta_{10} & \theta_{20} \end{bmatrix}.$$

The derivative matrix is then

$$\mathbf{D} = \frac{\partial \kappa_{bin,2}}{\partial \theta} =$$
$$\begin{bmatrix} \theta_{11}\theta_{21} - \theta_{10}\theta_{20} & \bar{\theta}_{21}\theta_{11} - \bar{\theta}_{20}\theta_{10} & \theta_{11}\theta_{21} - \theta_{10}\theta_{20} \\ -p\theta_{21} & p\bar{\theta}_{21} & p\theta_{21} \\ p\bar{\theta}_{11} & -p\theta_{11} & p\theta_{11} \\ -(1-p)\theta_{20} & (1-p)\bar{\theta}_{20} & (1-p)\theta_{20} \\ (1-p)\bar{\theta}_{10} & -(1-p)\theta_{10} & (1-p)\theta_{10} \end{bmatrix},$$

where $\bar{\theta}_{jk} = 1 - \theta_{jk}$. Using Maple, the code for which is given in `latentvariable.mw`, the derivative matrix can be shown to have rank 3.

However there are five parameters, therefore by Theorem 3 this model is parameter redundant with deficiency 3. The Maple code also shows how using the exhaustive summary $\kappa_{bin,1}$ gives the same rank, and leads to the same conclusion.

The Maple code also gives the Fisher Information matrix, and shows its rank is also 3. Therefore by Theorem 4a this model can also be shown to be parameter redundant. However it is not possible to use a symbolic version of Theorem 4b, as it is not possible to find explicit maximum likelihood estimates. Instead a numerical version of Theorem 4b would need to be used. In any model where explicit maximum likelihood estimates are unavailable we need to use this numerical version, called the Hessian method, which is discussed in Section 4.1.1.

3.2.3   Three Compartment Model

The compartmental model introduced in Section 1.3.2 has exhaustive summary κ_{comp}, given by equation (3.4) in Section 3.1.4.3. As the parameters are $\theta = [\theta_{01}, \theta_{21}, \theta_{32}]$, the derivative matrix is

$$\mathbf{D} = \frac{\partial \kappa_{bin,2}}{\partial \theta} = \begin{bmatrix} 0 & 1 & \theta_{32} \\ \theta_{32} & 1 & \theta_{32} \\ \theta_{21} & 1 & \theta_{01} + \theta_{21} \end{bmatrix}.$$

This derivative matrix has rank 3 (see Maple code `compartmental.mw`), and there are three parameters, therefore by Theorem 3 this model is not parameter redundant and at least locally identifiable. Note that by finding the rank of the derivative matrix it is only possible to state that this model is identifiable. We cannot tell at this stage whether this model is locally or globally identifiable. Local versus global identifiability for this example is discussed in Section 3.3.2.1.

3.2.4   Checking for Extrinsic Parameter Redundancy

In the three examples above we are investigating intrinsic parameter redundancy. If we select an exhaustive summary that includes the data then we can also check for extrinsic parameter redundancy. For example we could use the log-likelihood components to form an exhaustive summary. However if we only check for parameter redundancy using an exhaustive summary that includes the data, we only know that the model with that particular data set is parameter redundant. It could be the model is always parameter redundant regardless of the data (intrinsic parameter redundancy) or it could be the data that is causing the parameter redundancy (extrinsic parameter redundancy). To distinguish between these we need to explore both the exhaustive summary with the specific data and an exhaustive summary that assumes perfect data. The exhaustive summary with perfect data can often be simplified to an exhaustive summary that does not include the data; see, for example, Section 3.1.4.4.

In Section 1.2.4 we discussed a parameter redundant mark-recovery model, suitable for data on animals that are uniquely marked in a particular year, and then recovered dead in that or a subsequent year. Here we consider a different version of the model where the probability of survival, ϕ_t, is dependent on the time point t, and the probability of recovering a dead animal, λ_a is dependent on the age of the animal, a. Consider a study which has three years of marking and three years of recovery. The probabilities that an animal marked in year i is recovered dead in year j, P_{ij}, are

$$\mathbf{P} = \begin{bmatrix} (1-\phi_1)\lambda_1 & \phi_1(1-\phi_2)\lambda_2 & \phi_1\phi_2(1-\phi_3)\lambda_3 \\ 0 & (1-\phi_2)\lambda_1 & \phi_2(1-\phi_3)\lambda_2 \\ 0 & 0 & (1-\phi_3)\lambda_1 \end{bmatrix}.$$

Let F_i correspond to the number of animals marked in year i and let N_{ij} denote the number of animals marked in year i is recovered dead in year j. The log-likelihood is formed from a multinomial distribution for each year of marking, giving

$$l \propto \sum_{i=1}^{3} \sum_{j=i}^{3} N_{ij} \log(P_{ij}) + \sum_{i=1}^{3} \left(F_i - \sum_{j=i}^{3} N_{ij} \right) \log \left(1 - \sum_{j=i}^{3} P_{ij} \right).$$

Below, we use the log-likelihood exhaustive summary to show the model itself is not parameter redundant, that is, there is not intrinsic parameter redundancy, but that certain data sets can lead to extrinsic parameter redundancy. Maple code for this example is given in `markrecovery.mw`.

We start with intrinsic parameter redundancy. In this case perfect data would occur if none of the N_{ij} are equal to 0. The log-likelihood exhaustive summary is

$$\kappa_1 = \begin{bmatrix} N_{1,1} \log\{(1-\phi_1)\lambda_1\} \\ N_{1,2} \log\{\phi_1(1-\phi_2)\lambda_2\} \\ N_{1,3} \log\{\phi_1\phi_2(1-\phi_3)\lambda_3\} \\ A_1 \log\{1 - (1-\phi_1)\lambda_1 - \phi_1(1-\phi_2)\lambda_2 - \phi_1\phi_2(1-\phi_3)\lambda_3\} \\ N_{2,2} \log\{(1-\phi_2)\lambda_1\} \\ N_{2,3} \log\{\phi_2(1-\phi_3)\lambda_2\} \\ (F_2 - N_{2,2} - N_{2,3}) \log\{1 - (1-\phi_2)\lambda_1 - \phi_2(1-\phi_3)\lambda_2\} \\ N_{3,3} \log\{(1-\phi_3)\lambda_1\} \\ (F_3 - N_{3,3}) \log\{1 - (1-\phi_3)\lambda_1\} \end{bmatrix},$$

where $A_1 = (F_1 - N_{1,1} - N_{1,2} - N_{1,3})$, and the parameters are

$$\boldsymbol{\theta} = [\phi_1, \phi_2, \phi_3, \lambda_1, \lambda_2, \lambda_3].$$

The derivative matrix,

$$
\mathbf{D}_1 = \frac{\partial \boldsymbol{\kappa}_1}{\partial \boldsymbol{\theta}} =
\begin{bmatrix}
-\frac{N_{1,1}}{1-\phi_1} & \frac{N_{1,2}}{\phi_1} & \frac{N_{1,3}}{\phi_1} & \cdots & 0 \\
0 & -\frac{N_{1,2}}{1-\phi_2} & \frac{N_{1,3}}{\phi_2} & & 0 \\
0 & 0 & -\frac{N_{1,3}}{1-\phi_3} & \frac{(F_3-N_{3,3})\lambda_1}{1-(1-\phi_3)\lambda_1} \\
\vdots & & & & \vdots \\
0 & 0 & \frac{N_{1,3}}{\lambda_3} & \cdots & 0
\end{bmatrix},
$$

using Maple, is found to have rank 6. As there are six parameters this model is not parameter redundant, so the model does not have intrinsic parameter redundancy.

Now suppose that no animals were recovered dead after the year of marking, that is, $N_{1,2} = N_{1,3} = N_{2,3} = 0$. In this case, ignoring any zero terms, the exhaustive summary becomes

$$
\boldsymbol{\kappa}_2 =
\begin{bmatrix}
N_{1,1}\log\{(1-\phi_1)\lambda_1\} \\
(F_1 - N_{1,1})\log\{1 - (1-\phi_1)\lambda_1 - \phi_1(1-\phi_2)\lambda_2 - \phi_1\phi_2(1-\phi_3)\lambda_3\} \\
N_{2,2}\log\{(1-\phi_2)\lambda_1\} \\
(F_2 - N_{2,2})\log\{1 - (1-\phi_2)\lambda_1 - \phi_2(1-\phi_3)\lambda_2\} \\
N_{3,3}\log\{(1-\phi_3)\lambda_1\} \\
(F_3 - N_{3,3})\log\{1 - (1-\phi_3)\lambda_1\}
\end{bmatrix}.
$$

The parameters remain the same and the derivative matrix becomes

$$
\mathbf{D}_2 = \frac{\partial \boldsymbol{\kappa}_2}{\partial \boldsymbol{\theta}} =
$$

$$
\begin{bmatrix}
-\frac{N_{1,1}}{1-\phi_1} & \frac{(F_1-N_{1,1})\{\lambda_1-(1-\phi_2)\lambda_2-\phi_2(1-\phi_3)\lambda_3\}}{1-(1-\phi_1)\lambda_1-\phi_1(1-\phi_2)\lambda_2-\phi_1\phi_2(1-\phi_3)\lambda_3} & \cdots & 0 \\
0 & \frac{(F_1-N_{1,1})\{\phi_1\lambda_2-\phi_1(1-\phi_3)\lambda_3\}}{1-(1-\phi_1)\lambda_1-\phi_1(1-\phi_2)\lambda_2-\phi_1\phi_2(1-\phi_3)\lambda_3} & & 0 \\
0 & \frac{(F_1-N_{1,1})\phi_1\phi_2\lambda_3}{1-(1-\phi_1)\lambda_1-\phi_1(1-\phi_2)\lambda_2-\phi_1\phi_2(1-\phi_3)\lambda_3} & & \frac{(F_3-N_{3,3})\lambda_1}{1-(1-\phi_3)\lambda_1} \\
\frac{N_{1,1}}{\lambda_1} & \frac{(F_1-N_{1,1})(-1+\phi_1)}{1-(1-\phi_1)\lambda_1-\phi_1(1-\phi_2)\lambda_2-\phi_1\phi_2(1-\phi_3)\lambda_3} & & \frac{(F_3-N_{3,3})(-1+\phi_3)}{1-(1-\phi_3)\lambda_1} \\
0 & -\frac{(F_1-N_{1,1})\phi_1(1-\phi_2)}{1-(1-\phi_1)\lambda_1-\phi_1(1-\phi_2)\lambda_2-\phi_1\phi_2(1-\phi_3)\lambda_3} & & 0 \\
0 & -\frac{(F_1-N_{1,1})\phi_1\phi_2(1-\phi_3)}{1-(1-\phi_1)\lambda_1-\phi_1(1-\phi_2)\lambda_2-\phi_1\phi_2(1-\phi_3)\lambda_3} & \cdots & 0
\end{bmatrix}.
$$

Using Maple, we can show this derivative matrix has rank 5, which is less than the number of parameters. Therefore this data set causes extrinsic parameter redundancy.

The results for this mark-recovery model, and other mark-recovery models, are generalised to any number of years of marking and recovery and different patterns of $N_{ij} = 0$ in Cole et al. (2012).

3.2.5   Finding Estimable Parameter Combinations

Whilst the derivative matrix can be used to detect parameter redundancy, it can also be used to determine which parameters and parameter combinations can be estimated in a parameter redundant model. If a model, with a particular parameterisation, is parameter redundant, it is possible to find a reparameterisation of the model that is no longer parameter redundant. This reparameterisation will be a combination of the original parameters and have a unique maximum likelihood estimate for every parameter, for some region of the parameter space, even though the original parameterisation does not. Therefore, this reparameterisation is known as a set of estimable parameter combinations in Cole et al. (2010) and Catchpole et al. (1998a). It is also described as a minimal parameter set in Catchpole et al. (1998a) and a locally identifiable reparameterisation in Chappell and Gunn (1998) and Evans and Chappell (2000).

Finding a locally identifiable reparameterisation or a set of estimable parameter combinations involves examining the null-space of the transpose of the derivative matrix. The positions of the zeros in all null-space vectors indicate that an individual parameter is identifiable. The other estimable parameter combinations can then be found by solving a set of partial differential equation (PDEs) derived from the null-space. Catchpole et al. (1998a) first used this method for finding estimable parameter combinations in parameter redundant exponential family models. Independently, Chappell and Gunn (1998) and Evans and Chappell (2000) used this method for finding locally identifiable reparameterisations for non-identifiable compartmental models. Cole et al. (2010) generalised this method to any exhaustive summary in Theorem 5a below. Alternatively Dasgupta et al. (2007) show that a reparameterisation can be found from the Fisher information matrix, $\mathbf{J}$, as laid out in Theorem 5b below.

Theorem 5 *a. If the derivative matrix $\mathbf{D}$ of a model with q parameters has rank r with deficiency $d = q - r > 0$, the estimable parameters (or a locally identifiable reparameterisation) can be determined by solving $\boldsymbol{\alpha}(\boldsymbol{\theta})^T \mathbf{D}(\boldsymbol{\theta}) = 0$, which has d solutions, labelled $\alpha_j(\boldsymbol{\theta})$ for $j = 1 \dots d$, with individual entries $\alpha_{ij}(\boldsymbol{\theta})$. Any $\alpha_{ij}(\boldsymbol{\theta})$ which are zero for all d solutions correspond to a parameter, θ_i, which is estimable or is locally identifiable. The solutions of the system of linear first-order PDEs,*

$$\sum_{i=1}^{q} \alpha_{ij} \frac{\partial f}{\partial \theta_i} = 0, \ j = 1 \dots d, \tag{3.6}$$

form a set of estimable parameters or a reparameterisation that is at least locally identifiable (Theorem 2b Cole et al., 2010).

b. If the Fisher information matrix $\mathbf{J}$ of a model with q parameters has rank r_J, with deficiency $d_J = q - r_J > 0$, the estimable parameters (or a locally identifiable reparameterisation) can be determined by solving $\mathbf{J}\boldsymbol{\alpha}_J(\boldsymbol{\theta}) = 0$, which has d_J solutions, labelled $\alpha_{Jj}(\boldsymbol{\theta})$ for $j = 1 \dots d$, with individual entries $\alpha_{Jij}(\boldsymbol{\theta})$. Any $\alpha_{Jij}(\boldsymbol{\theta})$ which are zero for all d_J solutions correspond to a parameter, θ_i, which is estimable or is locally identifiable. The solutions of the system of *linear first-order PDEs,*

$$\sum_{i=1}^{q} \alpha_{Jij} \frac{\partial f}{\partial \theta_i} = 0, \ j = 1 \dots d, \tag{3.7}$$

form a set of estimable parameters or a reparameterisation that is at least locally identifiable (Theorem 7 Dasgupta et al., 2007).

The solutions of the PDEs are found using Maple, as explained in Appendix A.2. We demonstrate this method using two examples in Sections 3.2.5.1 and 3.2.5.2 below.

Note that in Theorem 5, the resulting set of estimable parameter combinations is not unique, and could be reparameterised to form an alternative set of estimable parameter combinations. For example, for a three parameter model with $\boldsymbol{\theta} = [\theta_1, \theta_2, \theta_3]$, suppose the null space is

$$\boldsymbol{\alpha} = \left[-\frac{\theta_1}{\theta_3}, -\frac{\theta_2}{\theta_3}, 1 \right]^T.$$

The PDEs are therefore

$$-\frac{\partial f(\theta_1, \theta_2, \theta_3)}{\partial \theta_1} \frac{\theta_1}{\theta_3} - \frac{\partial f(\theta_1, \theta_2, \theta_3)}{\partial \theta_2} \frac{\theta_2}{\theta_3} + \frac{\partial f(\theta_1, \theta_2, \theta_3)}{\partial \theta_3} = 0.$$

The solution of the PDEs involves an arbitrary function, so the solution is not unique. However, one solution gives us the estimable parameter combinations $\beta_1 = \theta_2/\theta_1$, $\beta_2 = \theta_1\theta_3$. We could reparameterise using $\beta_1\beta_2$ and an alternative set of estimable parameter combinations is $\theta_1\theta_3$ and $\theta_2\theta_3$. This illustration is actually the estimable parameter combinations for the multiple-input multiple-output model discussed in Section 1.3.3; Maple code for this example is given in `MIMO.mw`. The Maple code also demonstrates some alternative functions that are solutions of these PDEs.

3.2.5.1 Basic Occupancy Model

Consider the basic occupancy model with exhaustive summary $\boldsymbol{\kappa}_{occ,1}$, given by equation (3.1). The derivative matrix, given by equation (3.5) in Section Section 3.2.1, has null space

$$\boldsymbol{\alpha} = \left[-\frac{\psi}{p}, 1 \right]^T.$$

As neither entry is zero, neither parameter is identifiable. The corresponding partial differential equation is

$$-\frac{\partial f(\psi, p)}{\partial \psi} \frac{\psi}{p} + \frac{\partial f(\psi, p)}{\partial p} = 0$$

which has a solution ψp. By Theorem 5a the estimable parameter combination is ψp. The solution to this partial differential equation was found using Maple, see `basicoccupancy.mw`. This Maple file also finds the estimable parameter combinations for exhaustive summary $\kappa_{occ,2}$ (equation 3.2), which has identical null space and hence identical partial differential equations and solutions.

The Fisher information matrix, from Section 3.2.1,

$$\mathbf{J} = \begin{bmatrix} \frac{np}{\psi(1-p\psi)} & \frac{n}{1-p\psi} \\ \frac{n}{1-p\psi} & \frac{n\psi}{p(1-p\psi)} \end{bmatrix},$$

also has null space

$$\boldsymbol{\alpha} = \left[-\frac{\psi}{p}, 1 \right]^T.$$

As the null space is identical to the null space of the derivative matrix, the resulting partial differential equations and solution will be identical. Therefore by Theorem 5b we can also show an estimable parameter combination is ψp. This is also shown in Maple code `basicoccupancy.mw`.

3.2.5.2 Binary Latent Class Model

Here we continue with the binary latent class model from Section 3.2.2. The deficiency of this model is 2 (the number of parameters, 5, minus the rank of 3). There are two vectors that span the null space of the derivative matrix:

$$\boldsymbol{\alpha}_1 = \begin{bmatrix} \dfrac{1-p}{\theta_{20}-\theta_{21}} & \dfrac{(\theta_{10}-\theta_{11})(1-p)}{(\theta_{20}-\theta_{21})p} & 0 & 0 & 1 \end{bmatrix}$$

$$\boldsymbol{\alpha}_2 = \begin{bmatrix} \dfrac{1-p}{\theta_{10}-\theta_{11}} & 0 & \dfrac{(\theta_{20}-\theta_{21})(1-p)}{(\theta_{10}-\theta_{11})p} & 1 & 0 \end{bmatrix}.$$

As there is no $\alpha_{i,j}$ where both $\alpha_{1,j}$ and $\alpha_{2,j}$ equal zero, none of the original parameters are identifiable. The two corresponding partial differential equations are

$$\frac{\partial f}{\partial p}\frac{1-p}{\theta_{10}-\theta_{11}} + \frac{\partial f}{\partial \theta_{11}}\frac{(\theta_{10}-\theta_{11})(1-p)}{(\theta_{20}-\theta_{21})p} + \frac{\partial f}{\partial \theta_{20}} = 0$$

$$\frac{\partial f}{\partial p}\frac{1-p}{\theta_{10}-\theta_{11}} + \frac{\partial f}{\partial \theta_{21}}\frac{(\theta_{20}-\theta_{21})(1-p)}{(\theta_{10}-\theta_{11})p} + \frac{\partial f}{\partial \theta_{10}} = 0.$$

Using Maple (see `latentvariable.mw`) it can be shown that the estimable parameter combinations are $p\theta_{11} + (1-p)\theta_{10}$, $p(1-p)(\theta_{20}-\theta_{21})(\theta_{10}-\theta_{11})$ and $\theta_{20} + p(\theta_{21} - \theta_{20})$.

3.2.6 Generalising Parameter Redundancy Results to Any Dimension

In some cases, interest lies in the rank for a specific model with fixed dimensions. In other cases we may be interested in generalising parameter redundancy results to any dimension. For example, an ecology experiment may be carried out over several years; we wish to find out general parameter redundancy results for a generic n years rather than a fixed number of years. It is possible to generalise results to any dimension as demonstrated in Section 3.2.6.1 below.

3.2.6.1   The Extension Theorem

We start by considering the smallest structural version of a model, where a general result holds, and then extend the size of that model whilst maintaining the model structure, using a proof by induction argument to generalise the result to higher dimensions. This idea was introduced in Catchpole and Morgan (1997) for full rank product-multinomial models. Results for parameter redundant models are given in Catchpole and Morgan (2001) for product-multinomial models under certain conditions. This was generalised to any exhaustive summary in Cole et al. (2010) as set out in Theorem 6 below. In particular, Theorem 6b allows the generalisation of the results to any parameter redundant model, by first reparameterising the model so it is no longer parameter redundant.

Suppose the smallest dimension of the model that is being examined has exhaustive summary $\kappa_1(\theta_1)$, with parameters θ_1. The derivative matrix is $D_1(\theta_1) = [\partial\kappa_1/\partial\theta_1]$. This model is then extended, adding extra parameters, θ_2, and new exhaustive summary terms $\kappa_2(\theta')$, with $\theta' = [\theta_1, \theta_2]$. The extended model's exhaustive summary is $\kappa(\theta') = [\kappa_1(\theta_1), \kappa_2(\theta')]^T$. That is, increasing the dimension retains the existing exhaustive summary terms, as well as adding new exhaustive summary terms. The derivative matrix of the extended model is

$$D = \begin{bmatrix} D_1(\theta_1) & D_{2,1}(\theta_1) \\ 0 & D_{2,2}(\theta_2) \end{bmatrix} \text{ with } D_{2,1} = \left[\frac{\partial\kappa_2}{\partial\theta_1}\right] \text{ and } D_{2,2} = \left[\frac{\partial\kappa_2}{\partial\theta_2}\right].$$

Theorem 6 *The Extension Theorem*
a. If D_1 is full rank and $D_{2,2}$ is full rank, then the extended model is will be full rank.
b. If D_1 is not full rank, first reparameterise the model in terms of the estimable parameter combinations. Then Theorem 6a can be applied to the reparameterised model. The rank of the original model will be identical to the rank of the reparameterised model (Cole et al., 2010).

If proof by induction is applicable, for example if there is a general pattern, then the result can be extended to any dimension.

If no new parameters are added then $D = \begin{bmatrix} D_1(\theta_1) & D_{2,1}(\theta_1) \end{bmatrix}$, which will be full rank if $D_1(\theta_1)$ is full rank. If one new parameter, θ_2 is added then $D_{2,2}$ will always have full rank 1 as long as κ_2 is a function of θ_2 and $\theta_2 \neq 0$. In these two possible cases Theorem 6 can be applied without evaluating the rank of $D_{2,2}$. This is known as a trivial application of the extension theorem noted in Catchpole and Morgan (2001) and Cole et al. (2010), and given in Remark 1 below.

Remark 1 *Trivial application of the extension theorem: if zero or one new parameter is added to a full rank model then the model be full rank.*

Theorem 6 is applied in example 3.2.6.2 below, as well as demonstrating a case where Remark 1 can be used.

3.2.6.2 Mark-Recovery Model

In Section 1.2.4 we introduced a mark-recovery model with two age-categories of first year animals and adults. Here we consider three alternative mark-recovery models which illustrate Theorem 6a and 6b, as well as Remark 1. The Maple code for these examples is given in `markrecovery.mw`.

Firstly we consider a mark-recovery model that also includes time dependence to demonstrate Theorem 6a. In this example there are separate survival probabilities for first year animals and adult animals as before, however these also depend on the occasion, which is known as time-dependence. Conversely, this example has one constant recovery probability.

The general result for this model starts at 3 years of marking and recovery, so we begin the extension theorem by examining this case. For three years of marking and recovery the probability matrix is given by

$$
\mathbf{P}_A = \begin{bmatrix} (1-\phi_{1,1})\lambda & \phi_{1,1}(1-\phi_{a,2})\lambda & \phi_{1,1}\phi_{a,2}(1-\phi_{a,3})\lambda \\ 0 & (1-\phi_{1,2})\lambda & \phi_{1,2}(1-\phi_{a,3})\lambda \\ 0 & 0 & (1-\phi_{1,3})\lambda \end{bmatrix},
$$

where $\phi_{1,t}$ is the survival probability for animals in their first year of life at occasion t, $\phi_{a,t}$ is the survival probability for animals older than a year at occasion t and λ is the reporting probability. Using the exhaustive summary,

$$
\kappa_{A,1} = \begin{bmatrix} (1-\phi_{1,1})\lambda \\ \phi_{1,1}(1-\phi_{a,2})\lambda \\ \phi_{1,1}\phi_{a,2}(1-\phi_{a,3})\lambda \\ (1-\phi_{1,2})\lambda \\ \phi_{1,2}(1-\phi_{a,3})\lambda \\ (1-\phi_{1,3})\lambda \end{bmatrix},
$$

with parameters $\boldsymbol{\theta}_{A,1} = [\lambda, \phi_{1,1}, \phi_{1,2}, \phi_{1,3}, \phi_{a,2}, \phi_{a,3}]$, the derivative matrix is given by

$$
\mathbf{D}_{A,1} = \frac{\partial \kappa_{A,1}}{\partial \boldsymbol{\theta}_{A,1}} = \begin{bmatrix} \bar{\phi}_{1,1} & \phi_{1,1}\bar{\phi}_{a,2} & \phi_{1,1}\phi_{a,2}\bar{\phi}_{a,3} & \bar{\phi}_{1,2} & \phi_{1,2}\bar{\phi}_{a,3} & \bar{\phi}_{1,3} \\ -\lambda & \bar{\phi}_{a,2}\lambda & \phi_{a,2}\bar{\phi}_{a,3}\lambda & 0 & 0 & 0 \\ 0 & 0 & 0 & -\lambda & \bar{\phi}_{a,3}\lambda & 0 \\ 0 & 0 & 0 & 0 & 0 & -\lambda \\ 0 & -\phi_{1,1}\lambda & \phi_{1,1}\bar{\phi}_{a,3}\lambda & 0 & 0 & 0 \\ 0 & 0 & -\phi_{1,1}\phi_{a,2}\lambda & 0 & -\phi_{1,2}\lambda & 0 \end{bmatrix},
$$

where $\bar{\phi} = 1 - \phi$. Maple can be used to show this derivative matrix has full rank 6.

If we extend the model to four years of marking and recapture, the probability matrix is

$$
\mathbf{P}_{Aex} = \begin{bmatrix}
\bar{\phi}_{1,1}\lambda & \phi_{1,1}\bar{\phi}_{a,2}\lambda & \phi_{1,1}\phi_{a,2}\bar{\phi}_{a,3}\lambda & \phi_{1,1}\phi_{a,2}\phi_{a,3}\bar{\phi}_{a,4}\lambda \\
0 & \bar{\phi}_{1,2}\lambda & \phi_{1,2}\bar{\phi}_{a,3}\lambda & \phi_{1,2}\phi_{a,3}\bar{\phi}_{a,4}\lambda \\
0 & 0 & \bar{\phi}_{1,3}\lambda & \phi_{1,3}\bar{\phi}_{a,4}\lambda \\
0 & 0 & 0 & \bar{\phi}_{1,4}\lambda
\end{bmatrix},
$$

where $\bar{\phi} = 1 - \phi$. The extra exhaustive summary terms are

$$
\boldsymbol{\kappa}_{A,2} = \begin{bmatrix}
\phi_{1,1}\phi_{a,2}\phi_{a,3}\bar{\phi}_{a,4}\lambda \\
\phi_{1,2}\phi_{a,3}\bar{\phi}_{a,4}\lambda \\
\phi_{1,3}\bar{\phi}_{a,4}\lambda \\
\bar{\phi}_{1,4}\lambda
\end{bmatrix},
$$

and the extra parameters $\boldsymbol{\theta}_{A,2} = [\phi_{1,4}, \phi_{a,4}]$. Differentiating the extra exhaustive summary terms with respect to the extra parameters gives the second derivative matrix

$$
\mathbf{D}_{A,2,2} = \frac{\partial \boldsymbol{\kappa}_{A,2}}{\partial \boldsymbol{\theta}_{A,2}} = \begin{bmatrix}
0 & 0 & 0 & -\lambda \\
-\phi_{1,1}\phi_{a,2}\phi_{a,3}\lambda & -\phi_{1,2}\phi_{a,3}\lambda & -\phi_{1,3}\lambda & 0
\end{bmatrix},
$$

which, again using Maple, has full rank 2. As both $\mathbf{D}_{A,1}$ and $\mathbf{D}_{A,2,2}$ are full rank by the extension theorem (Theorem 6a) this model is full rank for four years of marking and recovery. Then by induction, for N years of marking and recovery will always be full rank, with rank $2N$. This result is given in Cole et al. (2012).

To demonstrate Remark 1 we consider a similar model where only first year survival is dependent on the occasion. For 3 years of marking and recovery the probability matrix is

$$
\mathbf{P}_B = \begin{bmatrix}
(1 - \phi_{1,1})\lambda & \phi_{1,1}(1 - \phi_a)\lambda & \phi_{1,1}\phi_a(1 - \phi_a)\lambda \\
0 & (1 - \phi_{1,2})\lambda & \phi_{1,2}(1 - \phi_a)\lambda \\
0 & 0 & (1 - \phi_{1,3})\lambda
\end{bmatrix},
$$

where $\phi_{1,t}$ is the survival probability for animals in their first year of life at occasion t, ϕ_a is the survival probability for animals older than a year and λ is the reporting probability. The exhaustive summary is then

$$
\boldsymbol{\kappa}_{B,1} = \begin{bmatrix}
(1 - \phi_{1,1})\lambda \\
\phi_{1,1}(1 - \phi_a)\lambda \\
\phi_{1,1}\phi_a(1 - \phi_a)\lambda \\
(1 - \phi_{1,2})\lambda \\
\phi_{1,2}(1 - \phi_a)\lambda \\
(1 - \phi_{1,3})\lambda
\end{bmatrix},
$$

with parameters $\boldsymbol{\theta}_{B,1} = [\lambda, \phi_{1,1}, \phi_{1,2}, \phi_{1,3}, \phi_a]$. Using Maple, the derivative matrix $\partial \boldsymbol{\kappa}_{B,1} / \partial \boldsymbol{\theta}_{B,1}$ can be shown to have full rank 5. The extra exhaustive summary terms are

$$\boldsymbol{\kappa}_{B,2} = \begin{bmatrix} \phi_{1,1} \phi_a^2 \bar{\phi}_a \lambda \\ \phi_{1,2} \bar{\phi}_a \phi_a \lambda \\ \phi_{1,3} \bar{\phi}_a \lambda \\ \bar{\phi}_{1,4} \lambda \end{bmatrix}.$$

In this case there is only one extra parameter $\boldsymbol{\theta}_2 = [\phi_{1,4}]$. The extra derivative matrix is

$$\mathbf{D}_{B,2,2} = \frac{\partial \boldsymbol{\kappa}_2}{\partial \boldsymbol{\theta}_2} = \begin{bmatrix} 0 & 0 & 0 & -\lambda \end{bmatrix}$$

which will obviously have rank 1, as long as $\lambda \neq 0$. Rather than calculating the derivative matrix when there is one extra parameter we can use Remark 1 and state that by a trivial application of the extension theorem this model will always be full rank.

The final example here is a parameter redundant model to demonstrate Theorem 6b. In this example the first year survival probability is constant, the adult survival probability is time dependent and the reporting probability depends on age. The general result starts with four years of marking and recovery, where the probability matrix is

$$\mathbf{P}_C = \begin{bmatrix} \bar{\phi}_1 \lambda_1 & \phi_1 \bar{\phi}_{a,2} \lambda_2 & \phi_1 \phi_{a,2} \bar{\phi}_{a,3} \lambda_3 & \phi_1 \phi_{a,2} \phi_{a,3} \bar{\phi}_{a,4} \lambda_4 \\ 0 & \bar{\phi}_1 \lambda_1 & \phi_1 \bar{\phi}_{a,3} \lambda_2 & \phi_1 \phi_{a,3} \bar{\phi}_{a,4} \lambda_3 \\ 0 & 0 & \bar{\phi}_1 \lambda_1 & \phi_1 \bar{\phi}_{a,4} \lambda_2 \\ 0 & 0 & 0 & \bar{\phi}_1 \lambda_1 \end{bmatrix},$$

where ϕ_1 is the survival probability for animals in their first year of life, $\phi_{a,t}$ is the survival probability for animals older than a year at occasion t and λ_j is the reporting probability of an animal aged j. Using the exhaustive summary,

$$\boldsymbol{\kappa}_{C,1} = \begin{bmatrix} \bar{\phi}_1 \lambda_1 \\ \phi_1 \bar{\phi}_{a,2} \lambda_2 \\ \phi_1 \phi_{a,2} \bar{\phi}_{a,3} \lambda_3 \\ \phi_1 \phi_{a,2} \phi_{a,3} \bar{\phi}_{a,4} \lambda_4 \\ \bar{\phi}_1 \lambda_1 \\ \phi_1 \bar{\phi}_{a,3} \lambda_2 \\ \phi_1 \phi_{a,3} \bar{\phi}_{a,4} \lambda_3 \\ \bar{\phi}_1 \lambda_1 \\ \phi_1 \bar{\phi}_{a,4} \lambda_2 \\ \bar{\phi}_1 \lambda_1 \end{bmatrix},$$

with parameters $\boldsymbol{\theta}_{C,1} = [\lambda_1, \lambda_2, \lambda_3, \lambda_4, \phi_1, \phi_{a,2}, \phi_{a,3}, \phi_{a,4}]$. The derivative matrix, $\mathbf{D}_{C,1} = \partial \boldsymbol{\kappa}_{C,1} / \partial \boldsymbol{\theta}_{C,1}$ is found using Maple to have rank 7, but there are 8 parameters, so the deficiency is 1 and the model is parameter redundant. Solving the appropriate set of PDEs the estimable parameters are $\phi_{a,2}$, $\phi_{a,3}$, $\phi_{a,4}$, $(1 - \phi_1)\lambda_1$, $\phi_1 \lambda_2$, $\phi_1 \lambda_3$, $\phi_1 \lambda_4$.

Let $\beta_1 = (1 - \phi_1)\lambda_1$, $\beta_2 = \phi_1\lambda_2$, $\beta_3 = \phi_1\lambda_3$, $\beta_4 = \phi_1\lambda_4$. We can reparameterise the exhaustive summary as

$$
\boldsymbol{\kappa}_{Cre,1} = \begin{bmatrix} \beta_1 \\ \bar{\phi}_{a,2}\beta_2 \\ \phi_{a,2}\bar{\phi}_{a,3}\beta_3 \\ \phi_{a,2}\phi_{a,3}\bar{\phi}_{a,4}\beta_4 \\ \beta_1 \\ \bar{\phi}_{a,3}\beta_2 \\ \phi_{a,3}\bar{\phi}_{a,4}\beta_3 \\ \beta_1 \\ \bar{\phi}_{a,4}\beta_2 \\ \beta_1 \end{bmatrix},
$$

which has parameters $\boldsymbol{\theta}_{Cre,1} = [\beta_1, \beta_2, \beta_3, \beta_4, \phi_{a,2}, \phi_{a,3}, \phi_{a,4}]$. As we have reparameterised using the estimable parameter combinations, the derivative matrix, $\mathbf{D}_{Cre,1} = \partial\boldsymbol{\kappa}_{Cre,1}/\partial\boldsymbol{\theta}_{Cre,1}$ will have full rank 7. (This can be confirmed in Maple). Extending the reparameterised model by one year of marking and recovery gives extra exhaustive summary terms,

$$
\boldsymbol{\kappa}_{Cre,2} = \begin{bmatrix} \phi_{a,2}\phi_{a,3}\phi_{a,4}\bar{\phi}_{a,5}\beta_5 \\ \phi_{a,3}\phi_{a,4}\bar{\phi}_{a,5}\beta_4 \\ \phi_{a,4}\bar{\phi}_{a,5}\beta_3 \\ \bar{\phi}_{a,5}\beta_2 \\ \beta_1 \end{bmatrix}.
$$

where $\beta_5 = \phi_1\lambda_5$. The extra parameters are $\boldsymbol{\theta}_{re,2} = [\beta_5, \phi_{a,5}]$. The derivative matrix

$$
\mathbf{D}_{Cre,2,2} = \frac{\partial\boldsymbol{\kappa}_{re,2}}{\partial\boldsymbol{\theta}_{re,2}} = \begin{bmatrix} \phi_{a,2}\phi_{a,3}\phi_{a,4}\bar{\phi}_{a,5} & 0 & 0 & 0 & 0 \\ -\phi_{a,2}\phi_{a,3}\phi_{a,4}\beta_5 & -\phi_{a,3}\phi_{a,4}\beta_4 & -\phi_{a,4}\beta_3 & \beta_2 & 0 \end{bmatrix}
$$

is shown in the Maple code to have full rank 2. Therefore the reparameterised model with $N \geq 4$ years of marking and recovery will always have full rank $2N$. The original parameterisation by Theorem 6b will also have rank $2N$, but as there are $2N + 1$ parameters the model will always be parameter redundant with deficiency 1.

3.2.7   Essentially and Conditionally Full Rank

A full rank model, that is, a model that is not parameter redundant, may not be full rank for all points in the parameter space. In these cases it is termed a conditionally full rank model. Alternatively a full rank model can be full rank for all points in the parameter space, in which case it is termed essentially full rank. (See Definition 4.) The derivative matrix can also be used to distinguish between essentially and conditionally full rank models, and in particular find points in the parameter space where the model will be parameter redundant.

Gimenez et al. (2003) show that it is possible to determine whether a model is essentially full rank by first multiplying a full rank derivative matrix, $\mathbf{D}$, by full rank matrices, $\mathbf{A}$ and $\mathbf{B}$, so that $\mathbf{L} = \mathbf{ADB}$. As $\mathbf{A}$ and $\mathbf{B}$ are full rank, $\text{rank}(\mathbf{L}) = \text{rank}(\mathbf{D})$ (Graybill, 2001). If $\mathbf{L}$ has only entries that are constant then the model is essentially full rank.

This approach was generalised in Cole et al. (2010) by providing a general decomposition method that can be used to determine whether a model is essentially or conditionally full rank. The decomposition that we employ is a (modified) PLUR decomposition, or a Turing factorisation (see Corless and Jeffrey, 1997).

Consider the derivative matrix, $\mathbf{D}$, which has dimensions q by k, where q is the number of parameters and k is the number of exhaustive summary terms. $\mathbf{D}$ is written as the product of 4 matrices, with $\mathbf{D} = \mathbf{PLUR}$. The four matrices are:

- $\mathbf{P}$, with dimensions q by q, is a permutation matrix, that is, a matrix consisting of ones and zeros, with exactly 1 one in each row.

- $\mathbf{L}$, with dimensions q by q, is a lower triangular matrix with ones on the diagonal.

- $\mathbf{U}$, with dimensions q by q, is an upper triangular matrix, with general entries on the diagonal.

- $\mathbf{R}$, with dimensions q by k, is a matrix in reduced echelon form.

In a reduced echelon form matrix all the rows with non-zero values are above any rows of all zeros. The leading coefficient of any row is always 1 and strictly to the right of the leading coefficient of any row above it and it is the only non-zero entry in its column. Writing a matrix in reduced echelon form can be used to determine the rank of a matrix, where the rank is the number of leading coefficients that are equal to 1. For example the following two matrices are in reduced echelon form,

$$\mathbf{R}_1 = \begin{bmatrix} 1 & 0 & 0 & 1 & 0 \\ 0 & 1 & 0 & 0 & 1 \\ 0 & 0 & 1 & 0 & 0 \end{bmatrix}, \mathbf{R}_2 = \begin{bmatrix} 1 & 0 & \theta_1\theta_2 \\ 0 & 1 & \theta_2 \\ 0 & 0 & 0 \end{bmatrix}.$$

The matrix $\mathbf{R}_1$ has three leading coefficients that are equal to 1, and therefore has rank 3. The matrix $\mathbf{R}_2$ has two leading coefficients that are equal to 1, and therefore has rank 2. This **PLUR** decomposition can be found using Gaussian elimination. Maple has an intrinsic function that will obtain the decomposition, as explained in Appendix A.3.

We can use the **PLUR** decomposition to detect points in the parameter space that are not full rank by examining the determinant of the matrix $\mathbf{U}$, as laid out in Theorem 7 below.

Theorem 7 *If we write the derivative matrix as* $\mathbf{D} = \mathbf{PLUR}$, *where* $\mathbf{P}$ *is a permutation matrix,* $\mathbf{L}$ *is a lower triangular matrix with ones on the diagonal,* $\mathbf{U}$ *is an upper triangular matrix and* $\mathbf{R}$ *is a matrix in reduced echelon form, then the model is parameter redundant at* $\boldsymbol{\theta}$ *if and only if* $Det(\mathbf{U}) = 0$ *at a point* $\boldsymbol{\theta} \in \Omega$, *as long as* $\mathbf{L}$, $\mathbf{U}$ *and* $\mathbf{R}$ *are defined at* $\boldsymbol{\theta}$ *(Cole et al., 2010).*

Theorem 7 can also be deduced from Corless and Jeffrey (1997)'s results on Turing factorisations and a proof is given in Cole et al. (2010).

We can use Theorem 7 to identify if any models nested in a full rank model are parameter redundant. If a full rank model has any nested models that are parameter redundant, formed from taking linear constraints on the elements of $\boldsymbol{\theta}$, then they will appear as solutions of $Det(\mathbf{U}) = 0$. The deficiency of the sub-model is d_U, where d_U is the deficiency of $\mathbf{U}$ evaluated with the sub-model constraints applied. This is described in Theorem 8 below.

Theorem 8 *The model with constraint* $\boldsymbol{\theta} = g(\boldsymbol{\theta})$, *nested within a full rank model with derivative matrix* $\mathbf{D} = \mathbf{PLUR}$, *will be parameter redundant if* $Det(\mathbf{U})|_{\boldsymbol{\theta}=g(\boldsymbol{\theta})} = 0$. *The deficiency of the nested model is equal to the deficiency of* $\mathbf{U}|_{\boldsymbol{\theta}=g(\boldsymbol{\theta})}$ *(Cole et al., 2010).*

3.2.7.1   Mark-Recovery Model

Consider a mark-recovery example with only first year survival dependent on the occasion, constant adult survival and two age classes for the recovery rate for three years of marking and recovery. An exhaustive summary formed from the probabilities of marking and recovery is

$$
\boldsymbol{\kappa} = \begin{bmatrix}
(1 - \phi_{1,1})\lambda_1 \\
\phi_{1,1}(1 - \phi_a)\lambda_a \\
\phi_{1,1}\phi_a(1 - \phi_a)\lambda_a \\
(1 - \phi_{1,2})\lambda_1 \\
\phi_{1,2}(1 - \phi_a)\lambda_a \\
(1 - \phi_{1,3})\lambda_1
\end{bmatrix},
$$

where $\phi_{1,t}$ is the survival probability for animals in their first year of life at occasion t, ϕ_a is the survival probability for animals older than a year, λ_1 is the reporting probability for animals in their first year of life and λ_a is the reporting rate of animals older than a year. The parameters are $\boldsymbol{\theta} = [\lambda_1, \lambda_a, \phi_a, \phi_{1,1}, \phi_{1,2}, \phi_{1,3}]$. The derivative matrix formed from differentiating $\boldsymbol{\kappa}$ with respect to $\boldsymbol{\theta}$ is

$$
\mathbf{D} = \begin{bmatrix}
\bar{\phi}_{1,1} & 0 & 0 & \bar{\phi}_{1,2} & 0 & \bar{\phi}_{1,3} \\
0 & \phi_{1,1}\bar{\phi}_a & \phi_{1,1}\phi_a\bar{\phi}_a & 0 & \phi_{1,2}\bar{\phi}_a & 0 \\
0 & -\phi_{1,1}\lambda_a & (\phi_{1,1} - 2\phi_{1,1}\phi_a)\lambda_a & 0 & -\phi_{1,2}\lambda_a & 0 \\
-\lambda_1 & \lambda_a\bar{\phi}_a & \phi_a\bar{\phi}_a\lambda_a & 0 & 0 & 0 \\
0 & 0 & 0 & -\lambda_1 & \lambda_a\bar{\phi}_a & 0 \\
0 & 0 & 0 & 0 & 0 & -\lambda_1
\end{bmatrix},
$$

with $\bar{\phi} = 1 - \phi$.

In the Maple code `Markrecoverymodel.mw` we show the derivative matrix is of full rank 6 and can be decomposed into the matrices

$$
\mathbf{P} = \begin{bmatrix} 1 & 0 & 0 & 0 & 0 & 0 \\ 0 & 1 & 0 & 0 & 0 & 0 \\ 0 & 0 & 1 & 0 & 0 & 0 \\ 0 & 0 & 0 & 1 & 0 & 0 \\ 0 & 0 & 0 & 0 & 1 & 0 \\ 0 & 0 & 0 & 0 & 0 & 1 \end{bmatrix}, \quad
\mathbf{L} = \begin{bmatrix} 1 & 0 & 0 & 0 & 0 & 0 \\ 0 & 1 & 0 & 0 & 0 & 0 \\ 0 & -\dfrac{\lambda_a}{\bar{\phi}_a} & 1 & 0 & 0 & 0 \\ -\dfrac{\lambda_1}{\phi_{1,1}} & \dfrac{\lambda_a}{\phi_{1,1}} & 0 & 1 & 0 & 0 \\ 0 & 0 & 0 & \dfrac{\bar{\phi}_{1,1}}{\bar{\phi}_{1,2}} & 1 & 0 \\ 0 & 0 & 0 & 0 & 0 & 1 \end{bmatrix},
$$

$$
\mathbf{U} = \begin{bmatrix} \bar{\phi}_{1,1} & 0 & 0 & \bar{\phi}_{1,2} & 0 & \bar{\phi}_{1,3} \\ 0 & \phi_{1,1}\bar{\phi}_a & \phi_{1,1}\phi_a\bar{\phi}_a & 0 & \phi_{1,2}\bar{\phi}_a & 0 \\ 0 & 0 & \phi_{1,1}(1-\phi_a)\lambda_a & 0 & 0 & 0 \\ 0 & 0 & 0 & \dfrac{\bar{\phi}_{1,2}\lambda_1}{\phi_{1,1}} & \dfrac{\phi_{1,2}\bar{\phi}_a\lambda_a}{\phi_{1,1}} & \dfrac{\bar{\phi}_{1,3}\lambda_1}{\phi_{1,1}} \\ 0 & 0 & 0 & 0 & \dfrac{(\phi_{1,1}-\phi_{1,2})\bar{\phi}_a\lambda_a}{\phi_{1,1}\bar{\phi}_{1,2}} & \dfrac{\bar{\phi}_{1,3}\lambda_1}{\bar{\phi}_{1,2}} \\ 0 & 0 & 0 & 0 & 0 & -\lambda_1 \end{bmatrix}
$$

and

$$
\mathbf{R} = \begin{bmatrix} 1 & 0 & 0 & 0 & 0 & 0 \\ 0 & 1 & 0 & 0 & 0 & 0 \\ 0 & 0 & 1 & 0 & 0 & 0 \\ 0 & 0 & 0 & 1 & 0 & 0 \\ 0 & 0 & 0 & 0 & 1 & 0 \\ 0 & 0 & 0 & 0 & 0 & 1 \end{bmatrix}.
$$

The determinant of $\mathbf{U}$ is

$$
\det(\mathbf{U}) = \phi_{1,1}(\phi_{1,1} - \phi_{1,2})(1 - \phi_a)^3 \lambda_1^2 \lambda_a^2.
$$

As the $\det(\mathbf{U}) = 0$ when $\phi_{1,1} = \phi_{1,2}$ the model is parameter redundant when $\phi_{1,1} = \phi_{1,2}$. By Theorem 7 the model is parameter redundant when $\phi_{1,1} = \phi_{1,2}$. This model is therefore conditionally full rank.

When examining mark-recovery data, typically several different models are fitted to data. We may for example be interested in whether first year survival changes over time or whether it is constant. Interest may therefore also lie in the model with constant first year survival (but with other parameters unchanged). This nested model can be obtained with the constraint $\phi_{1,1} = \phi_{1,2} = \phi_{1,3}$. As $\det(\mathbf{U}) = 0$ with this constraint, by Theorem 8, this nested model is parameter redundant. The rank of $\mathbf{U}$ with the constraint $\phi_{1,1} = \phi_{1,2} = \phi_{1,3}$ is 5. As there are six parameters, in the original model, the deficiency is 1. This nested model has earlier been shown to be parameter redundant directly.

3.2.7.2   Checking the Rank using PLUR Decomposition

Cole and Morgan (2010a) and Choquet and Cole (2012) both observe instances where Maple returns the incorrect rank of a derivative matrix. Both these cases involve exponential functions, and are a direct result of Maple being unable to simplify the exponential functions, without further instruction. If Maple is first instructed to simplify the derivative matrix before calculating the rank, then both of these cases return the correct rank. Alternatively a **PLUR** decomposition can be applied to the derivative matrix. If Maple had returned the incorrect rank then the determinant of **U** will be zero everywhere.

The example from Choquet and Cole (2012) is a capture-recapture model, which is similar to mark-recovery models discussed above, but the animals are recaptured alive rather than recovered dead. This model includes trap-dependence to take account of the effect of the capture on an individual. To capture animals some traps are baited, which encourages an animal that has already been caught to return again to the trap. Other traps may have a negative effect on the animal making it less likely to be caught again. To take this effect into account, a trap-dependence model was developed in Pradel (1993). The model considered here has a survival probability that is dependent on the occasion and the capture probability is dependent on the occasion and whether the animal was caught on the previous occasion. We consider the model with $T = 5$ capture occasions. The probability of an animal surviving occasion t is

$$\phi_t = \frac{1}{1 + \exp(-\beta_t)},$$

for $t = 1, \ldots, T - 1$. To account for trap dependence, there are two different probabilities of recapture depending on whether the animal was caught at the previous occasion or not. The probability that an animal that was caught at occasion $t - 1$ and recaptured at time t is

$$p_t^* = \frac{1}{1 + \exp(-\beta_{t+3} + m)}$$

and the probability that an animal who was not caught at occasion $t - 1$ is recaptured at time t is

$$p_t = \frac{1}{1 + \exp(-\beta_{t+3})},$$

both for $t = 2, \ldots, T$. The parameters of this model are $\boldsymbol{\theta} = [\beta_1, \ldots, \beta_8, m]$.

The probabilities of being marked in year i and first recaptured in year j, $Q_{i,j}$, are summarised in the matrix,

$$\mathbf{Q} = \begin{bmatrix} \phi_1 p_2^* & \phi_1 \bar{p}_2^* \phi_2 p_3 & \phi_1 \bar{p}_2^* \phi_2 \bar{p}_3 \phi_3 p_4 & \phi_1 \bar{p}_2^* \phi_2 \bar{p}_3 \phi_3 \bar{p}_4 \phi_4 p_5 \\ 0 & \phi_2 p_3^* & \phi_2 \bar{p}_3^* \phi_3 p_4 & \phi_2 \bar{p}_3^* \phi_3 \bar{p}_4 \phi_4 p_5 \\ 0 & 0 & \phi_3 p_4^* & \phi_3 \bar{p}_4^* \phi_4 p_5 \\ 0 & 0 & 0 & \phi_4 p_5^* \end{bmatrix}.$$

The exhaustive summary used is the natural logarithm of the non-zero entries in $\mathbf{Q}$, giving

$$\kappa = \begin{bmatrix} \log(\phi_1 p_2^*) \\ \log(\phi_1 \bar{p}_2^* \phi_2 p_3) \\ \log(\phi_1 \bar{p}_2^* \phi_2 \bar{p}_3 \phi_3 p_4) \\ \log(\phi_1 \bar{p}_2^* \phi_2 \bar{p}_3 \phi_3 \bar{p}_4 \phi_4 p_5) \\ \log(\phi_2 p_3^*) \\ \log(\phi_2 \bar{p}_3^* \phi_3 p_4) \\ \log(\phi_2 \bar{p}_3^* \phi_3 \bar{p}_4 \phi_4 p_5) \\ \log(\phi_3 p_4^*) \\ \log(\phi_3 \bar{p}_4^* \phi_4 p_5) \\ \log(\phi_4 p_5^*) \end{bmatrix}.$$

Using Maple v6, Gimenez et al. (2003) found the rank of the derivative matrix, $\mathbf{D} = \partial \kappa / \partial \theta$, to have full rank 9. However Maple failed to simplify the expression $1/(1 + e^x) + 1/(1 + e^{-x})$ which is equal to 1. More recent versions of Maple, such as Maple v18, have this simplification automatically built in. We show in the Maple code, `capturerecapture.mw`, that the rank of this model is 8, so the model is parameter redundant with deficiency 1. However, we also show in the Maple code that if the derivative matrix is manipulated to give a similar derivative matrix to the one presented in Gimenez et al. (2003) a rank of 9 is returned by Maple. A **PLUR** decomposition gives

$$\det(\mathbf{U}) = \frac{e^{-(\beta_1 + \beta_2 + \beta_3 + \beta_4)}(2 + e^{\beta_5 + m} + e^{-\beta_5 - m})(e^{\beta_8 + m} - e^{\beta_8})u_p}{(1 + e^{-\beta_1})(1 + e^{-\beta_2})(1 + e^{-\beta_3})(1 + e^{-\beta_4 - m}) \ldots (1 + e^{\beta_8})},$$

where

$$u_p = 4e^{\beta_7} - e^{-\beta_7} - e^{2\beta_6 - \beta_7} - e^{2\beta_6 - \beta_7} - e^{-2\beta_6 - \beta_7 - 2m} + \ldots - 4e^{\beta_7 + m}.$$

However u_p simplifies to 0. As $\det(\mathbf{U}) = 0$ everywhere in the parameter space the model rank is not 9.

As any symbolic algebra package only has the ability to simplify algebra as far as they are programmed, this could potentially be a problem with any functions that are not rational. In general when the exhaustive summary involves functions that are not rational it is recommended that a **PLUR** decomposition is performed on full rank models, to ensure that Maple has returned the correct rank (Cole and Morgan, 2010a).

3.3 Distinguishing Local and Global Identifiability

The symbolic method described in Section 3.2 can only determine whether a model is identifiable or non-identifiable; it does not distinguish between local and global identifiability. A model that is not parameter redundant, described as full rank, will be at least locally identifiable, if we exclude any points in the parameter space that are parameter redundant. This can be defined formally using the terms essentially and conditionally full rank in Theorem 9 below.

Theorem 9 *If a model is essentially full rank it is at least locally identifiable. If a model is conditionally full rank, it is at least locally identifiable almost everywhere. If we exclude the points found in Theorem 7 in Section 3.2.7, the model is at least locally identifiable. (Catchpole and Morgan, 1997, Cole et al., 2010)*

In many problems we may be interested in global identifiability, which is a more difficult result to prove (Rothenberg, 1971). Whilst it is possible that a full rank model can be globally identifiable, in many examples this is not the case. Section 3.3.1 provides global identifiability results for exponential family models. In Section 3.3.2 we provide a general method for checking whether a model is locally or globally identifiable.

3.3.1   Global Identifiability Results for Exponential Family Models

Suppose that we have a set of data $\mathbf{y} = [y_1, y_2, \ldots y_n]$, with a probability density function, f, from the exponential family model so that

$$\log(f) = A(\mathbf{y}) + B(\boldsymbol{\theta}) + \sum_{i=1}^{p} \theta_i C_i(\mathbf{y}),$$

with parameters $\boldsymbol{\theta} = [\theta_1, \theta_2, \ldots, \theta_p]$, and for some functions, $A(.)$, $B(.)$ and $C_i(.)$. We then have the following Theorem on global identifiability in exponential family models.

Theorem 10 *If the Fisher information matrix, $\mathbf{J}$, is non-singular, and $B(\boldsymbol{\theta})$ is continuously differentiable, then the model is globally identifiable (Rothenberg, 1971, Theorem 3 and Prakasa Rao, 1992, Theorem 6.3.2).*

As stated in Section 3.2, Bekker et al. (1994) and Catchpole and Morgan (1997) show that Theorems 3 and 4a are equivalent for exponential family models, which means that if $\mathbf{J}$ is full rank, an appropriate derivative matrix, $\mathbf{D}$, for the same model, will also be full rank. Therefore if an exponential family model is essentially full rank then it is globally identifiable.

3.3.1.1   Linear Regression Model

The linear regression models discussed in Sections 1.2.3 and 1.3.1 can be generalised to a model with $q - 1$ covariates, and q parameters. The model can be written as $E(\mathbf{y}) = \mathbf{X}\boldsymbol{\beta}$, where $\mathbf{y}$ is a vector of n dependent variables, $\mathbf{X}$ is a matrix of covariates (also known as the design matrix) and $\boldsymbol{\beta}$ is a vector of q parameters. A suitable exhaustive summary is $\kappa = \mathbf{X}\beta$. The derivative matrix will then be

$$\mathbf{D} = \frac{\partial \kappa}{\partial \beta} = \mathbf{X}^T.$$

As $\text{rank}(\mathbf{X}^T) = \text{rank}(\mathbf{X})$, checking the rank of the derivative matrix is equivalent to checking the rank of the design matrix. If $\text{rank}(\mathbf{X}) = q$ then the model will not be parameter redundant. As the derivative matrix just consists of covariate values, which have constant values, a full rank $\mathbf{D}$ will always be essentially full rank. By Theorem 10 above, the model will be globally identifiable if $\mathbf{X}$ is full rank. This is the well known identifiability condition for linear regression models (see, for example, Silvey, 1970 page 50).

Suppose there is one covariate and a sample size of n, so that the matrix of covariates is

$$\mathbf{X} = \begin{bmatrix} 1 & x_{1,1} \\ 1 & x_{1,2} \\ \vdots & \\ 1 & x_{1,n} \end{bmatrix}.$$

If there are at least $n = 2$ observations with at least two different covariate values $x_{1,i}$, $\mathbf{X}$ will be full rank 2 and the model is globally identifiable.

3.3.2 General Methods for Distinguishing Local and Global Identifiability

A linear compartmental model is globally identifiable if the solution to the moment invariant equations in terms of the parameters is unique (Bellman and Astrom, 1970 and Godfrey and Distefano, 1987); this is demonstrated in the example in Section 3.3.2.1 below. The moment invariant equations form an exhaustive summary, and this approach can be extended to any exhaustive summary, as explained by Theorem 11 below.

Theorem 11 *A full rank model with an exhaustive summary κ is globally identifiable if and only if there is a unique solution to the set of equations $\mathbf{k} = \kappa(\theta)$ in terms of θ, where $dim(\mathbf{k}) = dim(\kappa)$ (Cole et al., 2010).*

This method works best for simpler exhaustive summaries which are the same length as the number of parameters.

Theorem 5 shows how to find locally identifiable reparameterisations in a non-identifiable models. Such a reparameterisation is at least locally identifiable. To distinguish between local and global identifiability, first reparameterise the original model. Then apply Theorem 11.

3.3.2.1 Three Compartment Model

Consider again the example in Section 1.3.2 from Godfrey and Chapman (1990). In Section 3.1.4.3 we show the model is at least locally identifiable.

Writing $\kappa_{comp}^T = [\kappa_1, \kappa_2, \kappa_3]$, we have:

$$\kappa_1 = \theta_{21}\theta_{32} \tag{3.8}$$

$$\kappa_2 = \theta_{01} + \theta_{2i} + \theta_{32} \tag{3.9}$$

$$\kappa_3 = \theta_{32}(\theta_{01} + \theta_{21}). \tag{3.10}$$

Solving equations (3.8) to (3.10) in terms of the parameters $\theta_{01}, \theta_{21}, \theta_{32}$ gives

$$\theta_{01} = \frac{2(\kappa_3 - \kappa_1)}{\kappa_2 \pm \sqrt{\kappa_2^2 - 4\kappa_3}} \tag{3.11}$$

$$\theta_{21} = \frac{2\kappa_1}{\kappa_2 \pm \sqrt{\kappa_2^2 - 4\kappa_3}} \tag{3.12}$$

$$\theta_{32} = \frac{1}{2}\left(\kappa_2 \pm \sqrt{\kappa_2^2 - 4\kappa_3}\right) \tag{3.13}$$

As there are generically two solutions this model is locally identifiable.

3.4 Proving Identifiability Directly

In some examples, identifiability results can be derived directly from the definition of identifiability (Definition 1 in Section 3.1.1), see for example Prakasa Rao (1992), Van Wieringen (2005) and Holzmann et al. (2006). Deriving results directly may be simpler than using the symbolic method described above, especially when non-identifiability is obvious. This is illustrated in Section 3.4.1 using the population growth model. However the algebra involved in deriving directly from the definition of identifiability will be problem specific. It is usually possible to use the symbolic method instead, which may be simpler, as we demonstrate in Section 3.4.2 below for the closed population capture-recapture model from Holzmann et al. (2006).

3.4.1 The Population Growth Model

Consider the population growth model, introduced in Section 1.1 and 1.2.2. The solution of the differential equation

$$\frac{dP(t)}{dt} = (\theta_b - \theta_d)P(t), \ P(0) = P_0,$$

is

$$P(t, \boldsymbol{\theta}) = P_0 \exp\{(\theta_b - \theta_d)t\}.$$

For simplicity assume that the initial population size P_0 is a fixed known quantity. The parameters, which are the birth and death rate respectively, are

$\boldsymbol{\theta} = [\theta_b, \theta_d]$. In Section 1.2.2 we stated that this model is obviously parameter redundant, as it can be reparameterised in terms of a smaller number of parameters with $\beta = \theta_b - \theta_d$. Non-identifiability follows directly. Using Definition 1, we define the model as

$$M(\boldsymbol{\theta}) = P(t, \boldsymbol{\theta}) = P_0 \exp\{(\theta_b - \theta_d)t\}.$$

As

$$P_0 \exp\{(\theta_{b,1} - \theta_{d,1})t\} = P_0 \exp\{(\theta_{b,2} - \theta_{d,2})t\}$$

if

$$\beta = \theta_{b,1} - \theta_{d,1} = \theta_{b,2} - \theta_{d,2},$$

$P(t, \boldsymbol{\theta}_1) = P(t, \boldsymbol{\theta}_2)$ does not imply that $\boldsymbol{\theta}_1 = \boldsymbol{\theta}_2$, so the model is non-identifiable.

Alternatively, for the symbolic method, we first need an exhaustive summary. This model could be put into a compartmental model format, and a transform function exhaustive summary derived, as described in Section 3.1.4.3. The matrices for this format are $\mathbf{A} = [\theta_b - \theta_d]$, $\mathbf{B} = [P_0]$, $\mathbf{C} = [1]$. The transform function is

$$\frac{P_0}{s - \theta_b + \theta_d}.$$

Therefore an exhaustive summary is $\boldsymbol{\kappa} = [\theta_d - \theta_b]$. The derivative matrix

$$\mathbf{D} = \frac{\partial \boldsymbol{\kappa}}{\partial \boldsymbol{\theta}} = \begin{bmatrix} -1 \\ 1 \end{bmatrix},$$

has rank 1, but there are two parameters, therefore the model is non-identifiable. Maple code for this example is `popgrowth.mw`.

3.4.2   Closed Population Capture-Recapture Model

In Section 3.2.7.2 we discussed a capture-recapture model which included modelling animal survival. This model can be used for what is known as an open population, where births and deaths can occur. If a shorter time period is examined, we may be able to assume that births and deaths do not occur, and this is known as a closed population (see, for example, Williams et al., 2001 Chapter 14). In a closed population, we are typically interested in estimating the unknown population size, N. The population is surveyed on n occasions, and we record the number of animals which are captured x times, f_x.

In the model examined here, we assume that f_x follows a binomial distribution, $\text{Bin}(n, p)$, where p is the probability an animal is captured. Note that we do not observe f_0, the number of animals which are captured zero times, therefore we condition on $i > 0$. The population size can then be estimated as $N = n/\{1 - \Pr(x = 0)\}$ (Link, 2003).

To account for heterogeneity in the capture probability, p, is assumed to follow a distribution, $g(p)$. The probability that an individual is sampled x times is then

$$\pi(x) = \binom{n}{x} \int_0^1 p^x(1-p)^{n-x}g(p)dp, \ x = 0, 1, \ldots, n. \tag{3.14}$$

As we do not observe the number of animals which are captured zero times we are interested in the conditional distribution

$$\pi^c(x) = \frac{\pi(x)}{1-\pi(0)}, \ x = 1, 2, \ldots, n. \tag{3.15}$$

Link (2003) discusses the problem with identifiability in such models. Holzmann et al. (2006) find conditions for identifiability of $g(p)$ directly from the definition of identifiability. Here we consider a particular example where p follows a beta distribution,

$$g(p) = \frac{\Gamma(\alpha+\beta)}{\Gamma(\alpha)\Gamma(\beta)}p^{\alpha-1}(1-p)^{\beta-1},$$

for parameters $\alpha > 0$ and $\beta > 0$. The model is globally identifiable for $n \geq 3$, but non-identifiable for $n = 2$.

Firstly, we consider proving this using the direct method from Holzmann et al. (2006). Let $\pi_1(x)$ and $\pi_1^c(x)$ represent $\pi(x)$ and $\pi^c(x)$ when $g(p)$ follows a beta distribution with parameters α_1 and β_1, and let $\pi_2(x)$ and $\pi_2^c(x)$ represent $\pi(x)$ and $\pi^c(x)$ when $g(p)$ follows a beta distribution with parameters α_2 and β_2. The model is identifiable if $\pi_1^c(x) = \pi_2^c(x)$ implies that $\alpha_1 = \alpha_2$ and $\beta_1 = \beta_2$. As

$$\pi_1^c(x) = \pi_2^c(x) \implies \frac{\pi_1(x)}{1-\pi_1(0)} = \frac{\pi_2(x)}{1-\pi_2(0)} \implies \frac{\pi_1(x)}{\pi_2(x)} = \frac{1-\pi_1(0)}{1-\pi_2(0)} = A,$$

it follows that $\pi_1^c(x) = \pi_2^c(x)$ implies $\pi_1(x) = A\pi_2(x)$ for some $A > 0$.

It is possible to write $\pi_i(x)$ as

$$\pi_i(x) = \binom{n}{x} \sum_{k=x}^n (-1)^{k-x} \binom{n-x}{k-x} m_i(k),$$

where $m_i(k)$ is the moment generating function with

$$m_i(k) = \frac{(\alpha_i+k-1)\ldots\alpha_i}{(\alpha_i+\beta_i+k-1)\ldots(\alpha_i+\beta_i)}, \ i = 1, 2.$$

Therefore $\pi_1^c(x) = \pi_2^c(x)$ implies $m_1(x) = Am_2(x)$ for some $A > 0$.

When $n = 3$ the moment generating functions for $x = 1, 2, 3$ give us the following equations

$$\frac{\alpha_1}{\alpha_1+\beta_1} = \frac{\alpha_2}{\alpha_2+\beta_2},$$

$$\frac{(\alpha_1+1)\alpha_1}{(\alpha_1+\beta_1+1)\alpha_1+\beta_1} = \frac{(\alpha_2+1)\alpha_2}{(\alpha_2+\beta_2+1)\alpha_2+\beta_2},$$

$$\frac{(\alpha_1+2)(\alpha_1+1)\alpha_1}{(\alpha_1+\beta_1+2)(\alpha_1+\beta_1+1)\alpha_1+\beta_1} = \frac{(\alpha_2+2)(\alpha_2+1)\alpha_2}{(\alpha_2+\beta_2+2)(\alpha_2+\beta_2+1)\alpha_2+\beta_2}.$$

The solution is $A = 1$, $\alpha_1 = \alpha_2$, $\beta_1 = \beta_2$, proving the model is globally identifiable for $n = 3$. However when $n = 2$ we only have the first 2 equations, which do not imply that $\alpha_1 = \alpha_2$ and $\beta_1 = \beta_2$, therefore the model is non-identifiable (see Maple code, `betabin.mw`).

Alternatively consider using the symbolic method. Evaluating equations 3.14 and then 3.15 when $g(p)$ follows a Beta distribution for $n = 2$ gives the exhaustive summary terms

$$\kappa_2 = \left[\begin{array}{c} \pi^c(1) \\ \pi^c(2) \end{array} \right] = \left[\begin{array}{c} \frac{2\beta}{2\beta+\alpha+1} \\ \frac{\alpha+1}{2\beta+\alpha+1} \end{array} \right].$$

The vector of parameters is $\boldsymbol{\theta} = [\alpha, \beta]$. The derivative matrix,

$$\mathbf{D} = \frac{\partial \kappa_2}{\partial \boldsymbol{\theta}} = \left[\begin{array}{cc} -\frac{2\beta}{(2\beta+\alpha+1)^2} & \frac{2\beta}{(2\beta+\alpha+1)^2} \\ \frac{2(\alpha+1)}{(2\beta+\alpha+1)^2} & -\frac{2(\alpha+1)}{(2\beta+\alpha+1)^2} \end{array} \right],$$

has rank 1, but there are 2 parameters, therefore the model is non-identifiable.

For $n = 3$ the exhaustive summary is

$$\kappa_3 = \left[\begin{array}{c} \pi^c(1) \\ \pi^c(2) \\ \pi^c(3) \end{array} \right] = \left[\begin{array}{c} \frac{3\beta(\beta+1)}{\alpha^2+3\alpha\beta+3\beta^2+3\alpha+6\beta+2} \\ \frac{3\beta(\alpha+1)}{\alpha^2+3\alpha\beta+3\beta^2+3\alpha+6\beta+2} \\ \frac{(\alpha+1)(\alpha+2)}{\alpha^2+3\alpha\beta+3\beta^2+3\alpha+6\beta+2} \end{array} \right].$$

The derivative matrix, $\mathbf{D} = \partial\kappa_2/\partial\boldsymbol{\theta}$, is shown in the Maple code to have rank 2, therefore the model is at least locally identifiable. Global identifiability, and the generalisation that if $n \geq 3$ the beta family of distributions is identifiable, are both confirmed in the Maple code, `betabin.mw`.

3.5 Near-Redundancy, Sloppiness and Practical Identifiability

In this Section we discuss the concepts of near-redundancy (see Definition 6), sloppiness (see Definition 7) and practical non-identifiability (see Definition 8).

Cole et al. (2010) use the symbolic method to investigate near-redundancy. Theorem 7 uses a decomposition of the derivative matrix to show points in the parameter space that are parameter redundant within a full rank model. If a full rank model is parameter redundant at $\boldsymbol{\theta} = \boldsymbol{\theta}_{PR}$ then it will be near-redundant for parameter estimates close to $\boldsymbol{\theta}_{PR}$.

Catchpole et al. (2001) investigate near-redundancy by examining the eigenvalues of the Fisher Information matrix or the Hessian matrix. If the model is parameter redundant the Fisher Information matrix and the Hessian matrix evaluated at the maximum likelihood estimate have at least one zero eigenvalue. In a near-redundant model these matrices will have eigenvalues close to zero. This is the identical method used to determine sloppiness, which by the definition compares the smallest and largest eigenvalues of these matrices. Chis et al. (2016) state that a model is sloppy if $\lambda_{\min}/\lambda_{\max} < 0.001$, where $\lambda_{\min}$ is the smallest eigenvalue and $\lambda_{\max}$ is the largest eigenvalue. The same ratio of smallest and largest eigenvalues has been used to investigate near-redundancy (Zhou et al., 2019 and Cole, 2019).

Raue et al. (2009) show how practical identifiability is examined by finding the likelihood-based confidence region for each parameter and checking whether the values are finite. For a model with q parameters, $\boldsymbol{\theta}$, log-likelihood $l(\boldsymbol{\theta})$ and maximum likelihood estimate $\hat{\boldsymbol{\theta}}$, a $100(1-\alpha)\%$ likelihood based confidence region is defined by

$$\{\boldsymbol{\theta}|l(\hat{\boldsymbol{\theta}}) - l(\boldsymbol{\theta}) < \frac{1}{2}\chi^2_{\alpha,q}\},$$

where $\chi^2_{\alpha,q}$ is the α percentage point of the chi-squared distribution with q degrees of freedom. That is, $\Pr(X > \chi^2_{\alpha,q}) = \alpha$ where X follows a chi-squared distribution with q degrees of freedom. The boundaries of the confidence region give a likelihood based confidence interval. In practice, a log-likelihood profile can be used to obtain confidence limits. The log-likelihood profile for a parameter, θ_i, is found by fixing θ_i and maximising $l(\boldsymbol{\theta})$ with respect to the other parameters, that is

$$l_{PF}(\theta_i) = \max_{\theta_j \neq \theta_i}\{l(\boldsymbol{\theta})\}.$$

As stated in Section 2.1 a non-identifiable parameter will have a flat log-likelihood profile. A model that is practically non-identifiable will have a maximum but will be relatively flat. In a practically identifiable model the likelihood base confidence limits exists; that is, there will be a value of θ_i less than the maximum likelihood estimate $\hat{\theta}_i$ and a value of θ_i greater than the maximum likelihood estimate $\hat{\theta}_i$ where $l_{PF}(\theta_i) - l(\hat{\boldsymbol{\theta}}) = \frac{1}{2}\chi^2_{\alpha,q}$. Whereas for a model that is practically non-identifiable, these confidence limits do not exist. This is illustrated in Figure 3.1.

Whilst it is usual that a model will be both sloppy and practically non-identifiable, this is not always the case (see Dufresne et al., 2018).

3.5.1   Mark-recovery Model

We return again to the mark-recovery example in Section 3.2.7.1. The mark-recovery model, with time-dependent first year survival, $\phi_{1,t}$, constant adult survival, ϕ_a and two classes of recovery, λ_1 and λ_a, can be near-redundant. The model is full rank, but when $\phi_{1,1} = \phi_{1,2} = \phi_{1,3}$ it is parameter redundant. Using Cole et al. (2010)'s method we can say the model will be near-redundant when $\phi_{1,1} \approx \phi_{1,2} \approx \phi_{1,3}$.

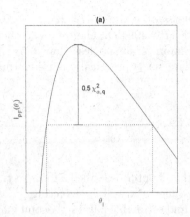
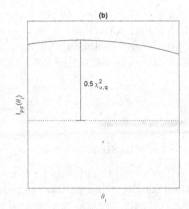

FIGURE 3.1

An illustration of practical identifiability showing profile log-likelihood plots. Figure (a) is practically identifiable; likelihood confidence limits (indicated by the dotted lines) exist. Figure (b) is practically non-identifiable. Whilst the profile log-likelihood is not completely flat and a maximum likelihood estimate exists, the likelihood confidence limits do not exist. The profile log-likelihood is always less than $0.5\chi^2_{\alpha,q}$ from its maximum.

Consider a data set where 1000 animals are marked each year and the numbers recovered in year j that were marked in year i, $N_{i,j}$, are summarised in the matrix

$$\mathbf{N} = \begin{bmatrix} 60 & 40 & 20 \\ 0 & 59 & 41 \\ 0 & 0 & 58 \end{bmatrix}.$$

The maximum likelihood estimates for this data set are $\hat{\phi}_{1,1} = 0.4$, $\hat{\phi}_{1,2} = 0.41$, $\hat{\phi}_{1,3}$, $\hat{\lambda}_1 = 0.1$, $\hat{\lambda}_a = 0.2$. In the Maple code `markrecovery.mw` we show that the Hessian matrix is

$$\mathbf{H} = \begin{bmatrix} 544.51 & 0 & 0 & -204.55 & -965.91 & 767.05 \\ 0 & 413.39 & 0 & -200.00 & -1000.00 & 500.00 \\ 0 & 0 & 183.03 & 0 & -1061.57 & 0 \\ -204.55 & -200.00 & 0 & 338.74 & -108.30 & -855.95 \\ -965.91 & -1000.00 & -1061.57 & -108.30 & 18852.98 & 338.93 \\ 767.05 & 500.00 & 0 & -855.95 & 338.93 & 2673.97 \end{bmatrix}.$$

The eigenvalues of this matrix are $19022.12, 3283.38, 473.31, 176.24, 51.54, 0.030$. Using the method from Catchpole et al. (2001) the model for this data set is near-redundant as the smallest eigenvalue is close to 0. We also classify the model with this data as sloppy as $\lambda_{\min}/\lambda_{\max} = 0.0000016 < 0.001$.

Using Raue et al. (2009)'s method the maximum likelihood estimate is $l(\hat{\boldsymbol{\theta}}) = -1102.490947$. We show in the Maple code that the profile log-likelihood for $\phi_{1,1}$ is relatively flat, for example $l_{PF}(\phi_{1,1} = 0.1) = -1102.494042$ and $l_{PF}(\phi_{1,1} = 0.9) = -1102.496459$. The difference between the profile log-likelihood and its maximum is always less than $0.5\chi^2_{0.05,6} = 6.295$. Therefore the model with this data set is practically non-identifiable.

3.6   Discussion

As well as giving the main definitions, this chapter described a formal symbolic method for detecting parameter redundancy. As well as detecting parameter redundancy the method can be used to find estimable parameter combinations, generalise results and distinguish between global and local identifiability. The method is described by Bailey et al. (2010) as the gold standard for detecting parameter redundancy. There are alternatives to the symbolic method for detecting parameter redundancy, which are are discussed in Chapter 4.

In a complex model the derivative matrix may be large and/or consist of algebraically complex terms. In these situations, a symbolic algebra package could run out of memory calculating the rank. In Section 5.2 we discuss an extended version of the symbolic method which can overcome this problem.

4

Practical General Methods for Detecting Parameter Redundancy and Identifiability

---

Different methods for detecting parameter redundancy are discussed in Gimenez et al. (2004). This chapter discusses the same types of methods, but also adds more recent developments. Firstly, we consider different methods for detecting parameter redundancy or non-identifiability. These methods are divided into three categories: numerical methods, symbolic methods and a hybrid of both numeric and symbolic methods, discussed in Sections 4.1, 4.2 and 4.3 respectively. We show how to use each of the methods.

Sometimes models will behave as a parameter redundant model, when they are not parameter redundant, because the model is in some way close to a parameter redundant model, or models will be theoretically identifiable but not identifiable in practice. These problems are known as near-redundancy and practical identifiability, and discussed in Section 4.4.

Each section demonstrates the limitations of each method, then the different methods are compared in Section 4.5.

4.1 Numerical Methods

Parameter redundancy can be detected numerically by examining the rank (or effective rank) of the Hessian matrix, using simulation or using profile log-likelihoods (see for example Viallefont et al., 1998 or Gimenez et al., 2004). Another numerical method is data cloning (Lele et al., 2010). Section 4.1.1 reviews the Hessian Method, Section 4.1.2 demonstrates how simulation can be used to detect parameter redundancy, and Section 4.1.3 explains data cloning for investigating parameter redundancy. Section 4.1.4 investigates using the profile log-likelihood to detect parameter redundancy and distinguish between local and global identifiability.

4.1.1 Hessian Method

As discussed in Section 2.3, if a model is parameter redundant then the Fisher information matrix will be singular, so that it is not possible to find the inverse of the matrix, and therefore standard errors do not exist. It follows that the Hessian matrix at the maximum likelihood estimate will also be singular. This is explained more formally in Theorem 4b of Section 3.2. Typically the Hessian matrix is simpler to use the Fisher information matrix (Little et al., 2009).

The Hessian matrix is a square matrix with dimensions identical to the number of parameters. The entries of the matrix correspond to the second differentials of the log-likelihood with respect to the parameters, that is

$$
\mathbf{H} = \begin{bmatrix}
\frac{\partial^2 l}{\partial \theta_1 \partial \theta_1} & \frac{\partial^2 l}{\partial \theta_1 \partial \theta_2} & \cdots & \frac{\partial^2 l}{\partial \theta_1 \partial \theta_q} \\
\frac{\partial^2 l}{\partial \theta_2 \partial \theta_1} & \frac{\partial^2 l}{\partial \theta_2 \partial \theta_2} & \cdots & \frac{\partial^2 l}{\partial \theta_2 \partial \theta_q} \\
\vdots & & & \vdots \\
\frac{\partial^2 l}{\partial \theta_q \partial \theta_1} & \frac{\partial^2 l}{\partial \theta_q \partial \theta_2} & \cdots & \frac{\partial^2 l}{\partial \theta_q \partial \theta_q}
\end{bmatrix},
$$

where l is the log-likelihood function and θ_i $(i = 1, \ldots, q)$ are the parameters. If the square matrix is singular then the rank of the matrix will be less than the dimensions of the matrix. Therefore, theoretically, parameter redundancy can be investigated by finding the rank of the Hessian matrix (see Theorem 4b).

It is frequently difficult to find the exact Hessian matrix, therefore in practice the Hessian matrix is found numerically. In Appendix B.1 we show a general method of finding an approximate Hessian matrix.

When the Hessian matrix is found numerically the rank may be different from the rank of the exact Hessian. The alternative involves finding the eigenvalues of the matrix. As the Hessian matrix at the maximum likelihood estimate will be singular for a parameter redundant model the matrix will have at least one zero eigenvalue. Due to numerical error, the eigenvalue may be close to zero when the Hessian matrix is found numerically. Viallefont et al. (1998) use this fact to create the Hessian method for investigating parameter redundancy, which involves checking the eigenvalues of a Hessian matrix found numerically. Example of the use of the Hessian method include Gimenez et al. (2004), Reboulet et al. (1999) and Little et al. (2010).

The steps to conduct the Hessian method (when the approximate Hessian matrix is used) are outlined below:

1. Calculate the approximate Hessian matrix.

2. Find the eigenvalues of the Hessian matrix.

3. Standardise the eigenvalues. Take the modulus of the eigenvalues. Then divide each eigenvalue by the largest value eigenvalue.

4. If any of the standardised eigenvalues are smaller than a threshold, τ, the model is parameter redundant. The number of estimable parameters and rank of the Hessian matrix is equal to the number of standardised eigenvalues greater than τ.

5. For parameter redundant models, examine the eigenvectors associated with the small eigenvalues. Any entries close to zero for all of

these eigenvalues correspond to parameters that are at least locally identifiable.

Appendix B.1 provides R code to execute this Hessian method. We demonstrate this method through a series of examples in Section 4.1.1.1 to 4.1.1.3.

The problem with the Hessian method is determining how close to zero an eigenvalue must be to denote parameter redundancy. The threshold used by Viallefont et al. (1998) was $\tau = q\epsilon$, where q is the number of parameters, and $\epsilon = 1 \times 10^{-9}$. However this value of ϵ does not work in all cases. In some cases it could be too strict and misclassify parameter redundant models. Alternative thresholds are discussed in Konstantinides and Yao (1988) and Little et al. (2010). The threshold varies depending on the computer package used, the machine precision, the error used in the Hessian approximation, as well as the model being fitted. The biggest source of this variation in the approximation, discussed in Appendix B.1, is the step length, δ, so a possible general threshold is $\tau = p\delta$. However, we demonstrate in Section 4.1.1.2 that using this threshold can also misclassify models.

The number of parameters that can be estimated in a model is needed for model selection using information criteria, such as the AIC, as discussed in Section 2.4.1. For example the AIC is calculated as $AIC = -2l + 2k$, where l is the log-likelihood, and k is the number of estimable parameters in the model. As given in step 4, the number of estimable parameters can be calculated as the number of eigenvalues that exceed the threshold τ. Some computer packages, such as the computer package Mark for marked populations (Cooch and White, 2017) use the Hessian method with an automatic threshold. This should always be used with caution, as using a general threshold can lead to incorrect computation of the number of parameters, and therefore the wrong AIC ranking of the models (Gimenez et al., 2004).

The Hessian method must have a specific data set to check for parameter redundancy. It therefore will not be clear whether the parameter redundancy was caused by the data set (extrinsic parameter redundancy) or the model is parameter redundant (intrinsic parameter redundancy). Intrinsic parameter redundancy can be investigated using a simulated data set, that would need to represent the best possible quality data, which is typically achieved using a very large sample size (Gimenez et al., 2004). We demonstrate the use of simulated data in example 4.1.1.3 below.

The big advantage of this method is that it can be calculated easily as the Hessian matrix is normally calculated to find the standard errors. Therefore it can be easily added to an existing computer package, for example the computer package Mark (Cooch and White, 2017).

4.1.1.1 Basic Occupancy Model

Consider again the basic occupancy model discussed in Section 1.2.1, as well as in Chapters 2 and 3. The log-likelihood is given in Section 3.1.4.1 of

Chapter 3 as

$$l = c + n_1 \log(\psi p) + n_2 \log(1 - \psi p),$$

for some constant c, where ψ and p are parameters representing the probability a site is occupied and the probability of detection respectively, and where n_1 and n_2 are the number of sites observed and unoccupied respectively.

In Section 3.2.1, we demonstrated that the exact Hessian matrix can be found for general data values of n_1 and n_2. Suppose that $n_1 = n_2 = 45$. The maximum likelihood estimates satisfy the equation $\hat{\psi}\hat{p} = 0.5$. For example one set of maximum likelihood estimates found is $\hat{\psi} = 0.75$ and $\hat{p} = 2/3$. The exact Hessian matrix is then

$$\mathbf{H} = \begin{bmatrix} 160 & 180 \\ 180 & 202.5 \end{bmatrix}.$$

This Hessian matrix has rank 1. (See Maple code `basicoccupancy.mw`.) However, as in most cases the exact Hessian matrix cannot be found, we demonstrate how the Hessian method works. Step 1 involves finding the approximate Hessian matrix. Using the R code `bassicocc.R` the approximate Hessian matrix, with step length $\delta = 0.00001$, is

$$\mathbf{H}_A = \begin{bmatrix} 160.0000132 & 180.0000149 \\ 180.0000149 & 202.5000256 \end{bmatrix}.$$

We note that the rank of this approximation of the Hessian matrix is 2, demonstrating why we cannot use the rank to check for parameter redundancy when the Hessian is found numerically. Instead step 2 of the Hessian method involves calculating the eigenvalues which are 362.50 and 0.0000039. Step 3 involves finding the modulus of the eigenvalues and dividing though by the largest value of 362.50. This gives standardised eigenvalues of 1 and 0.000000011. For step 4, the smallest eigenvalue value is smaller than the threshold of $\tau = q\delta = 2 \times 0.00001 = 0.00002$ therefore we conclude that this model is parameter redundant. There is one eigenvalue larger than τ so the rank and number of estimable parameters is 1.

In Step 5 we examine the eigenvector corresponding to the smallest eigenvalue. The eigenvector is $\begin{bmatrix} -0.74 & 0.66 \end{bmatrix}$. As neither entry is close to zero, neither of the parameters are individually identifiable.

4.1.1.2   Mark-Recovery Model

Here we demonstrate that the threshold for deciding on whether an eigenvalue is close to zero can misclassify parameter redundancy even in the same example. The example used is a parameter redundant mark-recovery model, first discussed in Section 1.2.4. This model has separate survival probabilities for first years, ϕ_1 and adults, ϕ_a, as well as separate recovery probabilities for first years, λ_1, and adults, λ_a. Table 4.1 shows two sets of simulated data from this mark-recovery model.

TABLE 4.1

Two simulated mark-recovery data sets. All data
sets have 1000 animals marked each year over 10
years of marking and recovery. The rows (i)
correspond to the years of marking, the columns (j)
correspond to the number of years marking. The
body of the table gives $N_{i,j}$, the number of animals
marked in year i and recovered dead in year j.

Data Set 1

i/j	1	2	3	4	5	6	7	8	9	10
1	58	29	12	5	3	3	0	0	2	0
2	0	66	22	15	15	3	3	1	0	0
3	0	0	58	29	13	7	7	2	0	1
4	0	0	0	68	25	18	9	7	0	2
5	0	0	0	0	66	33	10	3	1	1
6	0	0	0	0	0	61	24	16	7	5
7	0	0	0	0	0	0	66	28	7	11
8	0	0	0	0	0	0	0	59	26	8
9	0	0	0	0	0	0	0	0	53	27
10	0	0	0	0	0	0	0	0	0	63

Data Set 2

i/j	1	2	3	4	5	6	7	8	9	10
1	56	37	9	5	3	0	0	0	0	1
2	0	67	31	28	4	5	3	3	1	0
3	0	0	60	27	14	9	2	0	2	1
4	0	0	0	57	27	16	8	0	1	1
5	0	0	0	0	64	24	10	8	4	0
6	0	0	0	0	0	66	35	17	8	3
7	0	0	0	0	0	0	57	32	15	8
8	0	0	0	0	0	0	0	71	34	23
9	0	0	0	0	0	0	0	0	64	36
10	0	0	0	0	0	0	0	0	0	65

i	y_i	$x_{1,i}$	$x_{2,i}$
1	5.8	3.2	1.1
2	5.9	3.2	1.1
3	7.4	4.5	2.2
4	7.5	4.5	2.2

TABLE 4.2

A small regression data set with $n = 4$ observations, y_i and two covariates $x_{1,i}$ and $x_{2,i}$

The mark-recovery model with parameters ϕ_1, ϕ_a, λ_1 and λ_a is fitted to each of the three data sets and then parameter redundancy investigated using the Hessian Method. For the first data set, at a maximum likelihood estimate of $\hat{\phi}_1 = 0.22$, $\hat{\phi}_a = 0.53$, $\hat{\lambda}_1 = 0.08$, $\hat{\lambda}_a = 0.29$, the Hessian matrix has standardised eigenvalues of $1, 0.7918, 0.2106$ and 0.00001390. We again used a step-length $\delta = 0.00001$ to find the approximate Hessian matrix. As there are four parameters this results in a threshold of $\tau = 4 \times 0.00001 = 0.00004$. The smallest eigenvalue is smaller than the threshold, therefore we conclude the model with this data set is parameter redundant. The eigenvector for the only eigenvalue below the threshold is $\begin{bmatrix} 0.6619 & 0.00004273 & 0.1568 & -0.7330 \end{bmatrix}^T$. As the second entry is close to zero the second parameter, which is ϕ_a, is individually identifiable, whereas the other parameter are not.

However for the second data set the threshold fails. At a maximum likelihood estimate of $\hat{\phi}_1 = 0.54$, $\hat{\phi}_a = 0.48$, $\hat{\lambda}_1 = 0.15$, $\hat{\lambda}_a = 0.13$, the standardised eigenvalues of the Hessian matrix are $1, 0.63218, 0.2028$ and 0.00004162. Whilst the last eigenvalue is close to zero it is larger than the threshold of 0.00004. If we used an automatic threshold then we may incorrectly conclude this model is not parameter redundant. Close inspection of the smallest eigenvalue may lead us to suspect the model may be parameter redundant, as it is very close to the threshold.

4.1.1.3   Linear Regression Model

This example demonstrates how to distinguish between extrinsic and intrinsic parameter redundancy using the Hessian method. In Section 1.3.1 we introduced a simple linear regression model with a very small sample size. Here we also consider a linear regression model with a small sample size that results in a parameter redundant model. We then compare results with a large simulated data set to show that the model is not parameter redundant.

Firstly, we consider the small data set, which is given in Table 4.2. This data set is unrealistically small and it would not normally be sensible to fit a linear regression model in this case. We consider the linear regression model with $E(y_i) = \beta_0 + \beta_1 x_{1,i} + \beta_2 x_{2,i}$.

The design matrix, which is a matrix of covariates, is

$$\mathbf{X} = \begin{bmatrix} 1 & 3.2 & 1.1 \\ 1 & 3.2 & 1.1 \\ 1 & 4.5 & 2.2 \\ 1 & 4.5 & 2.2 \end{bmatrix}$$

This matrix has rank 2, but there are three regression parameters, β_0, β_1, β_2, therefore this model with this data set is parameter redundant. Examining the rank of the design matrix to check parameter redundancy is the well known identifiability condition for linear regression models (see, for example, Silvey, 1970 page 50). This is discussed in more detail in Section 3.3.1.1.

We can use the Hessian method to show this data set causes parameter redundancy. There are four parameters in this model, the three regression parameters $(\beta_0, \beta_1, \beta_2)$ and the standard deviation of the error term (σ). The R code `linear.R` demonstrates how the Hessian method can be used in this example. At a maximum likelihood estimate the standardised eigenvalues of the Hessian matrix are $1, 0.0049, 0.00026, 0.00000000015$ As the smallest eigenvalue is smaller than $\tau = 4 \times 0.00001 = 0.00004$ we conclude that this model with this data set is parameter redundant, as expected. This demonstrates the standard use of the Hessian method to check extrinsic parameter redundancy.

If we wish to test the parameter redundancy of the model in general (intrinsic parameter redundancy), we need to simulate a large set of ideal data. In this case, a data set is simulated with 1000 observations. The covariates are simulated from a continuous uniform distribution, the first covariate, $x_{1,i}$, has a range of $(0, 5)$ the second covariate, $x_{2,i}$ has a range of $(0, 2)$. The observations are simulated from a normal distribution with mean $\mu_i = 0.4 + 4.5x_{1,i} + 2.2x_{2,i}$ and variance $\sigma^2 = 1$. The standardised eigenvalues are $1, 0.21, 0.063, 0.011$. As the smallest eigenvalue is larger than $\tau = 0.00004$ we correctly conclude that the model is not parameter redundant.

4.1.2 Simulation Method

Simulation can be used to check for parameter redundancy by simulating a large data set from the model of interest and then finding the maximum likelihood estimates for that simulated data (see, for example, Kendall and Nichols, 2002, Gimenez et al., 2004, Bailey et al., 2010, Lavielle and Aarons, 2016, Auger-Méthé et al., 2016, Mosher et al., 2018). This method is also known as the analytic-numeric method (Burnham et al., 1987).

In simulating the data set, specific values of the parameters are used, which here we term the true values. The difference between the true values and the maximum likelihood estimates is known as the bias, that is

$$\text{bias}(\theta) = |\hat{\theta} - \theta_T|,$$

where θ_T is the true value of the parameter and $\hat{\theta}$ is the maximum likelihood estimate. As discussed in Section 2.2, an identifiable model should have a bias of approximately zero, however for a parameter redundant model the bias could be larger (Gimenez et al., 2004). This was discussed in Section 2.2. Starting values are used to initiate the algorithm which finds the maximum likelihood estimates. If the starting values are identical to the true values it is possible for the bias to be close to zero for a parameter redundant model.

The coefficient of variation is the ratio of the standard error of a parameter estimate to the actual parameter estimate, that is

$$\mathrm{CV}(\theta) = \frac{\mathrm{SE}(\theta)}{\hat{\theta}}.$$

As we explained in Section 2.3 the standard errors do not exist in parameter redundant models. In practice, when found numerically, the standard errors may be large, so in some cases undefined as the variance is negative. We would therefore expect the coefficient of variation to be small for an identifiable parameter, but be large or undefined for a non-identifiable parameter. Kendall and Nichols (2002) state that a parameter can be estimated, that is, the parameter is identifiable, if the parameter is unbiased to the 5th decimal place and the coefficient of variation is less than 1, using sample sizes of 3000 to 3500.

The simulated data is typically derived from expected values, as we demonstrate in Section 4.1.2.1. It is also possible to use random numbers to generate the simulated data set. As a single simulation generated from random numbers can be non-standard, it is recommended that multiple data sets are simulated and then the average bias and average coefficient of variation is examined. Regardless of the simulation method used the results will be dependent on the true values of the parameters used. Therefore Gimenez et al. (2004) recommend trying several different set of parameter values.

This method can be inaccurate, because the bias is highly dependent on the starting values used, and because the numerically found standard errors may be undefined or inaccurate.

As this method uses simulated data, it can only be used to check intrinsic parameter redundancy. It cannot be used to check extrinsic parameter redundancy.

4.1.2.1 Basic Occupancy Model

We revisit the basic occupancy model first discussed in Section 1.2.1 of Chapter 1, as well as Section 4.1.1.1 above. The data for this example consist of the number of site where the species was observed to be present, n_1, and the number of sites where the species was not observed to be present, n_2. If n sites were visited the expected data set comes from the expected values which are

$$E(n_1) = n\psi p, \ E(n_2) = n(1 - \psi p).$$

For example, if $n = 3000$, $\psi = 0.6$ and $p = 0.4$ the expected data set would be $n_1 = 720$, $n_2 = 2280$. Table 4.3 shows results from the simulation method for

TABLE 4.3

Simulation method for detecting parameter redundancy for the basic occupancy model example. The original model has parameters ψ and p and is parameter redundant. The reparameterised model has parameter $\beta = \psi p$ and is not parameter redundant. The column labelled True gives the true value used to generated the expected data set with $n = 3000$. The columns labelled Starting give the starting values, and the columns labelled CV give the coefficient of variation.

		Original model					
Parameter	True	Starting	Bias	CV	Starting	Bias	CV
ψ	0.6	0.6	0.00000	10.73	0.5	0.11010	NaN
p	0.4	0.4	0.00000	10.73	0.5	0.08990	NaN

		Reparameterised model					
Parameter	True	Starting	Bias	CV	Starting	Bias	CV
β	0.24	0.24	0.00000	0.032	0.5	0.00000	0.032

detecting parameter redundancy. We compare the parameter redundant basic occupancy model with parameters p and ψ with the reparameterised basic occupancy model with parameter $\beta = p\psi$. The reparameterised model is not parameter redundant.

In the original model the bias is dependent on the starting value used. When the starting value is identical to the true values the bias is close to zero, therefore from the bias alone the model would be classified incorrectly as identifiable. When the starting values are both 0.5 there is a large bias and the model is correctly identified as parameter redundant. In the first case the coefficient of variation is much larger than 1, and in the second case the standard error is undefined, therefore the coefficient of variation is also undefined.

For the reparameterised model, the model the bias is small regardless of the starting value used and the coefficient of variation is much smaller than 1. Both statistics correctly show the model is not parameter redundant.

We recommend that the true values are not used as starting values, if using simulation to detect parameter redundancy.

4.1.3 Data Cloning

Data cloning is a technique that uses Bayesian inference to obtain maximum likelihood estimates in complex models where the likelihood is intractable (Lele et al., 2007). As well as obtaining maximum likelihood estimates and standard errors, the method can also be used to detect identifiability (Lele et al., 2010). The Bayesian methodology is discussed in more detail in

Chapter 6, with Section 6.5 also discussing data cloning. However, as the method can be used to detect identifiability in classical models it is also discussed in this section.

Bayesian inference involves using both the likelihood, which we have already encountered, as well as prior information on the parameters. A prior distribution represents this prior knowledge on the parameters. Rather than finding maximum likelihood estimates of parameters, Bayesian methodology involves finding a distribution for the parameters called the posterior distribution. In most cases the posterior distribution is intractable and an MCMC method can be used to generate samples from the posterior distribution. For example, the program Winbugs (Lunn et al., 2000) will find MCMC samples from the posterior distribution. The summary statistics produced for each parameter include the mean of the generated samples, which is an approximation of the marginal posterior mean of that parameter, and the variance of the generated samples which is an approximation for the marginal posterior variance of that parameter. Below we refer to these as the posterior mean and posterior variance. More detail on Bayesian inference is given in Section 6.1.

Data cloning involves using Bayesian inference with a likelihood based on K clones or copies of the data, rather than the data itself (Lele et al., 2007). For approximate classical inference, uninformative priors are used on the parameters. For example, for a parameter $0 < \theta < 1$, an uninformative prior is the uniform distribution on the range 0 to 1. For a large number of clones, K, the maximum likelihood estimate of a parameter will be approximately the posterior mean for that parameter. The variance of the maximum likelihood estimate for a parameter will be approximately K times the posterior variance for the parameter.

The variance can be used to investigate parameter redundancy. The scaled variance is defined as the posterior variance for K clones divided by the posterior variance when $K = 1$. Lele et al. (2010) show that if a parameter is estimable, (that is, not parameter redundant for the given data set), then the scaled variance will be approximately $1/K$. However, if a parameter is not estimable, (that is, the parameter is non-identifiable for the given data set), then the scaled variance will be much larger than $1/K$. Examples of use of this method include Ponciano et al. (2012) and Mosher et al. (2018).

The data cloning method involves the following steps below.

1. Perform Bayesian inference with a likelihood based on the original data $(K = 1)$, using uninformative priors on all parameters.

2. Record the posterior variance for the parameters.

3. Create a data set consisting of K clones, that is, the data repeated K times.

4. Perform Bayesian inference with a likelihood based on the cloned data set of step 3, using uninformative priors on all parameters.

5. Record the posterior variance for the parameters. Scale this variance by dividing through by the posterior variance for $K = 1$, found in step 2.

6. Repeat steps 3 to 5 for successively larger values of K.

7. If the scaled variance is approximately equal to $1/K$ for a parameter then that parameter is identifiable and can be estimated. If the standardised variance is much larger than $1/K$ then that parameter is non-identifiable and cannot be estimated. If there is at least one non-identifiable parameter then the model with that set of data is parameter redundant.

Examples of utilising this method are given in Section 4.1.3.1 and 4.1.3.2 below, also demonstrating suitable values of K.

This is primarily a method of detecting extrinsic parameter redundancy, that is, this method is used to detect parameter redundancy for a specific data set. In data cloning literature, extrinsic parameter redundancy is called non-estimability of parameters. If real data are used it is not possible to tell whether the parameter redundancy is caused by the specific data set or the model (or both). This method could also be used to check intrinsic parameter redundancy, by using a large simulated data set. However as a large data set is needed, to guarantee the data are not causing the parameter redundancy, the cloned data sets will be extremely large which could make this method very slow to test intrinsic parameter redundancy, making it infeasible in some complex models.

In Section 4.1.3.2 we demonstrate that this method can give the incorrect results.

Campbell and Lele (2014) demonstrate how an ANOVA test can be used with data cloning to find estimable parameter combinations.

Data cloning can be used in a classical framework too. Rather than performing Bayesian inference, the maximum likelihood estimates are found using a standard classical method. If the parameter is identifiable we should have the following relationship between the standard error for the original data ($SE_{K=1}$) and the standard error for K clones (SE_K)

$$SE_{K=1} = SE_K \times K^{0.5}$$

(Cooch and White 2017). However note that the actual standard error is undefined in a non-identifiable model, numerically found standard errors may be undefined or inaccurate.

4.1.3.1 Basic Occupancy Model

Consider again the basic occupancy model first discussed in Section 1.2.1 and discussed in Sections 4.1.1.1 and 4.1.2.1 above.

The first parameterisation of the model we use in this section has log-likelihood

$$l = c + n_1 \log(\psi p) + n_2 \log(1 - \psi p),$$

TABLE 4.4
Results for data cloning of occupancy example. The scale variance is the sample variance for the marginal posterior distribution divided by the sample variance when $K = 1$. The scaled variance is given for parameters p and ψ for the first parameterisation and β for the second parameterisation.

K	$1/K$	Scaled Variance p	Scaled Variance ψ	Scaled Variance β
1	1	1	1	1
5	0.2	0.9205	0.9221	0.2036
10	0.1	0.9256	0.9221	0.1006
20	0.05	0.8727	0.917	0.0515
40	0.025	0.8827	0.9285	0.0256
60	0.0167	0.9802	0.9568	0.0171

(as given in Section 3.1.4.1) for some constant c, where ψ and p are parameters representing the probability a site is occupied and the probability of detection respectively, and where $n_1 = 45$ is the number of sites where the species was observed and $n_2 = 45$ is the number of sites where the species was not observed. As the parameters are are both probabilities, the uninformative priors used for both parameters are the continuous uniform distribution on the range 0 to 1. The second parameterisation of the model we use is a reparameterisation with $\beta = p\psi$ so that the log-liklihood is

$$l = c + n_1 \log(\beta) + n_2 \log(1 - \beta).$$

Again the prior used for the parameter β is the continuous uniform distribution on the range 0 to 1. The first parameterisation is obviously parameter redundant, and the second parameterisation is not parameter redundant.

Winbugs is used to perform data cloning. The Winbugs code used is given in Appendix B.2. Here we used a burn-in of 1000 and 10000 updates. The results are summarised in Table 4.4, which gives the scaled sample variance for parameters p and ψ for the first parameterisation and β for the second parameterisation. Clearly for the first parameterisation the scaled variance is not equal to $1/K$, so as expected we conclude neither parameter can be estimated and the model with this data set is parameter redundant. For the second parameterisation the scaled variance is approximately equal to $1/K$, therefore we can estimate β and the model with this data set is not parameter redundant.

4.1.3.2 Population Growth Model

Consider the population growth example introduced in Sections 1.1 and 1.2.2. Suppose that the initial population size is known to be 100, and the population is then measured, with potential error, at 10 time points, as given in Table 4.5, (note this is simulated data). The population size at time t is

$$y(t) = 100 \exp\{(\theta_b - \theta_d)t\},$$

TABLE 4.5

Simulated population growth data. t is the time point and $y(t)$ is the population size at time t.

t	1	2	3	4	5	6	7	8	9	10
$y(t)$	111	124	133	150	165	181	201	223	250	275

where θ_b is the birth rate and θ_d is the death rate. Model fitting typically uses the method of least squares, which is equivalent to assuming that the error on the observation is normally distributed with mean 0 and variance σ^2 and using maximum likelihood to find the parameter estimates (Raue et al., 2009). Data cloning can therefore be used to check for identifiability by assuming the the population size at time t follows a normal distribution with mean $100\exp\{(\theta_b - \theta_d)t\}$ and variance σ^2. The R code pop.R uses Winbugs (Lunn et al., 2000), which is called from R using the package R2WinBUGS (Sturtz et al., 2005) to perform data cloning. The results are given in Figure 4.1 for the three parameters θ_b, θ_d and σ^2.

This example is obviously parameter redundant, as explained in Section 1.2.2, and this is also shown formally in Section 3.4.1. However in Figure 4.1 the standardised variance is approximately equal to $1/K$, which wrongly suggests the model is not parameter redundant. A possible cause for the method giving the wrong result is the numerical algorithm, MCMC, is not sampling correctly from the posterior distribution, as the posterior decreases very sharply from the flat ridge in the posterior surface. Generally there are issues with convergence of MCMC in non-identifiable models which is discussed in Section 6.3.

The R code, pop.R, also demonstrates how to use the classical version of data cloning. However this method fails, as the inverse of the Hessian matrix cannot be found due to ill-conditioning, that is the smallest eigenvalue is very close to zero.

4.1.4 Profile Log-Likelihood Method

As explained in Chapter 2, if a model is parameter redundant then there will be a flat ridge in the likelihood surface (Catchpole and Morgan, 1997). Plotting the profile log-likelihood for a parameter can therefore be used to detect parameter redundancy. A flat profile indicates that a model is parameter redundant. This method is generally used for a given data set on a given model, but can also be used with a large simulated data set to consider parameter redundancy caused by the model (Gimenez et al., 2004). Examples of use of this profile method includes Freeman et al. (1992), Lebreton and Pradel (2002), Raue et al. (2009) and Tonsing et al. (2018).

We demonstrate this method in Section 4.1.4.1. Profile plots can also be used to check for local identifiability, as explained in Section 4.1.4.2.

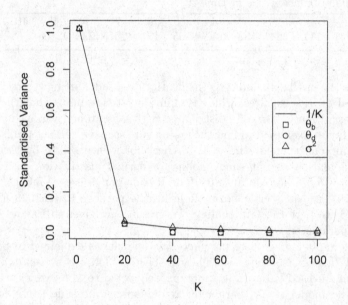

FIGURE 4.1

Data Cloning Population Growth plot of the scaled variance. The x-axis gives K, the number of data clones. The y-axis gives the standardised variance. The solid line, -, is $1/K$. The symbols represent the scaled variance for each of the different parameters at different values of K.

4.1.4.1 Mark-Recovery Model

Consider again the mark-recovery model discussed in Section 4.1.1.2 above, as well as in earlier chapters. Here, consider the model with time dependent first year survival probability ($\phi_{1,t}$), constant adult survival probability (ϕ_a) and two age classes for the recovery probabilities (λ_1 and λ_a). Suppose there are $n_1 = 5$ years of marking and $n_2 = 5$ years of recovery. The probabilities of being marked in year i and recovered dead in year j, $P_{i,j}$, are summarised in the matrix below.

$$
\mathbf{P} = \begin{bmatrix}
\bar{\phi}_{1,1}\lambda_1 & \phi_{1,1}\bar{\phi}_a\lambda_a & \phi_{1,1}\phi_a\bar{\phi}_a\lambda_a & \phi_{1,1}\phi_a^2\bar{\phi}_a\lambda_a & \phi_{1,1}\phi_a^3\bar{\phi}_a\lambda_a \\
0 & \bar{\phi}_{1,2}\lambda_1 & \phi_{1,2}\bar{\phi}_a\lambda_a & \phi_{1,2}\phi_a\bar{\phi}_a\lambda_a & \phi_{1,2}\phi_a^2\bar{\phi}_a\lambda_a \\
0 & 0 & \bar{\phi}_{1,3}\lambda_1 & \phi_{1,3}\bar{\phi}_a\lambda_a & \phi_{1,3}\phi_a\bar{\phi}_a\lambda_a \\
0 & 0 & 0 & \bar{\phi}_{1,4}\lambda_1 & \phi_{1,4}\bar{\phi}_a\lambda_a \\
0 & 0 & 0 & 0 & \bar{\phi}_{1,5}\lambda_1
\end{bmatrix},
$$

where $\bar{\phi} = 1 - \phi$.

To examine the parameter redundancy of this model using the profile log-likelihood method, a data set with 1000 animals marked each year is simulated. The number marked in year i and recovered in year j, $N_{i,j}$, is given by

$$
N = \begin{bmatrix}
348 & 42 & 30 & 11 & 4 \\
 & 309 & 39 & 34 & 17 \\
 & & 325 & 47 & 16 \\
 & & & 376 & 28 \\
 & & & & 255
\end{bmatrix}.
$$

The log-likelihood is then

$$
l = \sum_{i=1}^{5} \sum_{j=i}^{5} N_{i,j} \log(P_{i,j}) + \sum_{i=1}^{5} \left(1000 - \sum_{j=i}^{5} N_{i,j} \right) \log \left(1 - \sum_{j=i}^{5} P_{i,j} \right).
$$

Figure 4.2 shows the log-likelihood profile plots for each of the eight parameters. The profile log-likelihoods are all flat for part of the parameter space, suggesting the model is parameter redundant. In fact the model is not parameter redundant, as shown in Cole et al. (2012). The nested model, discussed in Section 1.2.4, with $\phi_{1,t} = \phi_1$, is parameter redundant. Therefore the model behaves similarly to a parameter redundant model when $\hat{\phi}_{1,t} \approx \hat{\phi}_1$ for all t (Catchpole et al., 2001). This is known as near-redundancy and is discussed in Section 4.4 below.

The profile log-likelihood method should be used with caution when examining parameter redundancy caused by the model, as it cannot distinguish between near-redundancy and actual parameter redundancy.

4.1.4.2   Profiles to check Locally Identifiability

Profile plots can also be used to check locally identifiability. As explained in Section 2.1, a locally identifiable model would have more than one maximum likelihood estimate, so a profile plot would have two or more peaks of the same value. As explained in Section 2.1.4 it is also possible to examine a profile plot for alternative functions other than the log-likelihood, for example the least squares function can replace the log-likelihood. In the case of a least squares profile a locally identifiable model would have two or more minimums with the same value.

An example is given by Figure 2.6 in Section 2.1.4. This shows the least squares profile plot for a compartment model. All three parameters have two minimums with identical values. Therefore the model is locally identifiable.

Note that the profile plot has only explored local identifiability in the region plotted. For example, suppose for the parameter θ_{01} the profile plot was only considered up to 0.3, then there would have only been one minimum, so local identifiability would be missed.

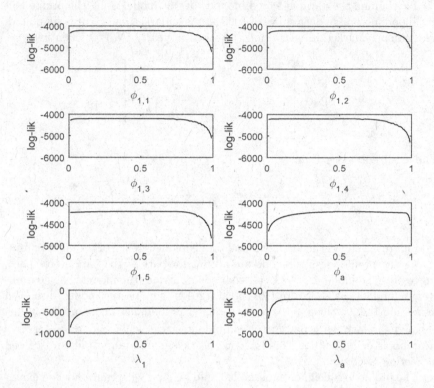

FIGURE 4.2
Profile log-likelihood plots for the mark-recovery model with time dependent first year survival probability ($\phi_{1,t}$), constant adult survival probability (ϕ_a), and two classes of recovery probability (λ_1 and λ_a). This example is based on simulated data with 5 years of marking and 5 years of recovery.

4.2   Symbolic Method

The symbolic method, or symbolic differentiation method, has been discussed in detail in Chapter 3. In this section we offer a step by step approach of how to apply the method, with reference to the appropriate section in Chapter 3 for more detail. This is the general method described in Cole et al. (2010), although the same method is also used in earlier papers, as explained in Section 3.2.

Step 1   Choose a suitable vector that uniquely represents the model, this
         is known as the exhaustive summary (Section 3.1.4).

Step 2   Form a derivative matrix by differentiating the exhaustive summary with respect to the parameters and calculate its rank. If the rank is less than the number of parameters the model is parameter redundant. If the rank is equal to the number of parameters the model is full rank (Section 3.2).

Step 3   a. If a model is parameter redundant, find which parameters and parameter combinations can be estimated solving a set of PDEs (Section 3.2.5).
b. If a model is full rank check whether it is full rank for all parameter values using a decomposition of the derivative matrix (Section 3.2.7).

Step 4   If applicable, generalise the result to any dimension of the model using the extension theorem (Section 3.2.6).

The approach usually needs to be executed in a symbolic algebra package such as Maple; see Appendix A. Instructions on executing step 2 in Maple are giving in Section A.1. Step 3a instructions are given in Section A.2 and step 3b instructions are given in Section A.3. (Step 4 involves the same code as step 2.)

We illustrate the use of these steps using the mark recovery model introduced in Section 1.2.4. It is recommended that this is read alongside the Maple code mrmsteps.mw, which demonstrates how each result is obtained.

Step 1: Exhaustive Summary

Firstly a representation of the model is needed, which is called an exhaustive summary. This will be some combination of the parameters that uniquely express how the parameters appear in the model in the form of a vector, κ. For example this could be the components of the log-likelihood or probabilities of events occurring. For the mark-recovery model the log-likelihood for three years of marking and three years of recovery is

$$l = \sum_{i=1}^{3} \sum_{j=i}^{3} N_{i,j} \log(P_{i,j}) + \sum_{i=1}^{3} \left(F_i - \sum_{j=i}^{3} N_{i,j} \right) \log \left(1 - \sum_{j=i}^{3} P_{i,j} \right),$$

where F_i is the number of animals marked in year i and $N_{i,j}$ is the number of animals recovered in year j, which were marked in year i. The probabilities of marking in year i and recovery in year j, $P_{i,j}$, are summarised in the matrix below:

$$\mathbf{P} = \begin{bmatrix} (1-\phi_1)\lambda_1 & \phi_1(1-\phi_a)\lambda_a & \phi_1\phi_a(1-\phi_a)\lambda_a \\ 0 & (1-\phi_1)\lambda_1 & \phi_1(1-\phi_a)\lambda_a \\ 0 & 0 & (1-\phi_1)\lambda_1 \end{bmatrix},$$

where ϕ_1, ϕ_a, λ_1 and λ_a are the parameters representing survival and reporting probabilities respectively, with the subscript representing the age of the animal (either first year, 1 or adult, a). The independent terms of the log-likelihood form an exhaustive summary

$$\kappa_l = \begin{bmatrix} N_{1,1}\log(P_{1,1}) \\ \vdots \\ N_{3,3}\log(P_{3,3}) \\ \left(F_1 - \sum_{j=1}^{3} N_{1,j}\right)\log\left(1 - \sum_{j=1}^{3} P_{1,j}\right) \\ \left(F_2 - \sum_{j=2}^{3} N_{2,j}\right)\log\left(1 - \sum_{j=2}^{3} P_{2,j}\right) \\ (F_3 - N_{3,j})\log\left(1 - P_{1,3}\right) \end{bmatrix}.$$

Assuming that none of the $N_{i,j}$ entries are zero, a simpler exhaustive summary consists of the terms in $\mathbf{P}$. We can remove repeated terms to give:

$$\kappa_P = \begin{bmatrix} (1-\phi_1)\lambda_1 \\ \phi_1(1-\phi_a)\lambda_a \\ \phi_1\phi_a(1-\phi_a)\lambda_a \end{bmatrix}.$$

We recommend using the simplest exhaustive summary, so continue with κ_P below, although the exhaustive summary κ_l is also shown in the Maple code.

Step 2: Derivative Matrix

The next step involves forming a matrix of partial derivative terms, differentiating each of the exhaustive summary terms, κ_i, with respect to each of the q parameters, θ_j. In the mark-recovery model the parameters are

$$\boldsymbol{\theta} = [\phi_1, \phi_a, \lambda_1, \lambda_a].$$

Maple is used to find the derivative matrix as

$$\mathbf{D} = \frac{\partial \kappa_P}{\partial \boldsymbol{\theta}} = \begin{bmatrix} -\lambda_1 & (1-\phi_a)\lambda_a & \phi_a(1-\phi_a)\lambda_a \\ 0 & -\phi_1\lambda_a & \phi_1(1-\phi_a)\lambda_a - \phi_1\phi_a\lambda_a \\ 1-\phi_1 & 0 & 0 \\ 0 & \phi_1(1-\phi_a) & \phi_1\phi_a(1-\phi_a) \end{bmatrix}.$$

Maple can be used to show that the rank of the derivative matrix is $r = 3$. As there are four parameters the deficiency is $d = r - q = 4 - 3 = 1$. As the deficiency is greater than zero the model is parameter redundant. If the deficiency is $d = 0$ then the model is not parameter redundant.

The rank, $r = 3$, is also the number of estimable parameters, which is used in some model selection methods such as AIC; see Section 2.4.1.

Step 3: Finding Estimable Parameter Combinations

If a model is parameter redundant it is possible to find which, if any, of the original parameters can be estimated, as well as which combinations of parameters can be estimated. These estimable parameter combinations can be found from further information available in the derivative matrix. This involves finding the left null space of the derivative matrix, that is, finding the solutions to $\boldsymbol{\alpha}^T\mathbf{D} = 0$, and then forming a set of partial differential equations using the terms in $\boldsymbol{\alpha}^T$. The full mathematical explanation of this method is given in Section 3.2.5. There is a Maple procedure that will find the estimable parameter combination (see Section A.2). For the mark-recovery model the estimable parameter combinations are ϕ_a, $(1 - \phi_1)\lambda_1$, $\phi_1\lambda_a$. The estimable parameter combinations give a reparameterisation, that allows the model to be utilised, for example in maximum likelihood estimation. If we wanted to use this parameter redundant model we could reparameterise with $\beta_1 = (1-\phi_1)\lambda_1$ and $\beta_2 = \phi_1\lambda_a$, to give a model that is not parameter redundant.

Step 4: Generalising Parameter Redundancy Results

In the mark-recovery models we have examined results for a fixed number of years of marking and recovery. Parameter redundancy results can be generalised to any dimension of model, which could be any number of years or any number of states in an ecological model. The extension theorem, given in Section 3.2.6.1, provides a method of obtaining general results. We demonstrate how this method works using the mark-recovery model.

Firstly if the model is parameter redundant it should be reparameterised in terms of the estimable parameter combinations. Therefore we use the exhaustive summary

$$\boldsymbol{\kappa}_{re} = \begin{bmatrix} \beta_1 \\ \beta_2(1 - \phi_a) \\ \beta_2\phi_a(1 - \phi_a) \end{bmatrix}.$$

The parameters are now $\boldsymbol{\theta}_{re} = [\phi_a, \beta_1, \beta_2]$. This reparameterised model with $q = 3$ parameters has rank $r = q = 3$.

Once we have a model with q parameters that has rank q, we consider any new exhaustive summary terms that are added when the dimension is increased. In this case adding an extra year of recovery adds an extra term $\kappa_2 = [\beta_2\phi_a^2(1 - \phi_a)]$. (Adding an extra year of marking adds no new distinct exhaustive summary terms). If there were additional parameters we would form another derivative matrix by differentiating the n_2 extra exhaustive summary terms with respect to the q_2 extra parameters, giving

$$\mathbf{D}_2 = \frac{\partial \boldsymbol{\kappa}_2}{\partial \boldsymbol{\theta}_2} = \begin{bmatrix} \frac{\partial \kappa_{2,1}}{\partial \theta_{2,1}} & \frac{\partial \kappa_{2,2}}{\partial \theta_{2,1}} & \cdots & \frac{\partial \kappa_{2,n_2}}{\partial \theta_{2,1}} \\ \frac{\partial \kappa_{2,1}}{\partial \theta_{2,2}} & \frac{\partial \kappa_{2,2}}{\partial \theta_{2,2}} & \cdots & \frac{\partial \kappa_{2,n_2}}{\partial \theta_{2,2}} \\ \vdots & & & \vdots \\ \frac{\partial \kappa_{2,1}}{\partial \theta_{2,q_2}} & \frac{\partial \kappa_{2,2}}{\partial \theta_{2,q_2}} & \cdots & \frac{\partial \kappa_{2,n_2}}{\partial \theta_{2,q_2}} \end{bmatrix}.$$

If the rank of $\mathbf{D}_2$, r_2, is equal to q_2 then the extended model will have rank $q + q_2$.

For the mark-recovery model, as there are no additional parameters in this example we can say that the extended model will always have rank q, therefore the original model will always have rank 3 for any number of years of data.

A more substantial example is given in Section 3.2.6.2.

4.3 Hybrid Symbolic-Numerical Method

The hybrid symbolic-numerical method of Choquet and Cole (2012) combines both the symbolic and numeric methods. As in the symbolic method the derivative matrix, $\mathbf{D} = \partial\boldsymbol{\kappa}/\partial\boldsymbol{\theta}$, is found symbolically. Then rather than evaluate the rank of the derivative matrix symbolically the derivative matrix is evaluated at a random set of parameter values, and then the rank is found numerically. This process is repeated n times, and the model rank is the maximum rank found numerically. If the model is parameter redundant it is possible to check whether any of the original parameters are individually identifiable by examining the left null space of the derivative matrix.

The algorithm for executing the hybrid symbolic-numerical method is explained by the steps below:

Step 1   Find the derivative matrix, $\mathbf{D}$, using symbolic algebraic.

Step 2   Evaluate the derivative matrix at a random set of parameter values to give $\mathbf{D}_i^\star$.

Step 3   Find the rank of $\mathbf{D}_i^\star$, r_i.

Step 4   If $\mathbf{D}_i^\star$ is not full rank, that is if $r_i < q$, find the left null space of $\mathbf{D}_i^\star$, i.e. solve $(\boldsymbol{\alpha}_{i,j}^\star)^T \mathbf{D}_i^\star = \mathbf{0}$. There will be $d = p - r_i$ vectors $\boldsymbol{\alpha}_{i,j}^\star$.

Step 5   Repeat steps 2 to 4 a total of n times.

Step 6   The model's rank is the maximum of the n ranks, $r = \max(r_i)$. The model is parameter redundant if the model rank is less than the number of parameters, $r < p$. The model is full rank, and at least locally identifiable, if the model rank is equal to the number of parameters, $r = p$.

Step 7   If the model is parameter redundant, consider any of the null space vectors for $\mathbf{D}_i^\star$ that obtained the maximum rank, that is any $\boldsymbol{\alpha}_{i,j}^\star$ for which $r_i = r$. The positions of any entries in $\boldsymbol{\alpha}_{i,j}^\star$ that are very close to zero for all j indicate the parameter with the same position is individually identifiable.

It is necessary to repeat steps 2 to 4 several times because some particular points in the parameter space may exist at which the rank is lower than the model rank. Repeating the process means that the rank is calculated at different points in the parameter space; not all of them will return a rank lower than the model rank. Choquet and Cole (2012) recommend approximately 5 random points.

Typically the random parameter values are chosen from an appropriate uniform distribution. However, the recommendation of 5 points is only valid with careful consideration of the boundaries of the parameter space. Make sure not to over-sample from regions close to the boundaries, which can have a lower rank than the model rank. If a parameter is a probability, sampling random numbers from a uniform distribution on the range 0 to 1 is recommended. If the parameter has an upper and/or a lower bound of infinity then it may be sensible to restrict the random sampling of the parameter space.

For example, suppose two parameters in a model are a and b, such that a probability for part of the model p_t is written as

$$p_t = \frac{1}{1 + \exp\{-(a + bt)\}}.$$

As a and b can take on any value, suppose the random values of a and b are drawn from a uniform distribution on the range -100 to 100, nearly all values of p_t will be close to 0 or close to 1, which could lead to boundary value problems. It is better in this case to draw a and b from a uniform distribution on the range -1.5 to 1.5, or to use an alternative distribution with a smaller probability of choosing higher and lower values, such as the normal distribution.

As with the symbolic method, the rank, r, gives the exact number of estimable parameters, which is needed in some model selection methods, such as AIC (see Section 2.4.1).

Step 1 (finding the derivative matrix symbolically) can be replaced with any method that finds the derivative matrix to machine precision. Alternative methods include the forward-backward algorithm (Lystig and Hughes, 2002), automatic differentiation tools (Hunter and Caswell, 2009), or using an explicit formula for the derivatives of products of matrices (Choquet et al., 2004, 2009). Using one of these methods allows the hybrid symbolic-numerical method to be added to software packages, for example M-Surge (Choquet et al. 2004) for multi-state ecology models and E-Surge (Choquet et al. 2009) for multi-event ecology models.

Evaluating the rank of the derivative matrix symbolically can be a computationally expensive. In more complex models symbolic algebra packages can run out of memory trying to calculate the rank of the derivative matrix (see for example Jiang et al., 2007, Forcina, 2008, Hunter and Caswell, 2009). The hybrid symbolic-numerical method is one method to deal with this problem.

Maple code for executing the hybrid symbolic-numerical method is given in Appendix A.3. Then in Section 4.3.1 and 4.3.2 we demonstrate how to use this method using two examples.

4.3.1   Basic Occupancy Model

Consider again the basic occupancy model of Section 1.2.1, as well as in Chapters 2 and 3 and in Sections 4.1.1.1, 4.1.2.1 and 4.1.3.1 above. We utilise the exhaustive summary consisting of the probabilities of the two events that can occur in the basic occupancy model: either the species is detected or it is not detected. This exhaustive summary is

$$\kappa = \left[\begin{array}{c} \psi p \\ 1 - \psi p \end{array} \right],$$

which is also given in Section 3.1.4.1 and labelled equation (3.1).

As shown in Section 3.2.1 the derivative matrix is

$$\mathbf{D} = \left[\begin{array}{cc} p & -p \\ \psi & -\psi \end{array} \right].$$

Although it is obvious in this case than the rank is 1, in Appendix A.4.1 and Maple code basicoccupancy.mw we show that the rank at 5 randomly chosen points is 1. This gives a model rank of 1, but there are two parameters, therefore we have again shown that this basic occupancy model is parameter redundant.

From Step 7 of the method we can also show that neither of the original parameters are individually identifiable, as the entries in α are not close to zero. However unlike the symbolic method the hybrid symbolic-numerical method cannot be used to show that ψp is the estimable parameter combination.

4.3.2   Complex Compartmental Model

In this section we consider a complex linear compartmental model where Maple runs out of memory calculating the rank of the derivative matrix. Audoly et al. (1998) present a linear compartment model which has

$$\mathbf{y}(t, \boldsymbol{\theta}) = \mathbf{C}\mathbf{x}(t, \boldsymbol{\theta}), \ \text{with} \ \frac{\partial}{\partial t}\mathbf{x}(t, \boldsymbol{\theta}) = \mathbf{A}(\boldsymbol{\theta})\mathbf{x}(t, \boldsymbol{\theta}) + \mathbf{B}(\boldsymbol{\theta})\mathbf{u}(t),$$

where $\mathbf{x}$ is the state-variable function, $\boldsymbol{\theta}$ is a vector of unknown parameters, $\mathbf{u}$ is the input function and t is the time recorded on a continuous time scale. The matrices $\mathbf{A}$, $\mathbf{B}$, $\mathbf{C}$ are the compartmental, input and output matrices respectively, which take on values

$$\mathbf{A} = \left[\begin{array}{cccc} -(k_{21} + k_{41}) & k_{12} & 0 & k_{14} \\ k_{21} & -(k_{12} + k_{32}) & k_{23} & 0 \\ 0 & k_{32} & -(k_{23} + k_{43} + k_{03}) & k_{34} \\ k_{41} & 0 & k_{43} & -(k_{14} + k_{34} + k_{04}) \end{array} \right],$$

$$\mathbf{B} = \left[\begin{array}{cccc} 1 & 0 & 0 & 0 \end{array} \right]^{T}, \mathbf{C} = \left[\begin{array}{cccc} 1/V_1 & 0 & 0 & 0 \\ 0 & 1 & 0 & 0 \\ 0 & 0 & 1 & 0 \end{array} \right],$$

The model has 11 parameters

$$\boldsymbol{\theta} = [k_{03}, k_{04}, k_{12}, k_{14}, k_{21}, k_{23}, k_{32}, k_{34}, k_{41}, k_{43}, V_1],$$

which are all assumed to be real values greater than zero.

As discussed in Section 3.1.4.3 an exhaustive summary can be formed from the terms in the transfer function. The transfer function is

$$Q(s) = \mathbf{C}(s\mathbf{I} - A)^{-1}\mathbf{U} =$$

$$
\begin{bmatrix}
\dfrac{\begin{array}{l}s^3+(k_3+k_4+k_{12}+k_{14}+k_{23}+k_{32}+k_{34}+k_{43})s^2\\+(k_3k_4+...+k_{32}k_{43})s+k_3k_4k_{12}+...+k_{14}k_{32}k_{43}\end{array}}{\begin{array}{l}V_1\big\{s^4+(k_3+...+k_{43})s^3+(k_3k_4+...+k_{41}k_{43})s^2\\+(k_3k_4k_{12}+...+k_{32}k_{41}k_{43})s+k_3k_4k_{12}k_{41}+...+k_4k_{32}k_{41}k_{43}\big\}\end{array}} \\[2em]
\dfrac{k_{21}s^2+(k_3k_{21}+...+k_{21}k_{43})s+k_3k_4k_{21}+...+k_{23}k_{34}k_{41}}{\begin{array}{l}s^4+(k_3+...+k_{43})s^3+(k_3k_4+...+k_{41}k_{43})s^2\\+(k_3k_4k_{12}+...+k_{32}k_{41}k_{43})s+k_3k_4k_{12}k_{41}+...+k_4k_{32}k_{41}k_{43}\end{array}} \\[2em]
\dfrac{(k_{21}k_{32}+k_{34}k_{41})s+k_4k_{21}k_{32}+...+k_{32}k_{34}k_{41}}{\begin{array}{l}s^4+(k_3+...+k_{43})s^3+(k_3k_4+...+k_{41}k_{43})s^2\\+(k_3k_4k_{12}+...+k_{32}k_{41}k_{43})s+k_3k_4k_{12}k_{41}+...+k_4k_{32}k_{41}k_{43}\end{array}}
\end{bmatrix}.
$$

The non-constant coefficients of s in each numerator and denominator form the exhaustive summary:

$$\boldsymbol{\kappa} = \begin{bmatrix} k_{03}k_{04}k_{12} + \ldots + k_{14}k_{32}k_{43} \\ k_{03}k_{04} + \ldots + k_{32}k_{43} \\ \vdots \\ k_{21}k_{32} + k_{34}k_{41} \end{bmatrix}.$$

Note we can ignore several repeated terms in the denominators. The full exhaustive summary terms are given in Audoly et al. (1998) and the Maple code for the example `complexcomp.mw`. Maple runs out of memory trying to calculate the symbolic rank of the derivative matrix $\mathbf{D} = \partial\boldsymbol{\kappa}(\boldsymbol{\theta})/\partial\boldsymbol{\theta}$. The hybrid symbolic-numerical method is applied in `complexcomp.mw` and the rank is 11. As there are 11 parameters the model is at least locally identifiable. However this method cannot distinguish whether the model is locally or globally identifiable.

There are two alternatives to the hybrid symbolic-numerical method, the first is given in Audoly et al. (1998) and is a specific method for compartmental models, the second is given in Cole et al. (2010). The latter is a more general method that involves extending the symbolic method; this is discussed in Section 5.2.2.2. The extended symbolic method can distinguish between local and global identifiability.

4.4 Near-Redundancy and Practical Identifiability

Models that are not parameter redundant can behave as though they are parameter redundant in practice, and we refer to this as near parameter redundancy. This is a consequence of model-fitting; when the parameter estimates are close to a parameter redundant nested model.

For example, suppose the mark-recovery example in Sections 3.2.7.1 and 3.5.1, was fitted to a certain set of data and the maximum likelihood parameter estimates were

$$\hat{\lambda}_1 = 0.1, \ \hat{\lambda}_a = 0.2, \ \hat{\phi}_a = 0.5, \ \hat{\phi}_{1,1} = 0.4, \ \hat{\phi}_{1,2} = 0.41, \ \hat{\phi}_{1,3} = 0.42.$$

Although the model is not parameter redundant, the parameter estimates are close to the nested model with $\phi_1 = \phi_{1,1} = \phi_{1,2} = \phi_{1,3}$. This nested model is parameter redundant, therefore the model with this data set will behave like the parameter redundant model and is termed near-redundant.

A similar and related problem is that of practical identifiability, where likelihood based confidence intervals are infinite. In the above example the parameters $\phi_{1,i}$, λ_1 and λ_a are all probabilities, with $0 \leq \theta_i \leq 1$. The likelihood based confidence interval has limits that are less than 0 and greater than 1, therefore the model with this data set is practically non-identifiable.

Formal definitions of near-redundancy and practical identifiability are given in Section 3.1.3.

A near-redundant model is one that is formally full rank, but might be classed as parameter redundant by an inevitably-imprecise numerical method, because the model is very similar to a model that is parameter redundant for a particular data set. In Section 4.4.1 near-redundancy is explored using the Hessian matrix method. The likelihood profile can also be used to investigate practical identifiability, which is discussed in Section 4.4.2. Near-redundancy can also be explored using the symbolic method, discussed in Section 3.5. However this method is not included in this chapter as it does not tell you when a model with a specific data set is near redundant only the conditions under which a model might be near redundant.

4.4.1   Detecting Near-Redundancy using the Hessian Matrix

In Section 3.5 we discussed detecting near-redundancy by examining small eigenvalues from the Hessian matrix (Catchpole et al., 2001). This gives rise to the following method for checking for near-redundancy:

Step 1   Find the Hessian matrix using a numerical method.

Step 2   Find the eigenvalues of the Hessian matrix.

Step 3   Standardise the eigenvalues by taking the modulus of the eigenvalues and dividing through by the largest eigenvalue.

Step 4   The model with that data set is near-redundant if the smallest eigenvalue is less than 0.001.

The value of 0.001 is used in Chis et al. (2016) to define a model as sloppy. The same value was used to investigate near-redundancy in Zhou et al. (2019) and Cole (2019).

This method is essentially applying the Hessian method, with a less stringent threshold. If only the Hessian method is used, it would be difficult to distinguish between near-redundancy and parameter redundancy. However if we use the Hessian method on a model with a specific data set that has formally been classified as not parameter redundant we can identify near-redundancy.

We demonstrate using this method in Section 4.4.1.1 below.

4.4.1.1 Mark-Recovery Model

In Section 3.5.1 we examined near-redundancy in a mark-recovery model, including using the Hessian method by finding the exact Hessian matrix in Maple. Here we revisit this same example to show how this can also be implemented numerically in R using the code in `markrecovery.R`.

In this mark-recovery example the first year survival probability is time dependent $(\phi_{1,t})$, the adult survival probability is constant (ϕ_a) and there are two classes of recovery for first year animals and adult animal (λ_1 and λ_a). We start by assuming that we have already established that this model is not parameter redundant (see Section 3.2.7.1).

We use the same data set as Section 3.5.1 which has 1000 animals marked each year and the numbers recovered in year j that were marked in year i, $N_{i,j}$, are summarised in the matrix

$$\mathbf{N} = \begin{bmatrix} 60 & 40 & 20 \\ 0 & 59 & 41 \\ 0 & 0 & 58 \end{bmatrix}.$$

The maximum likelihood estimates are $\hat{\phi}_{1,1} = 0.40$, $\hat{\phi}_{1,2} = 0.41$, $\hat{\phi}_{1,3} = 0.42$, $\hat{\phi}_a = 0.50$, $\hat{\lambda}_1 = 0.10$ and $\hat{\lambda}_a = 0.20$.

To execute Step 1 we use R to find the approximate Hessian matrix as

$$\mathbf{H} = \begin{bmatrix} 544.51 & 0.00 & 0.00 & -204.55 & -965.91 & 767.05 \\ 0.00 & 413.39 & 0.00 & -200.00 & -1000.00 & 500.00 \\ 0.00 & 0.00 & 183.03 & 0.00 & -1061.57 & 0.00 \\ -204.55 & -200.00 & 0.00 & 338.74 & -108.30 & -855.95 \\ -965.91 & -1000.00 & -1061.57 & -108.30 & 18852.98 & 338.93 \\ 767.05 & 500.00 & 0.00 & -855.95 & 338.93 & 2673.97 \end{bmatrix}.$$

At Step 2 we find the eigenvalues to be

$$19022.12, 3283.38, 473.31, 176.23, 51.54, 0.028.$$

Then at Step 3 we divide through by the largest eigenvalue of 19022.12, giving standardised eigenvalues

$$1, 0.170.025, 0.0093, 0.0027, 0.0000015$$

At Step 4 we classify the model with this data set as near redundant as $0.0000015 < 0.001$. (Note there are only slight differences between the results here and in Section 3.5.1. The slight differences stem from an approximate Hessian matrix being found in R, versus the exact Hessian matrix in Maple).

4.4.2 Practical Identifiability using Log-Likelihood Profiles

As discussed in Section 3.5, practical identifiability is used to describe a model that is identifiable but in practice for a specific data set has infinite confidence limits for one or more parameter (Raue et al., 2009). Practical identifiability can be investigated for a model with q parameters by plotting a likelihood profile along with a line at $l(\hat{\boldsymbol{\theta}}) - \frac{1}{2}\chi^2_{\alpha,q}$, where $l(\hat{\boldsymbol{\theta}})$ is the log-likelihood value evaulated at the maximum likelihood estimates $\hat{\boldsymbol{\theta}}$ and where $\chi^2_{\alpha,q}$ is the α percentage point of the chi-squared distribution with q degrees of freedom (Raue et al., 2009). The $\alpha\%$ confidence limits are the points at which the log-likelihood profile crosses this line. If a model is practically non-identifiable then it will only cross the line once or will not cross the line at all, so that the lower/upper or both confidence limits do not exist. It is important that the log-likelihood profile covers all values of the parameter space under consideration to avoid misclassification. We demonstrate this method in Section 4.4.2.1 below.

4.4.2.1 Mark-Recovery Model

Continuing with the mark-recovery model from Section 4.4.1.1, we consider two data sets with the first being the same as Section 4.4.1.1. The second data set has 100 000 animals marked each year and

$$\mathbf{N} = \begin{bmatrix} 9000 & 1000 & 500 \\ 0 & 6000 & 4000 \\ 0 & 0 & 7000 \end{bmatrix}.$$

The maximum likelihood estimates are $\hat{\phi}_{1,1} = 0.10$; $\hat{\phi}_{1,2} = 0.40$, $\hat{\phi}_{1,3} = 0.3$, $\hat{\phi}_a = 0.50$, $\hat{\lambda}_1 = 0.10$ and $\hat{\lambda}_a = 0.20$. This second data set is not realistic and just used to illustrate a case that is practically identifiable.

Figures 4.3 and 4.4 show plots of the profile log-likelihood for data sets 1 and 2 respectively. Data set 1 is practically non-identifiable as the upper confidence limits do not exist for all parameters and the lower confidence limit does not exist for some of the parameters. This can·be seen visually as the profile only crossing the dashed line representing $l(\hat{\boldsymbol{\theta}}) - \frac{1}{2}\chi^2_{5\%,6}$ to the left of the maximum, or not crossing the dashed line at all. Data set 2 is practically identifiable because both the lower and upper confidence limits exist for all parameters. This can be seen visually as the profile crossing the dashed line either side of the maximum. R code for producing these plots is given in `markrecovery.R`.

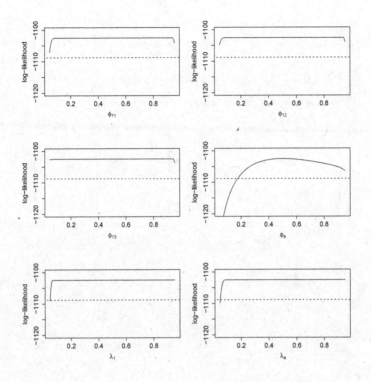

FIGURE 4.3
Practical Identifiability in Mark-Recovery Model $\phi_{1,t}$, ϕ_a, λ_1 and λ_a for data set 1. The figure shows profile log-likelihood plots for each parameter (solid line) and the line $l(\hat{\boldsymbol{\theta}}) - \frac{1}{2}\chi^2_{5\%,6}$ (dashed line). This example is practically non-identifiable.

4.5 Discussion and Comparison of Methods

There are several different methods for detecting parameter redundancy that have been discussed in this Chapter. We compare and contrast these different methods in Table 4.6, by examining their perform in the following categories:

- detecting intrinsic parameter redundancy (intrinsic);

- detecting extrinsic parameter redundancy (extrinsic);

- determining the number of estimable parameters, which is used in model selection methods such as AIC (AIC);

- whether the method is accurate (accurate);

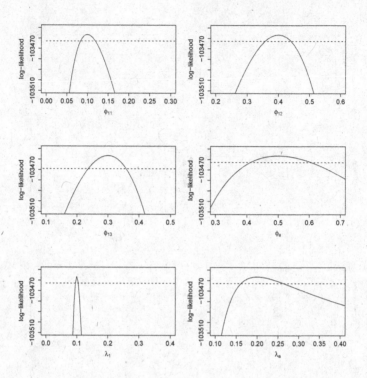

FIGURE 4.4

Practical identifiability in mark-recovery model $\phi_{1,t}$, ϕ_a, λ_1 and λ_a for data set 2. This figure shows profile log-likelihood plots for each parameter (solid line) and the line $l(\hat{\boldsymbol{\theta}}) - \frac{1}{2}\chi^2_{5\%,6}$ (dashed line). This example is practically identifiable.

- determining individual identifiable parameters (ident. pars.);

- finding estimable parameter combinations (est. par.);

- producing general results (general);

- whether the method is automatic (automatic); and

- whether the method works in a complex model (complex).

Firstly we consider whether the method can detect intrinsic and extrinsic parameter redundancy, that is parameter redundancy caused by the structure model and parameter redundancy caused by the data respectively. All methods can detect intrinsic parameter redundancy, though for some this requires the use of an ideal large simulated data set. All the methods except the simulation method can be used to detect extrinsic parameter redundancy.

In some model selection methods, such as AIC, we need to know the number of estimable parameters, the Hessian method, the symbolic and hybrid symbolic-numerical method are able to find this, whereas the other methods cannot.

The next category to compare methods is whether or not they are always accurate. All the numerical methods, due to their nature, have the possibility of returning the incorrect answer. This is commonly a problem of distinguishing between near-redundancy and actual parameter-redundancy in a full rank model. However numerical methods can also incorrectly identify a model as full rank, when it is in fact parameter redundant, as the threshold is usually subjective and may vary from example to example.

If a model is parameter redundant it is possible to find which of the original parameters can be estimated, which all methods can do, but with varying degrees of accuracy. In particular, data cloning can often return the wrong result about which parameters can be estimated. The Hessian method, profile method and simulation method also have the potential to be inaccurate, whereas the symbolic and hybrid symbolic-numerical method are accurate. It is also possible to find which combination of parameters can be estimated in a parameter redundant model using the symbolic method. Campbell and Lele (2014) provide a method for finding estimable parameter combinations in data cloning in some situations. In Section 5.3.1 we extend the hybrid symbolic-numerical method to give a method for finding estimable parameter combinations. It is also possible to generalise results to any dimension of the model using the symbolic method, and we extend the hybrid symbolic-numerical method to do this in Section 5.3.2.

Whilst the symbolic method has full functionality it needs to be executed using a symbolic algebra package so cannot be considered automatic; that is it cannot be easily added to an existing computer package. Simulation, data cloning and profile log-likelihood similarly are not considered automatic. However the Hessian and the hybrid symbolic-numerical method are automatic. The Hessian method is used in the computer program Mark (Cooch and White, 2017) and the hybrid symbolic-numerical method is used in the computer programs M-Surge (Choquet et al. 2004)and E-Surge (Choquet et al. 2009)

The final category is another problem with the symbolic method, which is dealing with complex models. Complex models may be structurally too complex for a computer to be able to execute part of the symbolic algebra. This problem is solved by extending the symbolic method is Section 5.2. The other methods work with complex models, although simulation and data cloning will get slower the more complex the model becomes.

Based on this comparison it is recommended that the symbolic method is used to detect intrinsic parameter redundancy where possible. If the rank of the derivative matrix cannot be calculated then switch to the hybrid symbolic-numerical method instead. The hybrid symbolic-numerical method is recommended for detecting extrinsic parameter redundancy. The Hessian method is recommended for investigating near-redundancy.

TABLE 4.6

Comparison of different methods for detecting parameter redundancy. Ident. pars. is short for individually identifiable parameters. Est. par. is short for estimable parmaeter combinations. Profile is short for profile log-likelihood, and hybrid is short for hybrid symbolic-numerical. Sim. is short for using simulated data. Ex. represents that it is possible in the extended version of the method in Chapter 5.

Method	Intrinsic	Extrinsic	AIC	Accurate	Ident. pars.	Est. par.	General	Automatic	Complex
Hessian	Sim.	Yes	Yes	No	Yes	No	No	Yes	Yes
Simulation	Yes	No	No	No	Yes	No	No	No	Slow
Data Cloning	Sim.	Yes	No	No	Yes	Yes	No	No	Slow
Profile	Sim.	Yes	No	No	Yes	No	No	No	Yes
Symbolic	Yes	Yes	Yes	Yes	Yes	Yes	Yes	No	Ex.
Hybrid	Yes	Yes	Yes	Yes	Yes	Ex.	Ex.	Yes	Yes

5

Detecting Parameter Redundancy and Identifiability in Complex Models

In some complex models it is not possible to use the symbolic method, as Maple or other symbolic algebra packages run out of memory calculating the rank of the appropriate derivative matrix. Examples of this problem occurring include complex ecological models (Jiang et al., 2007, Hunter and Caswell, 2009), latent class models with covariates (Forcina, 2008) and compartmental models (Chis et al., 2011a).

In this Chapter we examine methods for dealing with problems associated with complex models. The addition of covariates can, structurally, increase complexity of a model. A method for applying the symbolic method in models with covariates is discussed in Section 5.1. More generally for any complex model there is an extended symbolic method, discussed in Section 5.2.

The alternative method of checking parameter redundancy in complex models is to use the hybrid symbolic-numerical method. However, as we discussed in Section 4.5, this method is unable to find estimable parameter combinations and to create general results for any dimension of a method. In Section 5.3 we discuss the extended hybrid symbolic-numerical method that has this functionality.

5.1 Covariates

Adding covariates to certain parameters can separate out confounded parameters (see, for example, Forcina, 2008, Royle et al., 2012, Nater et al., 2019). However, adding covariates does not necessarily mean the model is identifiable; we still need to check. Models with covariates are structurally more complex particularly if a non-linear link is used. Applying the symbolic method to a such model will result in a more complex derivative matrix than the equivalent model without covariates. In some cases Maple may run out of memory trying to calculate the rank of more complex derivative matrices, for example, see Forcina (2008). Cole and Morgan (2010a) provide a method of examining parameter redundancy in model with covariates which involves checking the derivative matrix of the equivalent model without covariates.

Start by considering a model without covariates, with exhaustive summary κ and q parameters $\boldsymbol{\theta}$. For this model the derivative matrix $\mathbf{D} = \partial\kappa/\partial\boldsymbol{\theta}$ has rank r and estimable parameter combinations $\boldsymbol{\theta}_E = f(\boldsymbol{\theta})$, for some function f. If $r = p$ then $\boldsymbol{\theta}_E = \boldsymbol{\theta}$. Covariates are then added to the model, so that the exhaustive summary becomes κ_c and there are now q_c parameters, $\boldsymbol{\theta}_c$. Some of the parameters in $\boldsymbol{\theta}_c$ will be the same as the parameters in $\boldsymbol{\theta}$; other parameters will be new parameters. For instance we might set $\theta_i = \Psi_i(\beta_0 + \beta_1 x_{1,i} + \beta_2 x_{2,i})$, where $x_{1,i}$ and $x_{2,i}$ are two covariates and Ψ_i is some function such as the inverse logit function with $\Psi(w) = 1/\{1 + \exp(-w)\}$. The new parameters in $\boldsymbol{\theta}_c$ would be β_0, β_1, β_2.

We are interested in finding the rank of the derivative matrix $\mathbf{D}_c = \partial\kappa_c/\partial\boldsymbol{\theta}_c$. If $\partial\boldsymbol{\theta}_E/\partial\boldsymbol{\theta}_c$ is full rank the following Theorem applies:

Theorem 12 *The rank of the derivative matrices $\mathbf{D}_c = \partial\kappa_c/\partial\boldsymbol{\theta}_c$ is $min(p_c, r)$ (Cole and Morgan, 2010a).*

This means that rather than checking the rank of the more complex derivative matrix $\mathbf{D}_c$ it is only necessary to check the rank of the simpler derivative matrix to determine whether or not a model with covariates is parameter redundant. This method is illustrated using two examples in Sections 5.1.1 and 5.1.2 below.

Note we also need to consider whether $\partial\boldsymbol{\theta}_E/\partial\boldsymbol{\theta}_c$ is full rank. This will normally be the case for non-linear link functions. If it is not full rank, the rank of $\mathbf{D}_c = \partial\kappa_c/\partial\boldsymbol{\theta}_c$ will be equal to the rank of $\partial\boldsymbol{\theta}_E/\partial\boldsymbol{\theta}_c$. An example of a linear function where $\partial\boldsymbol{\theta}_E/\partial\boldsymbol{\theta}_c$ is not full rank is given in Section 5.1.1.

5.1.1 Conditional Mark-Recovery Model

In mark-recovery models, the total number of animals marked each year may be unknown or unreliable (McCrea et al., 2012). It is possible to fit a model by conditioning on an animal being recovered dead. Suppose that the probability of recovery is dependent on time λ_t and the probability of survival is dependent on age in two age classes: ϕ_1 the probability of surviving the first year of life and ϕ_a the probability of surviving subsequent years. The probability that an animal is marked in year i and recovered dead in year j, conditional on being found dead is

$$Q_{ij} = \begin{cases} (1 - \phi_1)\lambda_j/F_i & i = j \\ \phi_1\phi_a^{j-i-1}(1 - \phi_a)\lambda_j/F_i & i < j \end{cases} \text{ for } i = 1, \ldots, n_1, j = i, \ldots, n_2,$$

where

$$F_i = (1 - \phi_1)\lambda_i + \sum_{k=i+1}^{n_2} \phi_1\phi_a^{k-i-1}(1 - \phi_a)\lambda_k,$$

n_1 is the number of years of marking and n_2 is the number of years of recovery. An exhaustive summary consists of the probabilities Q_{ij}. For $n_1 = n_2 = 3$

years of marking and recovery the exhaustive summary is

$$
\kappa = \begin{bmatrix}
\dfrac{(1-\phi_1)\lambda_1}{-\phi_1\lambda_3\phi_a^2-\lambda_2\phi_1\phi_a+\phi_1\lambda_3\phi_a-\lambda_1\phi_1+\lambda_2\phi_1+\lambda_1} \\[2mm]
\dfrac{\phi_1(1-\phi_a)\lambda_2}{-\phi_1\lambda_3\phi_a^2-\lambda_2\phi_1\phi_a+\phi_1\lambda_3\phi_a-\lambda_1\phi_1+\lambda_2\phi_1+\lambda_1} \\[2mm]
\dfrac{\phi_1\phi_a(1-\phi_a)\lambda_3}{-\phi_1\lambda_3\phi_a^2-\lambda_2\phi_1\phi_a+\phi_1\lambda_3\phi_a-\lambda_1\phi_1+\lambda_2\phi_1+\lambda_1} \\[2mm]
\dfrac{(1-\phi_1)\lambda_2}{-\phi_1\lambda_3\phi_a-\lambda_2\phi_1+\phi_1\lambda_3+\lambda_2} \\[2mm]
\dfrac{\phi_1(1-\phi_a)\lambda_3}{-\phi_1\lambda_3\phi_a-\lambda_2\phi_1+\phi_1\lambda_3+\lambda_2}
\end{bmatrix}.
$$

($Q_{33} = 1$ is excluded from the exhaustive summary because it is a constant term that gives no information on identifiability.) The parameters are $\theta = [\phi_1, \phi_a, \lambda_1, \lambda_2, \lambda_3]$. The derivative matrix $\mathbf{D} = \partial\kappa/\partial\theta$ has rank $r = 3$, but there are five parameters therefore the model without covariates is parameter redundant with deficiency 2. The estimable parameter combinations are

$$
\theta_E = \left[\frac{\theta_a - \theta_1}{\theta_a(1-\theta_1)}, \frac{\lambda_2\phi_a}{\lambda_1}, \frac{\lambda_3\phi_a^2}{\lambda_1} \right].
$$

In this example the covariates are a time dependent trend on the reporting probabilities, so that

$$
\lambda_j = \frac{1}{1 + \exp\{-(\beta_0 + \beta_1 t)\}}.
$$

This model with covariates has parameters $\theta_c = [\phi_1, \phi_a, \beta_0, \beta_1]$.

For $n_1 = n_2 = 3$ years of marking and recovery the derivative matrix $\partial\theta_E/\partial\theta_c$ is of full rank 3, therefore we can check parameter redundancy using Theorem 12. As $\min(p_c, r) = \min(4, 3) = 3$, the model with covariates has rank 3, but there four parameters so is still parameter redundant.

For $n_1 = n_2 = 4$ years of marking and recovery the model without covariates now has 6 parameters $\theta = [\phi_1, \phi_a, \lambda_1, \lambda_2, \lambda_3, \lambda_4]$. The derivative matrix $D = \partial\kappa/\partial\theta$ has rank 4. Therefore the model without covariates is still parameter redundant with deficiency 2. The derivative matrix $\partial\theta_E/\partial\theta_c$ is of full rank 4, so again we can apply Theorem 12. The model with covariates still has four parameters, $\theta_c = [\phi_1, \phi_a, \beta_0, \beta_1]$. Then as $\min(p_c, r) = \min(4, 4) = 4$ the model with covariates model is not parameter redundant (Cole and Morgan, 2010a).

If $\lambda_j = \beta_0 + \beta_1 t$, for $n_1 = n_2 = 4$ years of marking and recovery, the rank of $\partial\theta_E/\partial\theta_c$ is 3. As the derivative matrix $\partial\theta_E/\partial\theta_c$ is not full rank, Theorem 12 cannot be used. However, we can deduce that the rank of $\mathbf{D}_c = \partial\kappa_c/\partial\theta_c$ is also 3, therefore the model is parameter redundant.

The Maple code for this example can be found in `conditional.mw`.

5.1.2   Latent Class Model with Covariates

The latent class model from Section 2.3.1 has two binary response variables, Y_1 and Y_2, which are conditionally independent given a binary latent variable Z. The probability function for a particular observation is

$$Pr(Y_1 = y_1 \text{ and } Y_2 = y_2) = p\theta_{1,1}^{y_1}(1 - \theta_{1,1})^{1-y_1}\theta_{2,1}^{y_2}(1 - \theta_{2,1})^{1-y_2} +$$
$$(1 - p)\theta_{1,0}^{y_1}(1 - \theta_{1,0})^{1-y_1}\theta_{2,0}^{y_2}(1 - \theta_{2,0})^{1-y_2}$$

where $\theta_{j,k} = Pr(Y_j = 1|Z = k)$ and $p = Pr(Z = 1)$. As discussed in Forcina (2008), as well as Sections 2.3.1 and 3.2.2, this model is parameter redundant with rank 3. This stems from the fact that whilst there are four possible options for $Pr(Y_1 = y_1 \text{ and } Y_2 = y_2)$, the probabilities sum to 1 so there are three independent probabilities.

Forcina (2008) presents an example that introduces two covariates to this model to change this non-identifiable model into an identifiable model. The covariate x_1 is assumed to only effect the latent variable and the covariate x_2 is assumed to only effect the observation, so that

$$p = \frac{1}{1 + \exp\{-(a_p + b_p x_1)\}} \text{ and } \theta_{j,k} = \frac{1}{1 + \exp\{-(a_{j,k} + b_j x_2)\}}.$$

The parameters are then $\boldsymbol{\theta}_c = [a_p, b_p, a_{1,1}, a_{2,1}, a_{1,0}, a_{2,0}, b_1, b_2]$. Forcina (2008) uses $20,000$ numerical evaluations of appropriate derivative matrices to conclude that the model with covariates is almost certainly full rank, because the symbolic method would be difficult to use in this case. Alternatively Cole and Morgan (2010a) show how the same result can be found using the symbolic method.

The probability that individual i has observation (y_1, y_2) is

$$Pr(Y_1 = y_1 \text{ and } Y_2 = y_2) = p_i\theta_{1,1,i}^{y_1}(1 - \theta_{1,1,i})^{1-y_1}\theta_{2,1,i}^{y_2}(1 - \theta_{2,1,i})^{1-y_2} +$$
$$(1 - p_i)\theta_{1,0,i}^{y_1}(1 - \theta_{1,0,i})^{1-y_1}\theta_{2,0,i}^{y_2}(1 - \theta_{2,0,i})^{1-y_2}$$

where

$$p_i = \frac{1}{1 + \exp\{-(\beta_0 + \beta_1 x_{1,i})\}} \text{ and } \theta_{j,k,i} = \frac{1}{1 + \exp\{-(a_{j,k} + b_j x_{2,i})\}}.$$

Each covariate value is assumed to be sampled on m separate occassions, (Forcina, 2008 uses $m = 5$). In the ideal situation, where at least three out of the four possible options of Y_1 and Y_2 are observed, then the exhaustive summary contribution for the ith covariate is

$$\boldsymbol{\kappa}_i = \begin{bmatrix} p_i(1 - \theta_{1,1,i})\theta_{2,1,i} + (1 - p_i)(1 - \theta_{1,0,i})\theta_{2,0,i} \\ p_i\theta_{1,1,i}(1 - \theta_{2,1,i}) + (1 - p_i)\theta_{1,0,i}(1 - \theta_{2,0,i}) \\ p_i\theta_{1,1,i}\theta_{2,1,i} + (1 - p_i)\theta_{1,0,i}\theta_{2,0,i} \end{bmatrix}.$$

First consider this model without covariates. The parameters for κ_i are $\boldsymbol{\theta}_i = [p_i, \theta_{1,1,i}, \theta_{2,1,i}, \theta_{1,0,i}, \theta_{2,0,i}]$. The derivative matrix for the ith part of the exhaustive summary,

$$
\mathbf{D} = \frac{\partial \boldsymbol{\kappa}_i}{\partial \boldsymbol{\theta}_i} =
\begin{bmatrix}
\bar{\theta}_{1,1,i}\theta_{2,1,i} - \bar{\theta}_{1,0,i}\theta_{2,0,i} & \theta_{2,1,i}\theta_{1,1,i} - \bar{\theta}_{2,0,i}\theta_{1,0,i} & -\theta_{1,0,i}\theta_{2,0,i} + \theta_{1,1,i}\theta_{2,1,i} \\
-p_i\theta_{2,1,i} & p_i\bar{\theta}_{2,1,i} & p_i\theta_{2,1,i} \\
p_i\bar{\theta}_{1,1,i} & -p\theta_{1,1,i} & p\theta_{1,1,i} \\
-(1-p_i)\theta_{2,0,i} & (1-p_i)\bar{\theta}_{2,0,i} & (1-p_i)\theta_{2,0} \\
(1-p_i)\bar{\theta}_{1,0,i} & -(1-p_i)\theta_{1,0,i} & (1-p_i)\theta_{1,0,i}
\end{bmatrix},
$$

has rank 3. By reparameterising in terms of the estimable parameter combinations and applying the extension theorem (Theorem 6b in Section 3.2.6.1) if there were m sets of covariates the model rank would be $3m$.

For the model with covariates we can apply Theorem 12. There are $p_c = 8$ parameters in the model with covariates. The covariates model rank is $\min(p_c, q) = \min(8, 3m)$. As long as there are at least three sets of covariates ($m \geq 3$) this model will not be parameter redundant (Cole and Morgan, 2010a). This demonstrates this model is full rank without the need for numerical investigation. The result is based on the ideal situation where at least three out of the four possible options of Y_1 and Y_2 are observed, so is the best result that can be obtained in terms of parameter redundancy. If the ideal situation did not occur the hybrid symbolic-numerical method could be used to check for parameter redundancy instead.

Maple code for this example is available in the file `latentcovariates.mw`. The Maple also shows that $\partial \boldsymbol{\theta}_E / \partial \boldsymbol{\theta}_c$ is full rank.

5.2   Extended Symbolic Method

As has already been observed, some models are structurally too complex for the symbolic method of testing for parameter redundancy to be of practical use. For example Jiang et al. (2007) and Hunter and Caswell (2009) found it was not possible to use the symbolic method to check parameter redundancy of certain models, due to Maple running out of memory trying to find the rank. Cole et al. (2010) show that it is still possible to use the symbolic method in complex models using a simpler exhaustive summary.

5.2.1   Simpler Exhaustive Summaries

For any model there is more than one exhaustive summary that will represent how the parameters appear in the model. The more structurally simple the exhaustive summary is, the structurally simpler the derivative matrix will be, and therefore Maple will need less memory to calculate the rank.

For example, in Section 3.1.4.4 we discussed different exhaustive summaries for the mark-recovery model with two age classes of survival and two ages classes for recovery. Three exhaustive summaries for this model are:

$$
\kappa_1 = \begin{bmatrix}
N_{1,1} \log\{(1-\phi_1)\lambda_1\} \\
N_{1,2} \log\{\phi_1(1-\phi_a)\lambda_a\} \\
N_{1,3} \log\{\phi_1\phi_a(1-\phi_a)\lambda_a\} \\
N_{2,2} \log\{(1-\phi_1)\lambda_1\} \\
N_{2,3} \log\{\phi_1(1-\phi_a)\lambda_a\} \\
N_{3,3} \log\{(1-\phi_1)\lambda_1\} \\
\left(F_1 - \sum_{j=1}^{3} N_{1,j}\right) \log\left(1 - \sum_{j=1}^{3} P_{1,j}\right) \\
\left(F_2 - \sum_{j=2}^{3} N_{2,j}\right) \log\left(1 - \sum_{j=2}^{3} P_{2,j}\right) \\
(F_3 - N_{3,3}) \log\{1 - (1-\phi_1)\lambda_1\}
\end{bmatrix},
$$

$$
\kappa_2 = \begin{bmatrix}
(1-\phi_1)\lambda_1 \\
\phi_1(1-\phi_a)\lambda_a \\
\phi_1\phi_a(1-\phi_a)\lambda_a \\
(1-\phi_1)\lambda_1 \\
\phi_1(1-\phi_a)\lambda_a \\
(1-\phi_1)\lambda_1
\end{bmatrix},
$$

and

$$
\kappa_3 = \begin{bmatrix}
\log\{(1-\phi_1)\lambda_1\} \\
\log\{\phi_1(1-\phi_a)\lambda_a\} \\
\log\{\phi_1\phi_a(1-\phi_a)\lambda_a\}
\end{bmatrix}.
$$

In the above exhaustive summaries $P_{i,j}$ are the probabilities in the matrix

$$
\mathbf{P} = \begin{bmatrix}
(1-\phi_1)\lambda_1 & \phi_1(1-\phi_a)\lambda_a & \phi_1\phi_a(1-\phi_a)\lambda_a \\
0 & (1-\phi_1)\lambda_1 & \phi_1(1-\phi_a)\lambda_a \\
0 & 0 & (1-\phi_1)\lambda_1
\end{bmatrix},
$$

$N_{i,j}$ is the number of animals recovered dead in year j, that were marked in year i, F_i is the number of animals marked in year i, ϕ_1 is the probability of survival for first year animals, ϕ_a is the probability of survival for adult animals, λ_1 is the recovery probability for first year animals and λ_a is the recovery probability for adult year animals.

The derivative matrices formed by differentiating respect to the parameters, $\boldsymbol{\theta} = [\phi_1, \phi_a, \lambda_1, \lambda_a]$, are:

$$
\mathbf{D}_1 = \frac{\partial \kappa_1}{\partial \boldsymbol{\theta}} = \begin{bmatrix}
-\frac{N_{1,1}}{1-\phi_1} & \frac{N_{1,2}}{\phi_1} & \cdots \\
\vdots & & \\
0 & \frac{N_{1,2}}{\lambda_a} & \cdots
\end{bmatrix},
$$

$$\mathbf{D}_2 = \frac{\partial \boldsymbol{\kappa}_2}{\partial \boldsymbol{\theta}} = \begin{bmatrix} -\lambda_1 & (1-\phi_a)\lambda_a & \phi_a(1-\phi_a)\lambda_a & -\lambda_1 & (1-\phi_a)\lambda_a & -\lambda_1 \\ 0 & -\phi_1\lambda_a & \phi_1(1-2\phi_a)\lambda_a & 0 & -\phi_1\lambda_a & 0 \\ 1-\phi_1 & 0 & 0 & 1-\phi_1 & 0 & 1-\phi_1 \\ 0 & \phi_1(1-\phi_a) & \phi_1\phi_a(1-\phi_a) & 0 & \phi_1(1-\phi_a) & 0 \end{bmatrix},$$

and

$$\mathbf{D}_3 = \frac{\partial \boldsymbol{\kappa}_3}{\partial \boldsymbol{\theta}} = \begin{bmatrix} -(1-\phi_1)^{-1} & \phi_1^{-1} & \phi_1^{-1} \\ 0 & -(1-\phi_a)^{-1} & \frac{1-2\phi_a}{\phi_a(1-\phi_a)} \\ \lambda_1^{-1} & 0 & 0 \\ 0 & \lambda_a^{-1} & \lambda_a^{-1} \end{bmatrix}.$$

In the Maple code, `markrecovery.mw`, we show that calculating the rank of $\mathbf{D}_1$ uses 399.84kb of memory, calculating the rank of $\mathbf{D}_2$ uses 158.18kb of memory and calculating the rank of $\mathbf{D}_3$ uses 111.08kb of memory (using Maple 2019). In all cases Maple code calculate the rank as 3, but the derivative matrix $\mathbf{D}_3$ is the simplest and uses the least memory.

Using the simplest exhaustive summary available could solve the issue the of lack of memory to calculate the rank. If this approach is not sufficient, then a new simpler exhaustive summary may be derived from any reparameterisation that simplifies the model structure. This method is discussed in Section 5.2.2 below.

5.2.2 Reduced-Form Exhaustive Summaries

A reduced-form exhaustive summary is an exhaustive summary of the smallest possible size. The smallest length of an exhaustive summary is the number of estimable parameters in the model. A formal definition is given below.

Definition 11 *For exhaustive summary $\boldsymbol{\kappa}_r$ with parameters $\boldsymbol{\theta}$, if the derivative matrix $\mathbf{D}_r = \partial \boldsymbol{\kappa}_r / \partial \boldsymbol{\theta}$ has rank r and $\dim(\boldsymbol{\kappa}_r) = r$ then $\boldsymbol{\kappa}_r$ is a reduced-form exhaustive summary (where $\dim(\boldsymbol{\kappa}_r)$ is the length of the vector $\boldsymbol{\kappa}_r$).*

In the same way that there is more than one exhaustive summary for any model, there can be more than one reduced-form exhaustive summary.

A method for creating a simpler reduced-form exhaustive summary involves reparameterising the exhaustive summary. This method can also be used to determine the rank of a complex model where Maple runs out of memory trying to calculate the rank using the standard symbolic method.

Suppose $\boldsymbol{\kappa}(\boldsymbol{\theta})$ is an exhaustive summary with $\dim\{\boldsymbol{\kappa}(\boldsymbol{\theta})\} = n$ and we choose a reparameterisation $\mathbf{s}(\boldsymbol{\theta})$ with $\dim\{\mathbf{s}(\boldsymbol{\theta})\} = q_s$. Typically the reparameterisation would be independent in terms of $\boldsymbol{\theta}$, so that $\text{rank}(\partial \mathbf{s}/\partial \boldsymbol{\theta}) = q_s$. Let $\boldsymbol{\kappa}(\mathbf{s})$ represent the exhaustive summary $\boldsymbol{\kappa}(\boldsymbol{\theta})$ reparameterised in terms of $\mathbf{s}$ rather than $\boldsymbol{\theta}$. A derivative matrix can now be formed by differentiating with respect to the reparameterisation rather than the original parameters, so that

$$
\mathbf{D}_s = \frac{\partial \boldsymbol{\kappa}(\mathbf{s})}{\partial \mathbf{s}} =
\begin{bmatrix}
\dfrac{\partial \kappa_1(\mathbf{s})}{\partial s_1} & \dfrac{\partial \kappa_2(\mathbf{s})}{\partial s_1} & \cdots & \dfrac{\partial \kappa_n(\mathbf{s})}{\partial s_1} \\
\dfrac{\partial \kappa_1(\mathbf{s})}{\partial s_2} & \dfrac{\partial \kappa_2(\mathbf{s})}{\partial s_2} & \cdots & \dfrac{\partial \kappa_n(\mathbf{s})}{\partial s_2} \\
\vdots & & & \vdots \\
\dfrac{\partial \kappa_1(\mathbf{s})}{\partial s_{q_s}} & \dfrac{\partial \kappa_2(\mathbf{s})}{\partial s_{q_s}} & \cdots & \dfrac{\partial \kappa_n(\mathbf{s})}{\partial s_{q_s}}
\end{bmatrix}.
$$

If rank $(\partial \mathbf{s}/\partial \boldsymbol{\theta}) = \dim(\mathbf{s})$, the derivative matrix $\mathbf{D}_s$ can be used to investigate parameter redundancy of the original parametrisation. Further, regardless of rank$(\partial \mathbf{s}/\partial \boldsymbol{\theta})$, a reduced form exhaustive summary can be formed from the reparameterisation $\mathbf{s}$. This is explained by the reparameterisation Theorem, given by Theorem 13 below.

Theorem 13 *Reparameterisation. Let $\mathbf{D}_s = \partial \boldsymbol{\kappa}(\mathbf{s})/\partial \mathbf{s}$, with $r_s = \mathrm{rank}(\mathbf{D}_s)$, and $q_s = \dim(\mathbf{s})$. Then the following holds:*

a. (i) If $r_s = q_s$, $\mathbf{s}$ is a reduced-form exhaustive summary.   (ii) If $r_s < q_s$ then $\mathbf{s}$ is not a reduced-form exhaustive summary. A reduced-form exhaustive summary may be found by first solving $\boldsymbol{\alpha}^T \mathbf{D}_s = 0$, which has $d = q_s - r_s$ solutions, labelled α_j for $j = 1 \ldots d$, with individual entries α_{ij}, and then solving the partial differential equations

$$
\sum_{i=1}^{q} \alpha_{ij} \frac{\partial f}{\partial \theta_i} = 0, \; j = 1 \ldots d. \tag{5.1}
$$

b. If $\mathrm{rank}(\partial \mathbf{s}/\partial \boldsymbol{\theta}) = q_s$, the number of estimable parameters is equal to r_s. If $r_s = p$ then the model in terms of $\boldsymbol{\theta}$ is not parameter redundant. If $r_s < p$ the model in terms of $\boldsymbol{\theta}$ is parameter redundant (Cole et al., 2010).

The reparameterisation Theorem is illustrated by three examples in Sections 5.2.2.1 to 5.2.2.3.

For the method to be useful, the reparameterisation needs to be chosen so that $\boldsymbol{\kappa}(\mathbf{s})$ is structurally simpler than $\boldsymbol{\kappa}(\boldsymbol{\theta})$. This would mean that $\mathbf{D}_s$ could be much simpler than original derivative matrix $\mathbf{D}$ and it may now be possible to find the rank symbolically.

Part a of Theorem 13 results in a reduced-form exhaustive summary that can be used to investigate local identifiability of a model. To ensure it can also be used to investigate global identifiability we also need to ensure there is a unique solution to the set of equations $\mathbf{k} = \boldsymbol{\kappa}(\mathbf{s})$ in terms of $\mathbf{s}$. This is demonstrated in example 5.2.2.2.

Part b of Theorem 13 only applies if rank$(\partial \mathbf{s}/\partial \boldsymbol{\theta}) = q_s$. If rank$(\partial \mathbf{s}/\partial \boldsymbol{\theta}) \neq q_s$, an exhaustive summary may be formed by solving $\boldsymbol{\alpha}^T \mathbf{D}_s = 0$ and then solving the appropriate PDEs, following part a of Theorem 13. The new exhaustive summary should then be used with Theorem 3 to determine whether or not the model is parameter redundant. An illustration of this is given in the example of Section 5.2.2.3.

5.2.2.1   Capture-Recapture Model

The Cormack Jolly Seber (CJS) (Cormack 1964, Jolly 1965 and Seber 1965) is a capture-recapture model which can be used to estimate the probability of survival. An extension of this model with trap dependence was discussed in Section 3.2.7.2. Capture-recapture models are similar to mark-recovery models, but animals are recaptured alive, rather than recovered dead. The CJS model has been widely applied to a variety of contexts (for example see Lebreton et al. 1992). The model assumes that not all animals are recaptured, therefore it also has parameters representing the probability of recapture.

Suppose R_i animals are released at occasion i; one way of summarising capture-recapture data is to count the number of individuals released at occasion i who are next recaptured at occasion j, $N_{i,j}$. Let ϕ_i denote the probability of surviving from occasion i to occasion $i+1$, and p_i denote the probability an animal alive at occasion i is recaptured. The probabilities of an animal being captured at occasion j, given they were last captured and released at occasion i, $P_{i,j}$, can be summarised by the matrix:

$$\mathbf{P} = \begin{bmatrix} \phi_1 p_2 & \phi_1(1-p_2)\phi_2 p_3 & \phi_1(1-p_2)\phi_2(1-p_3)\phi_3 p_4 \\ 0 & \phi_2 p_3 & \phi_2(1-p_3)\phi_3 p_4 \\ 0 & 0 & \phi_3 p_4 \end{bmatrix}.$$

The natural logarithm of the non-zero probabilities in the matrix $\mathbf{P}$ form an exhaustive summary with

$$\kappa = \begin{bmatrix} \log(\phi_1 p_2) \\ \log\{\phi_1(1-p_2)\phi_2 p_3\} \\ \log\{\phi_1(1-p_2)\phi_2(1-p_3)\phi_3 p_4\} \\ \log(\phi_2 p_3) \\ \log\{\phi_2(1-p_3)\phi_3 p_4\} \\ \log(\phi_3 p_4) \end{bmatrix}.$$

The parameters are $\boldsymbol{\theta} = [\phi_1, \phi_2, \phi_3, p_1, p_2, p_3]$. Observe the parameters ϕ_3 and p_4: they only ever appear in the product $\phi_3 p_4$. It is obvious that this model is parameter redundant, as we could reparameterise in terms of a smaller number of parameters.

To formally check identifiability, the standard symbolic approach will work for this model (see Cole et al., 2010 and the Maple code for this example). So in this case the reparameteration Theorem (Theorem 13) is not needed. However, the example is useful to demonstrate how the reparameterisation Theorem can be applied.

We start with forming a suitable reparameterisation. For the CJS model we note that the exhaustive summary consists of terms $\phi_i p_{i+1}$ and $\phi_i(1-p_{i+1})$, which leads to a possible reparameterisation of:

$$
\mathbf{s} = \begin{bmatrix} s_1 \\ s_2 \\ s_3 \\ s_4 \\ s_5 \end{bmatrix} = \begin{bmatrix} \phi_1 p_2 \\ \phi_1(1 - p_2) \\ \phi_2 p_3 \\ \phi_2(1 - p_3) \\ \phi_3 p_4 \end{bmatrix}.
\tag{5.2}
$$

We then check $\mathrm{rank}(\partial \mathbf{s}/\partial \boldsymbol{\theta}) = q_s = 5$. The derivative matrix

$$
\frac{\partial \mathbf{s}}{\partial \boldsymbol{\theta}} = \begin{bmatrix} p_2 & 1 - p_2 & 0 & 0 & 0 \\ 0 & 0 & p_3 & 1 - p_3 & 0 \\ 0 & 0 & 0 & 0 & p_4 \\ \phi_1 & -\phi_1 & 0 & 0 & 0 \\ 0 & 0 & \phi_2 & -\phi_2 & 0 \\ 0 & 0 & 0 & 0 & \phi_3 \end{bmatrix}
$$

has rank 5 as required to use part b of Theorem 13.

Next, we rewrite the exhaustive summary in terms of the reparameterisation, $\mathbf{s}$, which gives:

$$
\boldsymbol{\kappa}(\mathbf{s}) = \begin{bmatrix} \log(s_1) \\ \log(s_2 s_3) \\ \log(s_2 s_4 s_5) \\ \log(s_3) \\ \log(s_4 s_5) \\ \log(s_5) \end{bmatrix}.
$$

The derivative matrix is then

$$
\mathbf{D}_s = \frac{\partial \boldsymbol{\kappa}(\mathbf{s})}{\partial \mathbf{s}} = \begin{bmatrix} s_1^{-1} & 0 & 0 & 0 & 0 & 0 \\ 0 & s_2^{-1} & s_2^{-1} & 0 & 0 & 0 \\ 0 & s_3^{-1} & 0 & s_3^{-1} & 0 & 0 \\ 0 & 0 & s_4^{-1} & 0 & s_4^{-1} & 0 \\ 0 & 0 & s_5^{-1} & 0 & s_5^{-1} & s_5^{-1} \end{bmatrix}.
$$

The rank of $\mathbf{D}_s$ is $r_s = 5$. Therefore, by Theorem 13b the rank of the original model is 5. As there are six parameters this shows that this model is parameter redundant with deficiency 1. Further, as $r_s = q_s = 5$ by Theorem 13a the reparameterisation given by equation (5.2) is a reduced-form exhaustive summary (Cole et al., 2010).

The Maple code for this example is CJS.mw.

5.2.2.2   Complex Compartmental Model

In Section 4.3.2 we introduced a complex compartmental model from Audoly et al. (1998). There we used the hybrid symbolic-numerical method to show the model is at least locally identifiable. Audoly et al. (1998) show the model is globally identifiable using a topologically method. Here we use reduced-form exhaustive summaries as an alternative way of showing that this model is globally identifiable.

The model has 11 parameters

$$\boldsymbol{\theta} = [k_{03}, k_{04}, k_{12}, k_{14}, k_{21}, k_{23}, k_{32}, k_{34}, k_{41}, k_{43}, V_1].$$

In Section 4.3.2 we showed an exhaustive summary is

$$\boldsymbol{\kappa} = \begin{bmatrix} k_{03}k_{04}k_{12} + \ldots + k_{14}k_{32}k_{43} \\ k_{03}k_{04} + \ldots + k_{32}k_{43} \\ \vdots \\ k_{21}k_{32} + k_{34}k_{41} \end{bmatrix}.$$

The full exhaustive summary terms are given in the Maple code `complexcomp.mw`.

It is not possible to calculate the symbolic rank of the derivative matrix $\mathbf{D} = \partial \boldsymbol{\kappa}(\boldsymbol{\theta})/\partial \boldsymbol{\theta}$. Instead we need to use the reparameterisation Theorem so we can apply the symbolic method. A possible reparameterisation is

$$\mathbf{s} = \begin{bmatrix} s_1 \\ s_2 \\ s_3 \\ s_4 \\ s_5 \\ s_6 \\ s_7 \\ s_8 \\ s_9 \\ s_{10} \\ s_{11} \end{bmatrix} = \begin{bmatrix} -(k_{21} + k_{41}) \\ -(k_{12} + k_{32}) \\ -(k_{23} + k_{43} + k_{03}) \\ -(k_{14} + k_{34} + k_{04}) \\ k_{12}k_{21} \\ k_{14}k_{41} \\ k_{23}k_{32} \\ k_{34}k_{43} \\ k_{21} \\ k_{41}k_{34} \\ V_1 \end{bmatrix}.$$

This reparameterisation is based on the combinations of parameters that are given in Audoly et al. (1998) from the topological properties of the graph associated with this model.

We can check that $\text{rank}(\partial \mathbf{s}/\partial \boldsymbol{\theta}) = q_s = 11$, so Theorem 13b can be used. Next we write the exhaustive summary in terms of $\mathbf{s}$, which gives

$$\boldsymbol{\kappa}(\mathbf{s}) = \begin{bmatrix} s_2 s_8 + s_4 s_7 - s_2 s_3 s_4 \\ s_4 s_3 + s_2 s_4 + s_2 s_3 - s_7 - s_8 \\ \vdots \\ s_{10} - s_5 - s_2 s_9 \end{bmatrix}.$$

The rank of $\mathbf{D}_s = \partial \boldsymbol{\kappa}(\mathbf{s})/\partial \mathbf{s}$ is $r_s = 11$, so as $q_s = 11$ the reparameterisation $\mathbf{s}$ is a reduced-form exhaustive summary, by Theorem 13a. As the original model has 11 parameters, using Theorem 13b we can conclude this model is not parameter redundant (Cole et al., 2010).

We can also show the model is globally identifiable by first demonstrating that there is a unique solution to the equations $b_i = \kappa_i(\mathbf{s})$, for $i = 1, ..., 11$, of the form $s_1 = (b_3b_8 - b_7)/b_8$, $s_2 = (b_{10} - b_3b_{11})/b_{11}, \ldots$. Therefore $\mathbf{s}$ is a reduced-form exhaustive summary that can be used to distinguish between local and global identifiability. Then we need to solve the equations $b'_i = s_i(\boldsymbol{\theta})$, for $i = 1, ..., 11$, which has a unique solution $k_{21} = b'_9$, $k_{12} = b'_5/b'_9$, $k_{32} = -(b'_2b'_9 + b'_5)/b'_9, \ldots, V_1 = b'_{11}$. As there is a unique solution the model is globally identifiable. In both cases the full solution to the equations are provided in the Maple code `complexcomp.mw`.

5.2.2.3 Age-Dependent Fisheries Model

This example demonstrates how a reparameterisation that does not satisfy $\text{rank}(\partial\mathbf{s}/\partial\boldsymbol{\theta}) = q_s$ can be used to find a simpler exhaustive summary. Jiang et al. (2007) present an age-dependent model for data on fish that are tagged annually, and then the tags returned when the fish are harvested. The probability that a fish tagged at age k, released in year i is harvested and then the tagged returned in year j is

$$
P_{i,j,k} = \begin{cases} \left(1 - e^{-F_j\sigma_{k+j-i} - M}\right) \frac{F_j\sigma_k\lambda}{F_j\sigma_k + M} & j = i \\ \left(\prod_{\nu=i}^{j-1} e^{-F_\nu\sigma_{k+\nu-i} - M}\right) \left(1 - e^{-F_j\sigma_{k+j-i} - M}\right) \frac{F_j\sigma_{k+j-i}\lambda}{F_j\sigma_{k+j-i} + M} & j > i \end{cases}.
$$

The model has the following parameters:

- F_j, the instantaneous fishing mortality rate for fully recruited fish;

- σ_a, the selectivity coefficient for fish ages a, with $\sigma_a = 1$ for $a > a_c$;

- M, the instantaneous natural mortality rate;

- λ, the probability a tag is reported for a dead fish.

The parameters M and λ can depend on year and age, so that $M_{y,a} = M_y^Y M_a^A$ with $M_1^Y = 1$, and $\lambda_{y,a} = \lambda_y^Y \lambda_a^A$ with $\lambda_1^Y = 1$.

The Maple code for this example is provided in `fisheries.mw`.

Jiang et al. (2007) tried to use the symbolic method to investigate parameter redundancy but found that Maple ran out of memory trying to find the rank with, so instead used a numerical method. For example, consider a model with four years of tagging, four years of recovery, two age classes and $a_c = 2$. The full model has parameters:

$$
\boldsymbol{\theta} = [F_1, F_2, F_3, F_4, \sigma_1, \sigma_2, M_2^Y, M_3^Y, M_4^Y, M_1^A, M_2^A, \lambda_2^Y, \lambda_3^Y, \lambda_4^Y, \lambda_1^A, \lambda_2^A].
$$

Using an exhaustive summary consisting of the terms $P_{i,j,k}$, with

$$\boldsymbol{\kappa}(\boldsymbol{\theta}) = \begin{bmatrix} \dfrac{\left(1-e^{-F_1\sigma_1-M_1^A}\right)F_1\sigma_1\lambda_1^A}{F_1\sigma_1+M_1^A} \\[2ex] \dfrac{e^{-F_1\sigma_1-M_1}\left(1-e^{-F_2\sigma_2-M_2^Y M_2^A}\right)F_2\sigma_2\lambda_2^Y\lambda_2^A}{F_2\sigma_2+M_2^Y M_2^A} \\[1ex] \vdots \\[1ex] \dfrac{\left(1-e^{-F_4\sigma_2-M_4^Y M_2^A}\right)F_4\sigma_2\lambda_4^Y\lambda_2^A}{F_4\sigma_2+M_4^Y M_2^A} \end{bmatrix}$$

results in the derivative matrix $\mathbf{D} = \partial\boldsymbol{\kappa}/\partial\boldsymbol{\theta}$, where it is not possible to calculate the rank. Instead Cole and Morgan (2010b) use the reparameterisation

$$\mathbf{s} = \begin{bmatrix} s_1 \\ s_2 \\ s_3 \\ s_4 \\ s_5 \\ s_6 \\ s_7 \\ s_8 \\ s_9 \\ s_{10} \\ s_{11} \\ s_{12} \\ s_{13} \\ s_{14} \\ s_{15} \\ s_{16} \\ s_{17} \end{bmatrix} = \begin{bmatrix} F_1\sigma_1 + M_1^A \\ F_1\sigma_2 + M_2^A \\ F_2\sigma_1 + M_2^Y M_1^A \\ F_2\sigma_2 + M_2^Y M_2^A \\ F_2 + M_2^Y M_2^A \\ F_3\sigma_1 + M_3^Y M_1^A \\ F_3\sigma_2 + M_3^Y M_2^A \\ F_3 + M_3^Y M_2^A \\ F_4\sigma_1 + M_4^Y M_1^A \\ F_4\sigma_2 + M_4^Y M_2^A \\ F_4 + M_4^Y M_2^A \\ F_1\sigma_1\lambda_1^A \\ F_1\sigma_2\lambda_2^A \\ F_1\lambda_2^A \\ F_2\sigma_1\lambda_2^Y\lambda_1^A \\ F_3\sigma_1\lambda_3^Y\lambda_1^A \\ F_4\sigma_1\lambda_4^Y\lambda_1^A \end{bmatrix}$$

to finder a simpler exhaustive summary. The $\text{rank}(\partial\mathbf{s}/\partial\boldsymbol{\theta}) = 17$. However $q_s = 16$, therefore it is not possible to use Theorem 13b to determine that rank of the original model. Instead we create a reduced form exhaustive summary using Theorem 13a. Rewriting the original exhaustive summary in terms of $\mathbf{s}$ gives

$$\boldsymbol{\kappa}(\mathbf{s}) = \begin{bmatrix} \dfrac{s_{12}\left(1-e^{-s_1}\right)}{s_1} \\[2ex] \dfrac{s_{13}s_{15}e^{-s_1}\left(1-e^{-s_4}\right)}{s_4 s_{12}} \\[1ex] \vdots \\[1ex] \dfrac{s_{13}s_{17}\left(1-e^{-s_{10}}\right)}{s_{10}s_{12}} \end{bmatrix}.$$

The derivative matrix

$$\mathbf{D}_s = \frac{\partial \boldsymbol{\kappa}(\mathbf{s})}{\partial \mathbf{s}} =$$

$$\begin{bmatrix} \frac{\left(s_1 e^{-s_1} + e^{-s_1} - 1\right)s_{12}}{s_1^2} & -\frac{s_{13}s_{15}\left(1 - e^{-s_4}\right)e^{-s_1}}{s_4 s_{12}} & \cdots & 0 \\ 0 & 0 & & 0 \\ 0 & 0 & & 0 \\ 0 & \frac{\left(s_4 e^{-s_4} + e^{-s_4} - 1\right)s_{13}s_{15}e^{-s_1}}{s_4^2 s_{12}} & & 0 \\ \vdots & & & \vdots \\ 0 & 0 & \cdots & \frac{s_{13}\left(1 - e^{-s_{10}}\right)}{s_{10}s_{12}} \end{bmatrix}$$

has rank 16. Therefore the reparameterisation $\mathbf{s}$ is not an exhaustive summary, however we can find a new exhaustive summary applying Theorem 5 of Chapter 3. A reduced form exhaustive summary consists of the terms:

$$\boldsymbol{\kappa}_{re,1}(\mathbf{s}) = \begin{bmatrix} s_1 \\ s_2 \\ s_3 \\ s_4 \\ s_5 \\ s_6 \\ s_7 \\ s_8 \\ s_{12} \\ s_{13} \\ s_{14} \\ s_{15} \\ s_{16} \\ \frac{s_{17}\left(1 - e^{-s_9}\right)}{s_9} \\ \log\left\{\frac{s_{10}\left(1 - e^{s_9}\right)}{s_9\left(1 - e^{s_{10}}\right)}\right\} + s_{10} - s_9 \\ \log\left\{\frac{s_{11}\left(1 - e^{s_9}\right)}{s_9\left(1 - e^{s_{11}}\right)}\right\} + s_{11} - s_9 \end{bmatrix}$$

In terms of the original parameters this becomes:

$$\kappa_{re,1}(\boldsymbol{\theta}) = \begin{bmatrix} F_1\sigma_1 + M_1^A \\ F_1\sigma_2 + M_2^A \\ F_2\sigma_1 + M_2^Y M_1^A \\ F_2\sigma_2 + M_2^Y M_2^A \\ F_2 + M_2^Y M_2^A \\ F_3\sigma_1 + M_3^Y M_1^A \\ F_3\sigma_2 + M_3^Y M_2^A \\ F_3 + M_3^Y M_2^A \\ F_1\sigma_1\lambda_1^A \\ F_1\sigma_2\lambda_2^A \\ F_1\lambda_2^A \\ F_2\sigma_1\lambda_2^Y\lambda_1^A \\ F_3\sigma_1\lambda_3^Y\lambda_1^A \\ F_4\sigma_1\lambda_4^Y\lambda_1^A \dfrac{\left(1-e^{-F_4\sigma_1-M_4^Y M_1^A}\right)}{F_4\sigma_1 + M_4^Y M_1^A} \\ \log\left\{\dfrac{(F_4\sigma_2 + M_4^Y M_2^A)(1 - e^{F_4\sigma_1 + M_4^Y M_1^A})}{(F_4\sigma_1 + M_4^Y M_1^A)(1 - e^{F_4\sigma_2 + M_4^Y M_2^A})}\right\} + F_4\sigma_2 + M_4^Y M_2^A - F_4\sigma_1 - M_4^Y M_1^A \\ \log\left\{\dfrac{(F_4 + M_4^Y M_2^A)(1 - e^{F_4\sigma_1 + M_4^Y M_1^A})}{(F_4\sigma_1 + M_4^Y M_1^A)(1 - e^{F_4 + M_4^Y M_2^A})}\right\} + F_4 + M_4^Y M_2^A - F_4\sigma_1 - M_4^Y M_1^A \end{bmatrix}$$

This new reduced form exhaustive summary can then be used checked parameter redundancy by forming the derivative matrix

$$\mathbf{D}_{re} = \frac{\partial \kappa_{re,1}}{\partial \boldsymbol{\theta}} = \begin{bmatrix} \sigma_1 & \sigma_2 & 0 & 0 & 0 & \cdots \\ 0 & 0 & \sigma_1 & \sigma_2 & 1 & \\ 0 & 0 & 0 & 0 & 0 & \\ 0 & 0 & 0 & 0 & 0 & \\ F_1 & 0 & F_2 & 0 & 0 & \\ \vdots & & & & & \end{bmatrix},$$

which has rank 16. As there are 16 parameters this model is not parameter redundant.

5.2.3 General Results using Extension and Reparameterisation Theorems

In Section 3.2.6.1 we showed that it is possible to generalise results to any dimension of the model using the extension theorem (Theorem 6). In some examples, application of the extension theorem is straightforward using the original exhaustive summary. However in other instances, particularly in parameter redundant models, the reparsmeterisation Theorem (Theorem 13) and the extension theorem need to be used together to obtain general results.

This is demonstrated using the example in Section 5.2.3.1. Other examples of using the extension theorem and reparameterisation Theorem combined to obtain general results are given in Cole et al. (2014) and Cole (2012).

5.2.3.1 CJS Model

Consider again the CJS model. As shown in Section 5.2.2.1 a reduced-form exhaustive summary for 3 years of marking and recapture is

$$
\mathbf{s}_1 =
\begin{bmatrix}
s_1 \\
s_2 \\
s_3 \\
s_4 \\
s_5
\end{bmatrix}
=
\begin{bmatrix}
\phi_1 p_2 \\
\phi_1 (1 - p_2) \\
\phi_2 p_3 \\
\phi_2 (1 - p_3) \\
\phi_3 p_4
\end{bmatrix}.
$$

The derivative matrix $\mathbf{D}_s = \partial \kappa(\mathbf{s}_1)/\partial \mathbf{s}_1$ was shown to be of full rank $r_s = 5$. If we extend the CJS model by 1 extra year of marking and recovery the extra original exhaustive summary terms (the natural logarithm of the probabilities) are

$$
\mathbf{k}_2(\boldsymbol{\theta}_1, \boldsymbol{\theta}_2) =
\begin{bmatrix}
\log\{\phi_1(1 - p_2)\phi_2(1 - p_3)\phi_3(1 - p_4)\phi_4 p_5\} \\
\log\{\phi_2(1 - p_3)\phi_3(1 - p_4)\phi_4 p_5\} \\
\log\{\phi_3(1 - p_4)\phi_4 p_5\} \\
\log(\phi_4 p_5)
\end{bmatrix}.
$$

where the original parameters are $\boldsymbol{\theta}_1 = [\phi_1, \phi_2, \phi_3, p_2, p_3, p_4]$ and the extra parameters are $\boldsymbol{\theta}_3 = [\phi_4, p_5]$. The reparameterisation can be extended with the terms

$$
\mathbf{s}_2 =
\begin{bmatrix}
s_6 \\
s_7
\end{bmatrix}
=
\begin{bmatrix}
\phi_3(1 - p_4) \\
\phi_4 p_5
\end{bmatrix}.
$$

The extra exhaustive summary terms $\mathbf{k}_2(\boldsymbol{\theta}_1, \boldsymbol{\theta}_2)$ can be rewritten in terms of $\mathbf{s}_1$ and $\mathbf{s}_2$ to give

$$
\mathbf{k}_2(\mathbf{s}_1, \mathbf{s}_2) =
\begin{bmatrix}
\log(s_2 s_4 s_6 s_7) \\
\log(s_4 s_6 s_7) \\
\log(s_6 s_7) \\
\log(s_7)
\end{bmatrix}.
$$

The derivative matrix,

$$
\mathbf{D}_{s,2,2} = \frac{\partial \mathbf{k}_2(\mathbf{s}_1, \mathbf{s}_2)}{\mathbf{s}_2} =
\begin{bmatrix}
s_6^{-1} & s_6^{-1} & s_6^{-1} & 0 \\
s_7^{-1} & s_7^{-1} & s_7^{-1} & s_7^{-1}
\end{bmatrix},
$$

has full rank 2. Therefore by the extension theorem (Theorem 6) this reparameterisation will always be full rank. Then by the reparameterisation theorem (Theorem 13) we can say that the general reduced-form exhaustive summary for n years of marking and recapture is

$$\mathbf{s}_{re} = \begin{bmatrix} \phi_1 p_2 \\ \phi_1(1 - p_2) \\ \vdots \\ \phi_{n-1} p_n \\ \phi_{n-1}(1 - p_n) \\ \phi_n p_{n+1} \end{bmatrix}.$$

As this is a reduced-form exhaustive summary of length $2n - 1$ and there are $2n$ parameters, by the reparameterisation Theorem this model will always be parameter redundant with deficiency 1 (Cole et al., 2010). Although this result is obvious from the form of the model, this example demonstrates that applying the extension theorem to a reduced-form exhaustive summary, that increases in dimension, could be used to create a general reduced-form exhaustive summary for any dimension. The general reduced-form exhaustive summary can then be used to deduce general results about the parameter redundancy in a model of any dimension.

5.3 Extended Hybrid Symbolic-Numerical Method

As discussed in Section 4.5, the hybrid symbolic-numerical method, which was introduced in Section 4.3, is unable to find estimable parameter combinations and cannot be used to find general results. In this Section we outline extensions to the hybrid symbolic-numerical method, which we call the extended hybrid symbolic-numerical method. The first extension uses subset profiling, discussed in Section 5.3.1, to find estimable parameter combinations. Then in Section 5.3.2 we show how the hybrid symbolic-numerical method can be used to find general results.

5.3.1 Subset Profiling

As well as using a profile log-likelihood to decide whether or not a parameter is identifiable, there is also information in profile log-likelihoods on confounding between pairs of parameters. If a pair of parameters are confounded, for example $\beta_1 = \theta_1 \theta_2$, then plotting the fixed parameter against the maximum likelihood estimates of the other parameters will show the relationship between those two parameters. Multiple profile log-likelihood plots for each pair of parameters can be used to determine the estimable parameter combinations; that is, which combinations of parameters can be estimated in parameter redundant models. Not every pair of parameters should be compared. The choice of which parameters to compare is determined by examining the deficiency of subsets of parameters. This subset profiling method was developed in Eisenberg and Hayashi (2014).

The first stage of the method involves finding nearly full rank subsets. It is possible to examine parameter redundancy for a subset of parameters in a model. This involves fixing parameters that are not part of the subset to a known constant. Eisenberg and Hayashi (2014) define a nearly full rank subset as a subset of the parameters $\mathbf{A}$ which is parameter redundant but also has the following property: if each parameter in turn is removed from $\mathbf{A}$ then the subset becomes full rank. This needs to be true for all parameters in the subset of $\mathbf{A}$.

For example, suppose a model has four parameters, θ_1, θ_2, θ_3 and θ_4, and is parameter redundant. Consider the subset $\mathbf{A} = \{\theta_2, \theta_3, \theta_4\}$, which would be formed by treating the parameter θ_1 as a fixed constant. Suppose subset $\mathbf{A}$ is parameter redundant. For $\mathbf{A}$ to be nearly full rank we then also need the subsets with parameters $\{\theta_2, \theta_3\}$ and $\{\theta_2, \theta_4\}$ and $\{\theta_3, \theta_4\}$ to be full rank.

If a model is parameter redundant, it may still be possible to estimate some of the original parameters. Those parameters would be individually identifiable. It is simpler to start by removing any identifiable parameters before examining whether a subset is nearly full rank.

Once nearly full rank subsets of parameters are found, we fix one parameter in the subset and find the profile log-likelihood for that parameter. Rather than plotting the fixed parameter against the maximum likelihood values we plot the fixed parameter against the maximum likelihood estimates of each other parameter in the nearly full rank subset.

An example is shown in Figure 5.1. This corresponds to an artificial multinomial model with probabilities θ_1, $\theta_2 + \theta_3$, $\theta_3\theta_4$ and $(1 - \theta_1 - \theta_2 - \theta_3 - \theta_3\theta_4)$. This model is obviously parameter redundant with estimable parameter combinations θ_1, $\theta_2 + \theta_3$ and $\theta_3\theta_4$, but is just used here as an illustration. In this case a nearly full rank subset is $\mathbf{A} = \{\theta_2, \theta_3, \theta_4\}$. We consider a log-likelihood profile for each pair of parameters. Figure 5.1 shows the profile for the pair (θ_2, θ_3). The standard likelihood profile of θ_2 is shown in Figure 5.1a. The parameter is non-identifiable therefore the profile is flat. When we create the likelihood profile for θ_2 we fix θ_2 and maximise with respect to all the other parameters. This is then repeated for other fixed values of θ_2. In Figure 5.1b we plot the maximum likelihood estimates of θ_3 against the fixed values of θ_2. (Note we could have equally fixed θ_3; the choice is arbitrary.) If (θ_2, θ_3) is a pair from a nearly full rank subset then, as shown in Figure 5.1b, there will be a line representing the relationship between the two parameters. In this case the line is $\theta_3 = -\theta_2 + c$ for some constant c, which can be rearranged to $\theta_2 + \theta_3 = c$. This shows us that part of the confounding of θ_2 and θ_3 is of the form $\theta_2 + \theta_3$.

Once this procedure is repeated for every pair of parameters in a nearly full rank subset then we use the equations for each line, such as $\theta_3 + \theta_2 = c$, to find the estimable parameter combinations.

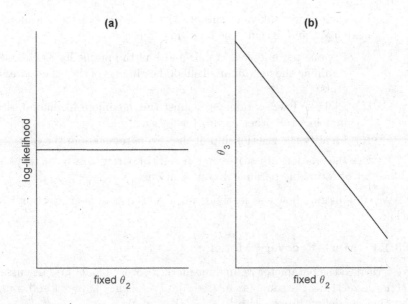

FIGURE 5.1
Example subset profile. Figure (a) shows the standard log-likelihood profile for non-identifiable parameter θ_2. The x-axis of figure (b) is identical, but the y-axis shows the maximum likelihood estimates of θ_3, corresponding to fixing θ_2. The line is $\theta_3 = -\theta_2 + c$ for some constant c.

The likelihood profiles would result in the following relationships:

$$\theta_2 + \theta_3 = c_1 \tag{5.3}$$
$$\theta_3\theta_4 = c_2 \tag{5.4}$$
$$(c_1 - \theta_2)\theta_4 = c_3 \tag{5.5}$$

where c_i are constants. We ignore equation (5.5), as this can be obtained from equations (5.3) and (5.4). Then it is clear from the relationships that the estimable parameter combinations are $\theta_2 + \theta_3$ and $\theta_3\theta_4$.

To summarise, the subset profile procedure is as follows:

1. Test whether a model is parameter redundant, using the hybrid symbolic-numerical method or another numerical method. If it is parameter redundant proceed to use subset profiling for finding estimable parameter combinations.

2. Exclude any individually identifiable parameters.

3. Find nearly full rank subsets from the non-identifiable parameters, which can be carried out using any of the methods for detecting parameter redundancy.

4. For each nearly full rank subset consider every pair of parameters in that set and execute the following steps:

   (a) Fix one parameter in a pair and find the profile log-likelihood, retaining the maximum likelihood estimates of the other parameters.

   (b) Plot the fixed parameter against the maximum likelihood estimates of the other parameter for each pair.

   (c) Identify the relationship of the two parameters in the plot.

5. Use the relationships between the pairs of parameters to recover the set of estimable parameter combinations.

We demonstrate how this methods works in the example in Section 5.3.1.1 below.

5.3.1.1 Mark-Recovery Model

We return to the parameter redundant mark-recovery model first discussed in Section 1.2.4 using the same (simulated) data set as in Chapter 1 and consider the same model with parameters ϕ_1, ϕ_a, λ_1 and λ_a.

Here we demonstrate the subset profiling method using the likelihood profile and the Hessian method combined. However, the use of the hybrid symbolic-numerical method of Section 4.3 is recommended, if more accuracy is required.

Although we know that this model is parameter redundant, the first step is to test whether or not the model is parameter redundant. Here we show this using the Hessian method. The standardised eigenvalues are 1, 0.30, 0.011 and -0.00000032. The smallest eigenvalue is close to zero suggesting this model is parameter redundant.

Then we examine a profile plot to determine if any of the original parameters are identifiable. This is similar to the profile plot produced in Figure 2.3 of Chapter 2, but with different data. Figure 5.2 reveals that ϕ_1, λ_1 and λ_a are non-identifiable as they have flat profiles, however the profile for ϕ_a is not flat suggesting that ϕ_a is identifiable. We exclude any identifiable parameters from the subset, so we exclude ϕ_a.

Next we consider the subset $\mathbf{A} = \{\phi_1, \lambda_1, \lambda_a\}$. Using the Hessian method the standardised eigenvalues are 1, 0.30 and 0.0000048. The smallest eigenvalue indicates that this model is parameter redundant. The following subsets and their eigenvalues are then considered:

- $A_1 = \{\lambda_1, \lambda_a\}$ has standardised eigenvalues, 1 and 0.22, so is not parameter redundant;

- $A_2 = \{\phi_1, \lambda_a\}$ has standardised eigenvalues, 1 and 0.14, so is not parameter redundant;

- $A_3 = \{\phi_1, \lambda_1\}$ has standardised eigenvalues, 1 and 0.10, so is not parameter redundant.

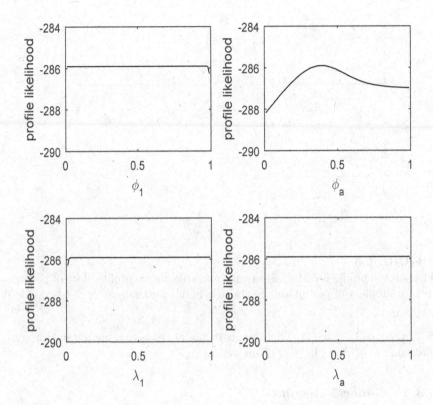

FIGURE 5.2
Profile log-likelihood for mark-recovery model with parameters ϕ_1, ϕ_a, λ_1 and λ_a.

Therefore $\mathbf{A} = \{\phi_1, \lambda_1, \lambda_a\}$ is a nearly full rank subset.

The last stage is to produce profile plots of each pair of parameters in this subset. This is given in Figure 5.3. The plots show the relationships

$$\lambda_1 = \frac{0.011}{1 - \phi_1} \tag{5.6}$$

$$\lambda_a = \frac{0.004}{\phi_1} \tag{5.7}$$

$$\lambda_a = \frac{0.004\lambda_1}{\lambda_1 - 0.011} \tag{5.8}$$

Equation (5.8) can be obtained from equations (5.6) and (5.7), so it is discounted. Equations (5.6) and (5.7) then specify the estimable parameter combinations. Rearranging (5.6) gives $(1 - \phi_1)\lambda_1 = 0.011$ and rearranging

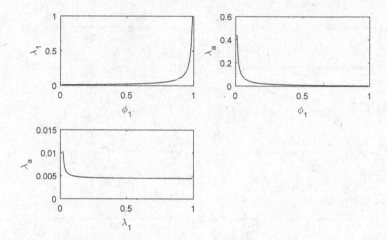

FIGURE 5.3

This subset profile for Mark-Recovery example shows profile plots of parameters for all pairs of parameters in the nearly full rank subset $\mathbf{A} = \{\phi_1, \lambda_1, \lambda_a\}$.

(5.7) gives $\phi_1 \lambda_a = 0.004$. Therefore this model has estimable parameter combinations $\phi_1 \lambda_a$ and $(1 - \phi_1)\lambda_1$, as well as ϕ_a.

5.3.2   General Results

The hybrid symbolic-numerical method can also be used to obtain general results. The extension theorem of Section 3.2.6.1 can be used in the same way as the symbolic method, but with the rank calculated numerically, as described below.

Suppose a simpler version of a model exists with exhaustive summary $\boldsymbol{\kappa}_1$ with n_1 terms and with p_1 parameters $\boldsymbol{\theta}_1$. This model is then extended to give exhaustive summary $\boldsymbol{\kappa} = \begin{bmatrix} \boldsymbol{\kappa}_1^T & \boldsymbol{\kappa}_2^T \end{bmatrix}^T$ with $n_1 + n_2$ terms and with $p_1 + p_2$ parameters $\boldsymbol{\theta} = [\boldsymbol{\theta}_1, \boldsymbol{\theta}_2]$. Let $\mathbf{D}_{11}$ be the derivative matrix with entries $D_{11ij} = \partial \kappa_{1i}/\partial \theta_{1j}$, $\mathbf{D}_{12}$ be the derivative matrix with entries $D_{12ij} = \partial \kappa_{2i}/\partial \theta_{1j}$ and $\mathbf{D}_{22}$ be the derivative matrix with entries $D_{22ij} = \partial \kappa_{2i}/\partial \theta_{2j}$. The extended model then has derivative matrix

$$\mathbf{D} = \begin{bmatrix} \mathbf{D}_{11} & \mathbf{D}_{12} \\ \mathbf{0} & \mathbf{D}_{22} \end{bmatrix}.$$

Theorem 14 *The Extension Theorem*
If $\mathbf{D}_{11}$ has full rank p_1 for at least one random point in the parameter space and $\mathbf{D}_{22}$ has full rank p_2 for at least one random point in the parameter space then $\mathbf{D}$ is of full rank $r = p_1 + p_2$ and the model is at least locally identifiable.

An example is given in Section 5.3.2.1 below.

5.3.2.1 Mark-Recovery Model

Consider the mark-recovery model with time dependent first year survival, $\phi_{1,t}$, constant adult survival, ϕ_a, and time dependent reporting probability, λ_t. For $n_1 = 4$ years of marking and $n_2 = 4$ years of recovery an exhaustive summary, κ_1, consists of all the non-zero terms of $\mathbf{P}$ with

$$
\mathbf{P} = \begin{bmatrix}
(1-\phi_{11})\lambda_1 & \phi_{11}(1-\phi_a)\lambda_2 & \phi_{11}\phi_a(1-\phi_a)\lambda_3 & \phi_{11}\phi_a^2(1-\phi_a)\lambda_4 \\
0 & (1-\phi_{12})\lambda_2 & \phi_{12}(1-\phi_a)\lambda_3 & \phi_{12}\phi_a(1-\phi_a)\lambda_4 \\
0 & 0 & (1-\phi_{13})\lambda_3 & \phi_{13}(1-\phi_a)\lambda_4 \\
0 & 0 & 0 & (1-\phi_{14})\lambda_4
\end{bmatrix}.
$$

The parameters are $\boldsymbol{\theta}_1 = [\phi_{11}, \phi_{1,2}, \phi_{1,3}, \phi_{1,4}, \phi_a, \lambda_1, \lambda_2, \lambda_3, \lambda_4]$. The derivative matrix is found symbolically and the rank calculated numerically at five random points in the parameter space; all have rank 9. Therefore by the hybrid symbolic-numerical method this mark-recovery model with $n_1 = n_2 = 4$ has full rank 9. The model is extended to $n_1 = 5$ years of marking and $n_2 = 5$ years of recovery. The extra exhaustive summary terms are

$$
\boldsymbol{\kappa}_2 = \begin{bmatrix}
\phi_{11}\phi_a^3(1-\phi_a)\lambda_5 \\
\phi_{12}\phi_a^2(1-\phi_a)\lambda_5 \\
\phi_{13}\phi_a(1-\phi_a)\lambda_5 \\
\phi_{14}(1-\phi_a)\lambda_5 \\
(1-\phi_{15})\lambda_5
\end{bmatrix}.
$$

There are two extra parameters $\boldsymbol{\theta}_2 = [\phi_{15}, \lambda_5]$. The derivative matrix, $\mathbf{D}_{22} = \partial\boldsymbol{\kappa}_2/\partial\boldsymbol{\theta}_2$, is found symbolically then the rank is calculated numerically at five random sets of parameters, each of which has rank 2. Therefore by Theorem 14 the model with $n_1 = n_2 = n_5$ has rank $r = 9+2 = 11$. There are 11 parameters in the extended model so the extended model is also not parameter redundant.

5.4 Conclusion

In complex models it is still possible to check identifiability using the symbolic method if simpler exhaustive summaries can be found. One way to do this is via reparameterisation. An alternative is to use the hybrid symbolic-numerical method. For the later method we show it is possible to find estimable parameter combinations using subset profiling and to create general results using the extension theorem.

6

Bayesian Identifiability

Lindley (1971) noted that identifiability does not present any difficulty when a Bayesian approach is used. His reasoning was that prior distributions can be used to provide extra information on non-identifiable parameters. In contrast Kadane (1975) stated that 'identification is a property of the likelihood function and is the same whether considered classically or from the Bayesian approach'. Bayesian identifiability has been discussed in many areas including generalised linear models (Gelfand and Sahu, 1999), categorical data (Swartz et al. 2004) and ecological statistics (Gimenez et al., 2009). San Martin and Gonzalez (2010) provide a detailed theoretical discussion of Bayesian identifiability. Here we consider a more applied approach to demonstrate that identifiability can still be a issue when a Bayesian approach is used, and therefore must be considered.

We start with a brief introduction to Bayesian statistics in Section 6.1, then give the definitions of identifiability used in this chapter in Section 6.2. In Section 6.3 we discuss the problems associated with identifiability when using a Bayesian approach. There are several methods for checking Bayesian identifiability. Section 6.4 is concerned with examining the overlap between the posterior and prior distribution. In Section 6.5 data cloning is used to investigate identifiability. Section 6.6 examines how the symbolic method can be used to investigate Bayesian identifiability.

6.1 Introduction to Bayesian Statistics

In classical inference, parameters are assumed to be unknown fixed constants. The parameters can be estimated using maximum likelihood, and the likelihood can form the starting point for determining whether the model is identifiable. In Bayesian statistics the parameters are assumed to be random variables that follow a posterior distribution, $f(\boldsymbol{\theta}|\mathbf{y})$, for parameters $\boldsymbol{\theta}$ and data $\mathbf{y}$. The posterior distribution is

$$f(\boldsymbol{\theta}|\mathbf{y}) \propto f(\mathbf{y}|\boldsymbol{\theta})f(\boldsymbol{\theta}),$$

where $f(\boldsymbol{\theta})$ is the prior distribution and $f(\mathbf{y}|\boldsymbol{\theta})$ is the sampling model describing the data (see for example Bernardo and Smith, 1994, Lee, 2012). The sampling model, $f(\mathbf{x}|\boldsymbol{\theta})$, is often referred to as the likelihood, as the form is identical to the likelihood used in classical inference. Here we also refer to $f(\mathbf{y}|\boldsymbol{\theta})$ as the likelihood, because identifiability in the Bayesian formation of a model can be easily linked to identifiability in a classical model via this identical function.

The prior distribution $f(\boldsymbol{\theta})$ provides information on the prior beliefs of the parameters. Here we use the term uninformative to represent a prior with $f(\boldsymbol{\theta}) \propto 1$, such as the uniform distribution. Although technically the prior still could provide some information, such as restricting the parameter space, in terms of identifiability this uninformative prior provides no additional information. Conversely, we use the term informative to refer to a prior that does provide information on the parameter, so that $f(\boldsymbol{\theta})$ is not proportional to 1.

Bayesian inference involves finding and reporting the posterior distribution. The posterior can also be summarised using statistics such as the posterior mean

$$\mathrm{E}(\boldsymbol{\theta}|\mathbf{y}) = \int_{-\infty}^{\infty} \boldsymbol{\theta} f(\boldsymbol{\theta}|\mathbf{y}) d\boldsymbol{\theta}$$

or the posterior variance

$$\mathrm{Var}(\boldsymbol{\theta}|\mathbf{y}) = \int_{-\infty}^{\infty} \{\boldsymbol{\theta} - \mathrm{E}(\boldsymbol{\theta}|\mathbf{y})\}^2 f(\boldsymbol{\theta}|\mathbf{y}) d\boldsymbol{\theta}.$$

In some cases where the posterior and any summary statistics can be found analytically, but in most examples numerical methods are required to approximate the integration involved. Typically a Markov Chain Monte Carlo (MCMC) method is used. MCMC methods are simulation algorithms for sampling from probability distributions, such as the posterior distribution (see for example Gamerman and Lopes, 2006 or Chapter 4 of Link and Barker, 2010). In this chapter we use Winbugs (Lunn et al., 2000) to find MCMC samples from the posterior distribution, which is called from R using the package R2WinBUGS (Sturtz et al., 2005). A long run of an MCMC algorithm is usually required before the chain of samples has converged to the posterior distribution. Therefore we typically disregard a certain number of the first samples, called the burn-in. For example we might use a burn-in of 1000, so disregard the first 1000 samples. Inference is then based on a large number of subsequent samples, where the algorithm is assumed to have converged to the posterior distribution; for example we might use the next 10 000 samples as a sample from the posterior distribution.

6.2   Definitions

The general definition of identifiability, Definition 1, from Section 3.1.1, is based on the specification of the model. If we consider the model to be the posterior distribution, then we have the following definition of identifiability of the posterior distribution, $f(\boldsymbol{\theta}|\mathbf{y})$:

Definition 12 *A Bayesian model, described by the posterior distribution, is globally identifiable if $f(\boldsymbol{\theta}_a|\mathbf{y}) = f(\boldsymbol{\theta}_b|\mathbf{y})$ implies that $\boldsymbol{\theta}_a = \boldsymbol{\theta}_b$. A model is locally identifiable if there exists an open neighbourhood of any $\boldsymbol{\theta}$ such that this is true. Otherwise the posterior is non-identifiable.*

Lindley (1971)'s comment, that by including priors with extra information a posterior can become identifiable, fits with this definition, which we demonstrate in Section 6.6.

However Dawid (1979) gives a different definition of Bayesian identifiability:

Definition 13 *If the parameters $\boldsymbol{\theta}$ are partitioned as $\boldsymbol{\theta} = [\boldsymbol{\theta}_1, \boldsymbol{\theta}_2]$ then the parameters $\boldsymbol{\theta}_2$ are non-identifiable if*

$$f(\boldsymbol{\theta}_2|\boldsymbol{\theta}_1, \mathbf{y}) = f(\boldsymbol{\theta}_2|\boldsymbol{\theta}_1).$$

Gelfand and Sahu (1999) note that Bayesian non-identifiability does not imply that $f(\boldsymbol{\theta}_2|\mathbf{y}) = f(\boldsymbol{\theta}_2)$. That is, Bayesian non-identifiability does not imply there is no Bayesian learning.

Gelfand and Sahu (1999) show that Bayesian non-identifiability is equivalent to non-identifiability of the likelihood. As $f(\boldsymbol{\theta}_2|\boldsymbol{\theta}_1, \mathbf{y}) \propto f(\mathbf{y}|\boldsymbol{\theta}_1, \boldsymbol{\theta}_2)f(\boldsymbol{\theta}_2|\boldsymbol{\theta}_1)f(\boldsymbol{\theta}_1)$, Bayesian non-identifiability will only occur if $f(\mathbf{y}|\boldsymbol{\theta}_1, \boldsymbol{\theta}_2)$ is free of $\boldsymbol{\theta}_2$. That is, if both $f(\mathbf{y}|\boldsymbol{\theta}_1, \boldsymbol{\theta}_2)$ and the parameters $\boldsymbol{\theta}_2$ are non-identifiable it will be possible to reparameterise $f(\mathbf{y}|\boldsymbol{\theta}_1, \boldsymbol{\theta}_2)$ so that it not a function of $\boldsymbol{\theta}_2$. This leads to the conclusion that identifiability of the likelihood, $f(\mathbf{y}|\boldsymbol{\theta}_1, \boldsymbol{\theta}_2)$, can be considered as equivalent of Dawid (1979)'s definition of Bayesian identifiability. If this equivalence is used, then Bayesian identifiability can be checked from just the likelihood, by applying methods discussed in early chapters. Bayesian identifiability does not then depend on the prior distribution (Gelfand and Sahu 1999), and Bayesian identifiability results will be identical to the identifiability results in the equivalent classical model. This fits with Kadane (1975)'s statement that 'identification is a property of the likelihood function and is the same whether considered classically or from the Bayesian approach'.

We deal with these two contrasting definitions of identifiability by considering identifiability of the posterior (Definition 12) and identifiability of the likelihood, which is equivalent to Bayesian identifiability (Definition 13).

It is recommended that both identifiability of the likelihood and identifiability of the posterior are considered. In some situations identifiability of the posterior might be the more applicable than identifiability of the likelihood. For example Swartz et al. (2004) discuss a hierarchical Bayesian model where $y_1, \ldots, y_n$ are a sample of size n from a normal distribution with mean μ and variance 1. The prior specification is: $\mu|\alpha$ follows a normal distribution with mean α and variance 1 and α follows a normal distribution with fixed mean α_0 and variance 1. The likelihood will only depend on μ, but not α, therefore the parameter α is non-identifiable in the likelihood. However the posterior can be shown to be a bivariate normal distribution and both parameters are identifiable in the posterior.

In the classical specification of a model, for specific data sets, it is possible for a parameter redundant model to behave as a model that is not parameter redundant. This is known as near redundancy, and was discussed in Section 4.4. Weak identifiability can be considered the Bayesian equivalent of near redundancy (Cole et al. 2010). There are two definitions of weak identifiability provided in Gelfand and Sahu (1999), as given below.

Definition 14 *a. If the parameters $\boldsymbol{\theta}$ are partitioned as $\boldsymbol{\theta} = [\boldsymbol{\theta}_1, \boldsymbol{\theta}_2]$, $\boldsymbol{\theta}_2$ is defined as weakly identifiable if*

$$f(\boldsymbol{\theta}_2|\boldsymbol{\theta}_1, \mathbf{y}) \approx f(\boldsymbol{\theta}_2|\boldsymbol{\theta}_1).$$

b. A parameter θ is weakly identifiable if $f(\theta|\mathbf{y}) \approx f(\theta)$.

Note these two definitions will not necessarily be equivalent. Garrett and Zeger (2000) utilise definition 14b and is discussed further in Section 6.4.

6.3   Problems with Identifiability in Posterior Distribution

Identifiability can be an issue in Bayesian models as well as classical models; we discuss some of these issues in the is Section.

Most Bayesian models do not have a closed form for the posterior distribution and MCMC is used to find samples from the posterior distribution. Performing the MCMC analysis can be more difficult in non-identifiable models due to problems with poor mixing in MCMC samples and slow convergence (Carlin and Louis, 2000, Rannala, 2002, Gimenez et al., 2009).

If the classical formation of a model is non-identifiable there will be a flat ridge in the likelihood surface (see Section 2.1). If the equivalent model is examined using Bayesian methodology and uninformative priors are used there will also be a flat ridge in the posterior surface. When informative priors are used this flat ridge may disappear. This is demonstrated for the basic occupancy model in Section 6.3.1 below.

Whilst the use of informative priors can result in a posterior that is technically identifiable, the results will be heavily dependent on the choice of prior, even with large sample sizes (Neath and Samaniego, 1997). Therefore results for non-identifiable parameters may be misleading (Garrett and Zeger, 2000,

Gimenez et al., 2009). We demonstrate that this is the case in two examples in Sections 6.3.2 and 6.3.3 below.

6.3.1 Basic Occupancy Model

The basic occupancy model, first encountered in Section 1.2.1, is an obviously non-identifiable model in the classical formation of the model. The model is also simple enough to be able to find an explicit distribution for the posterior distribution for certain prior distributions.

Suppose that only two sites were visited. At one site the species was observed and at the other site it was not observed. (Note this data is not realistic, it has been chosen to simplify the algebra required to obtain explicit results.) The sampling model (or likelihood) is then

$$f(\mathbf{y}|p, \psi) \propto p\psi(1 - p\psi),$$

where ψ is the probability the site is occupied, p is the probability the species is detected and $\mathbf{y}$ represents the data. Also suppose uninformative uniform priors are used on both parameters, with $f(p) = f(\psi) = 1$. The posterior distribution is then

$$f(p, \psi|\mathbf{y}) = \frac{36}{5}p\psi(1 - p\psi).$$

If instead the prior on both parameters was a beta distribution with both parameters equal to 2, so that $f(p) = 6p(1 - p)$ and $f(\psi) = 6\psi(1 - \psi)$, then the posterior distribution is

$$f(p, \psi|\mathbf{y}) = 225(p\psi)^2(1 - p\psi)(1 - p)(1 - \psi).$$

Contour plots for both posteriors are given in Figure 6.1. With uninformative uniform priors there is a flat ridge in the posterior surface, indicated by the dotted line. But when informative beta priors are used there is no longer a flat ridge in the posterior surface.

6.3.2 Mark-Recovery Model

Consider the mark-recovery model, which was introduced in Section 1.2.4. The mark-recovery model has constant survival probabilities for animals in their first year of life, ϕ_1, and as adults, ϕ_a, and constant reporting probabilities for animals in their first year of life, λ_1, and as adults, λ_a. In the classical formation of the model, it is obviously parameter redundant, because it can be reparameterised as $(1 - \phi_1)\lambda_1$, $\phi_1\lambda_a$. The parameters ϕ_1, λ_1 and λ_a are individually non-identifiable and the parameter ϕ_a is individually identifiable.

Suppose that the true values of the parameters were $\phi_1 = 0.6$, $\phi_a = 0.8$, $\lambda_1 = 0.3$ and $\lambda_a = 0.1$. The probabilities of marking in year i and being recovered in year j for 5 years of marking and recovery, $P_{i,j}$, (in matrix form)

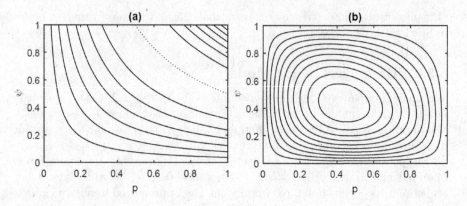

FIGURE 6.1

Contour plot of the posterior distribution for the basic occupancy model. Figure (a) has uniform priors on both parameters. The dotted line represents the ridge in the surface. Figure (b) has beta(2,2) priors on both parameters.

are,

$$
\mathbf{P} = \begin{bmatrix}
\bar{\phi}_1\lambda_1 & \phi_1\bar{\phi}_a\lambda_a & \phi_1\phi_a\bar{\phi}_a\lambda_a & \phi_1\phi_a^2\bar{\phi}_a\lambda_a & \phi_1\phi_a^3\bar{\phi}_a\lambda_a \\
0 & \bar{\phi}_1\lambda_1 & \phi_1\bar{\phi}_a\lambda_a & \phi_1\phi_a\bar{\phi}_a\lambda_a & \phi_1\phi_a^2\bar{\phi}_a\lambda_a \\
0 & 0 & \bar{\phi}_1\lambda_1 & \phi_1\bar{\phi}_a\lambda_a & \phi_1\phi_a\bar{\phi}_a\lambda_a \\
0 & 0 & 0 & \bar{\phi}_1\lambda_1 & \phi_1\bar{\phi}_a\lambda_a \\
0 & 0 & 0 & 0 & \bar{\phi}_1\lambda_1
\end{bmatrix}
$$

$$
= \begin{bmatrix}
0.12 & 0.012 & 0.0096 & 0.00768 & 0.006144 \\
0 & 0.12 & 0.012 & 0.0096 & 0.00768 \\
0 & 0 & 0.12 & 0.012 & 0.0096 \\
0 & 0 & 0 & 0.12 & 0.012 \\
0 & 0 & 0 & 0 & 0.12
\end{bmatrix},
$$

where $\bar{\phi} = 1 - \phi$. If 1000 animals were marked each year, the expected number of animals marked in year i and recovered in year j, $N_{i,j}$, (in matrix form) are

$$
\mathbf{N} = \begin{bmatrix}
120 & 12 & 10 & 8 & 6 \\
0 & 120 & 12 & 10 & 8 \\
0 & 0 & 120 & 12 & 10 \\
0 & 0 & 0 & 120 & 12 \\
0 & 0 & 0 & 0 & 120
\end{bmatrix}.
$$

Expected numbers have been rounded to the nearest whole number where applicable.

This gives an expected data set where we know what the true parameter values should be. However this model can be reparameterised with $\beta_1 = (1 - \phi_1)\lambda_1$ and $\beta_2 = \phi_1\lambda_a$. The true values of β_1 and β_2 are 0.12

TABLE 6.1

Priors used for the mark-recovery model. U(0,1) is uniform distribution with the range 0 to 1. Beta(a,b) is a beta distribution with parameters a and b. The variance for all beta priors is 0.01.

Parameter	Uninformative	Informative	Disinformative
ϕ_1	U(0,1)	Beta(13.8,9.2)	Beta(9.2,13.8)
ϕ_a	U(0,1)	Beta(12,3)	Beta(13.8,9.2)
λ_1	U(0,1)	Beta(6,14)	Beta(1.7625,9.9875)
λ_a	U(0,1)	Beta(0.8,7.2)	Beta(9.2,13.8)

TABLE 6.2

Results from Bayesian analysis for the mark-recovery example. The mean and standard deviation (SD) of the sample from the posterior distribution are given for each parameter for each prior. The column labelled true refers to the parameter values used to generated the expected data set.

Parameter	True	Uninformative Mean	SD	Informative Mean	SD	Disinformative Mean	SD
ϕ_1	0.6	0.48	0.26	0.60	0.08	0.26	0.08
ϕ_a	0.8	0.85	0.08	0.82	0.06	0.78	0.06
λ_1	0.3	0.33	0.22	0.31	0.06	0.16	0.02
λ_a	0.1	0.33	0.25	0.12	0.05	0.26	0.08

and 0.06 respectively. So in fact any values of ϕ_1, λ_1 and λ_a that are solutions of the equations $\beta_1 = (1 - \phi_1)\lambda_1 = 0.12$ and $\beta_2 = \phi_1\lambda_a = 0.06$ with $\phi_a = 0.8$ could have been used to create this expected data set.

In the Bayesian formation of the model, prior information is introduced. To investigate how this affects the identifiability of the Bayesian formation of the model we consider three different sets of priors. The first has uninformative uniform priors on all four parameters. This is to represent the case that there is no prior information available. The second set of priors represent the case that an expert has correct information on the parameters, so informative priors are developed from the same true parameters used to create the data set. A beta distribution is used with mean values based on the true parameter values with a variance of 0.01. The third set of priors demonstrate the effect of an expert giving incorrect information, so that disinformative priors are used. Again beta priors are used with a variance of 0.01, but different means to the true values are used (arbitrary chosen). The priors are given in Table 6.1.

Using Winbugs, a Bayesian analysis was performed with the three different priors on the parameters. In each case a burn-in of 100 000 and a sample of 100 000 was used. Table 6.2 gives the mean and standard deviation of the sample from the posterior distribution.

The identifiable parameter ϕ_a has a mean that is close to the true value for all three priors. The prior is not having a large effect on the result. However, when a uninformative prior or the disinformative prior is used, for the three non-identifiable parameters, the posterior mean values are now affected by the prior. As expected, if the informative prior is used the means are all close to the true values. We can see how sensitive the results are to the prior used.

If a classical formation of a model is non-identifiable it can lead to biased parameter estimates, as discussed in Section 2.2. If the equivalent Bayesian formation of the model is used it is also possible for the results to be biased for non-identifiable parameters, unless informative prior information is close to the true parameter values.

6.3.3 Logistic Growth Curve Model

The logistic growth curve model (also known as a logistic regression model) is an example used in Gelfand and Sahu (1999) and originates from Agresti (1990). It models the success rate of a new drug versus a standard drug, following two groups of people, who either have the disease mildly or severely, over three time points. The probability of success is

$$\text{logit}(p_{ijk}) = \log\left(\frac{p_{ijk}}{1 - p_{ijk}}\right) = \mu + \alpha_{D,i} + \alpha_{T,j} + \beta_0 x_k + \beta_j x_k,$$

where $j = 1$ represents the standard drug, $j = 2$ represents the new drug, $i = 1$ represents the diagnostic group who have the disease mildly and $i = 2$ represents the diagnostic group who have the disease severely. There are k occasions, with $k = 1, 2, 3$, which are assumed to be evenly spaced, so that $x_1 = -1$, $x_2 = 0$ and $x_3 = 1$. There are eight parameters,

$$\boldsymbol{\theta} = [\mu, \alpha_{D,1}, \alpha_{D,2}, \alpha_{T,1}, \alpha_{T,2}, \beta_0, \beta_1, \beta_2]^T,$$

where μ is the mean response, $\alpha_{D,i}$ is the drug effect, $\alpha_{T,j}$ is the disease severity effect and β_j are occasion effects. The design matrix is,

$$\mathbf{X} = \begin{bmatrix}
1 & 1 & 0 & 1 & 0 & -1 & -1 & 0 \\
1 & 1 & 0 & 1 & 0 & 0 & 0 & 0 \\
1 & 1 & 0 & 1 & 0 & 1 & 1 & 0 \\
1 & 1 & 0 & 0 & 1 & -1 & 0 & -1 \\
1 & 1 & 0 & 0 & 1 & 0 & 0 & 0 \\
1 & 1 & 0 & 0 & 1 & 1 & 0 & 1 \\
1 & 0 & 1 & 1 & 0 & -1 & -1 & 0 \\
1 & 0 & 1 & 1 & 0 & 0 & 0 & 0 \\
1 & 0 & 1 & 1 & 0 & 1 & 1 & 0 \\
1 & 0 & 1 & 0 & 1 & -1 & 0 & -1 \\
1 & 0 & 1 & 0 & 1 & 0 & 0 & 0 \\
1 & 0 & 1 & 0 & 1 & 1 & 0 & 1
\end{bmatrix},$$

so that

$$\text{logit}(p_{ijk}) = \mathbf{X}\boldsymbol{\theta}.$$

As stated in Gelfand and Sahu (1999), the likelihood is not identifiable because the design matrix has rank 5, but there are eight parameters.

The data for this example consists of the number of successes, y_{ijk}, out of the number of people in each category, n_{ijk}. The number of successes, y_{ijk}, follows a binomial distribution with index n_{ijk} and probability of success, p_{ijk}. Here we use the data set

$$
\mathbf{y} = \begin{bmatrix} y_{111} \\ y_{112} \\ y_{113} \\ y_{121} \\ y_{122} \\ y_{123} \\ y_{211} \\ y_{212} \\ y_{213} \\ y_{221} \\ y_{222} \\ y_{223} \end{bmatrix} = \begin{bmatrix} 31 \\ 21 \\ 12 \\ 44 \\ 27 \\ 8 \\ 41 \\ 33 \\ 24 \\ 48 \\ 38 \\ 18 \end{bmatrix}.
$$

This data set has been created by assuming that there are $n_{ijk} = 50$ people in each category, that the parameters are $\mu = 0.5$, $\alpha_{D,1} = -0.5$, $\alpha_{D,2} = 0.5$, $\alpha_{T,1} = -0.3$, $\alpha_{T,2} = 0.2$, $\beta_0 = -1$, $\beta_1 = 0.2$ and $\beta_2 = -0.8$. The number of successes is $50p_{ijk}$ rounded to nearest whole number. This creates an expected data set, where the approximate values of the parameters are known.

In a Bayesian MCMC analysis (shown in logisticgrowth.R), we investigate the effect that various prior distributions have on this model. The priors are given in Table 6.3a. Technically an uninformative prior would have limits of infinity in this case to match the limits of the parameters, however this would not be practical to apply in Winbugs (or any other computer package). Instead we choose very large lower and upper values for this example. The alternative to using an uniformative prior is to use a vague prior, one that provides little information on the parameter, by assuming a large variance. The vague prior used here is the same prior used in Gelfand and Sahu (1999). The informative priors have the same mean as the values the expected data set is created from, whereas the disinformative priors have been chosen to have incorrect means. Both these two priors have a smaller variance than the vague prior.

The Bayesian analysis was performed in Winbugs (called from R using the package R2WinBUGS) with a burn-in of 10 000 and the results in Table 6.3b give the mean and standard deviation of 10 000 samples from the posterior distribution.

The posterior distributions all vary considerably depending on the prior used. The data was generated from an expected data set, so we know what the correct parameter values should be. However the posterior distribution is only close to the correct parameter values when the informative prior is used. The

TABLE 6.3

Bayesian analysis of logistic growth curve example. (a) Priors used for the logistic growth curve with no constraints. U(-5,5) is uniform distribution with the range -5 to 5. N(a,b) is the normal distribution with mean a and variance b. (b) The mean and standard deviation (SD) of the sample from the posterior distribution are given for each parameter.

(a)				
Parameter	Uninformative	Vague	Informative	Disinformative
μ	U(-5,5)	N(0,1)	N(0.5,0.1)	N(-0.5,0.1)
α_1^D	U(-5,5)	N(0,1)	N(-0.5,0.1)	N(0.2,0.1)
α_2^D	U(-5,5)	N(0,1)	N(0.5,0.1)	N(-0.4,0.1)
α_1^T	U(-5,5)	N(0,1)	N(-0.3,0.1)	N(0.3,0.1)
α_2^T	U(-5,5)	N(0,1)	N(0.2,0.1)	N(-0.4,0.1)
β_0	U(-5,5)	N(0,1)	N(-1,0.1)	N(1,0.1)
β_1	U(-5,5)	N(0,1)	N(0.2,0.1)	N(-0.2,0.1)
β_2	U(-5,5)	N(0,1)	N(-0.8,0.1)	N(0.4,0.1)

(b)								
	Uninformative		Vague		Informative		Disinformative	
Parameter	Mean	SD	Mean	SD	Mean	SD	Mean	SD
μ	-0.17	3.01	0.15	0.79	0.51	0.23	-0.004	0.22
α_1^D	0.24	2.41	-0.41	0.69	-0.54	0.22	-0.23	0.22
α_2^D	1.28	2.41	0.62	0.68	0.51	0.22	0.51	0.22
α_1^T	-0.40	2.41	-0.07	0.65	-0.31	0.22	0.08	0.21
α_2^T	0.11	2.42	0.42	0.65	0.20	0.22	0.31	0.21
β_0	-0.55	2.14	-0.85	0.58	-1.01	0.20	-0.42	0.20
β_1	-0.26	2.15	0.03	0.58	0.20	0.21	-0.34	0.21
β_2	-1.28	2.14	-0.95	0.59	-0.82	0.22	-0.87	0.22

other priors all lead to incorrect conclusions, demonstrating it is not possible to accurately estimate parameters if the likelihood is not identifiable, unless you have accurate informative prior information.

Agresti (1990) imposes the following constraints on the model: $\alpha_{D,2} = \alpha_{T,2} = \beta_2 = 0$. Under these constraints the likelihood is now identifiable. The parameters used to create the same expected data set used above then become $\mu = 1.2$, $\alpha_{D,1} = -1$, $\alpha_{T,1} = -0.5$, $\beta_0 = -1.8$ and $\beta_1 = 1$. (See logisticgrow.mw for details.) The priors used are given in Table 6.4a. Note an alternative way to introduce constraints, which is used in Gelfand and Sahu (1999), would be to use priors with mean zero and very small variance on the parameters $\alpha_{D,2}$, $\alpha_{T,2}$ and β_2.

A similar Bayesian analysis was performed in Winbugs, with a burn-in of 10 000 and the results in Table 6.4b give the mean and standard deviation of 10 000 samples from the posterior distribution. Now that the likelihood is iden-tifiable there is little difference between using a uniform, vague or informative

TABLE 6.4

Bayesian analysis of the logistic growth model with constraints. (a) The priors used for the logistic growth curve model with constraints. U(-5,5) is uniform distribution with the range -5 to 5. N(a,b) is the normal distribution with mean a and variance b. (b) The mean and standard deviation (SD) of the sample from the posterior distribution are given for each parameter.

(a)				
Parameter	Uninformative	Vague	Informative	Disinformative
μ	U(-5,5)	N(0,1)	N(1.2,0.1)	N(0.2,0.1)
α_1^D	U(-5,5)	N(0,1)	N(-1,0.1)	N(-0.3,0.1)
α_1^T	U(-5,5)	N(0,1)	N(-0.5,0.1)	N(0.1,0.1)
β_0	U(-5,5)	N(0,1)	N(-1.8,0.1)	N(-0.8,0.1)
β_1	U(-5,5)	N(0,1)	N(1,0.1)	N(0.6,0.1)

(b)								
	Uninformative		Vague		Informative		Disinformative	
Parameter	Mean	SD	Mean	SD	Mean	SD	Mean	SD
μ	1.22	0.19	1.12	0.18	1.20	0.15	0.80	0.14
α_1^D	-1.05	0.20	-0.97	0.19	-1.03	0.16	-0.73	0.16
α_1^T	-0.51	0.20	-0.44	0.19	-0.50	0.16	-0.21	0.15
β_0	-1.85	0.21	-1.71	0.19	-1.82	0.15	-1.46	0.14
β_1	1.03	0.26	0.88	0.25	1.00	0.19	0.65	0.18

prior. However, the disinformative prior is still having some effect on the mean because of the relatively small variance of the prior. It should be noted that the smaller the variance the greater the effect the prior will have.

Whilst we have demonstrated that a lack of identifiability results in a strong dependence on the prior information and potential bias, it would be difficult to judge a lack of identifiability from considering different priors alone.

6.4 Detecting Weak Identifiability using Prior and Posterior Overlap

In this Section we use Garrett and Zeger (2000)'s definition of weak identifiability (Definition 14b). A parameter θ is weakly identifiable if $f(\theta|\mathbf{y}) \approx f(\theta)$, where $f(\theta|\mathbf{y})$ is the marginal posterior distribution for θ and $f(\theta)$ is the prior distribution for θ.

Garrett and Zeger (2000) propose checking for weak identifiability in latent class models by checking the overlap between the posterior and the prior distributions, either by displaying the posterior and the prior on the same figure or by calculating the percentage overlap between the prior and posterior,

τ. Examples of utilising this method include Gimenez et al. (2009), Abadi et al. (2010) and Koop et al. (2013).

Garrett and Zeger (2000) suggest that if $\tau > 35\%$ then the parameter is weakly identifiable. If at least one parameter is weakly identifiable then the model is weakly identifiable. However they also state that this critical value of 35% needs further investigation. Gimenez et al. (2009) investigated weak identifiability in mark-recovery and capture-recapture models and found the critical value of 35% works well with uniform priors but it is difficult to correctly calibrate the critical value if other priors are used.

As a non-identifiable model is within the class of weakly identifiable models (Gelfand and Sahu, 1999), we would expect that the method would always classify non-identifiable models as weakly identifiable. This expectation is investigated in Sections 6.4.1 to 6.4.3 with a series of examples, which are known to be non-identifiable.

It is sometimes possible to find the exact marginal posterior distribution and calculate the exact overlap between the prior and posterior distribution, as we demonstrate in Section 6.4.1. However in most cases, it is not possible to find the exact marginal posterior distribution. Therefore we use MCMC to generate a sample of parameter values from the posterior distribution, w_i, $i = 1, \ldots, m$. An estimate of the marginal posterior distribution, $\hat{f}$, can be found using a kernel density estimate with,

$$\hat{f}(w) = \frac{1}{m} \sum_{i=1}^{m} \frac{1}{h} K\left(\frac{w - w_i}{h}\right),$$

where K is the kernel function and h is the bandwidth. Here, a Gaussian kernel is used with

$$K(w) = \frac{1}{\sqrt{2\pi}} \exp\left(-\frac{w^2}{2}\right)$$

with a bandwidth of $h = 0.9Sm^{-1/5}$, where S is the minimum of the standard deviation of the sample and the interquartile range of the sample divided by 1.34 (using recommended values from, Silverman, 1986 and Morgan, 2009). An estimate of the overlap, $\hat{\tau}$, is then calculated using a Monte Carlo approximation as

$$\hat{\tau} = \frac{1}{m} \sum_{i=1}^{m} \min\{\hat{f}(w_i), g(w_i)\},$$

where $g(w)$ is the prior distribution (Gimenez et al., 2009). This is the method used in the example in Sections 6.4.2 and 6.4.3. The percentage overlap between the posterior and prior is also given in the R package MCMCvis, which is a package for visualising, manipulating and summarising MCMC ouput (Youngflesh, 2018).

Koop et al. (2013) show that the posterior for a non-identifiable parameter may differ considerably from its prior. We also note that the posterior for an identifiable parameter may have a large overlap with the prior, if the prior

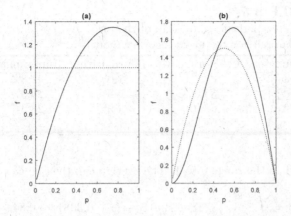

FIGURE 6.2
Weakly identifiable basic occupancy model graph showing the the posterior distribution (solid line) with the prior density (dotted line) for the parameter p, when (a) priors of $f(p) = f(\psi) = 1$ are used, and (b) priors of $f(p) = 6p(1 - p)$ and $f(\psi) = 6\psi(1 - \psi)$ are used.

happens to correspond with the posterior. Both these problems occur in the examples in Sections 6.4.2 and 6.4.3 below.

6.4.1   Basic Occupancy Model

We continue with the basic occupancy model from Section 6.3.1, where the sampling model or likelihood is

$$f(\mathbf{y}|p, \psi) \propto p\psi(1 - p\psi).$$

Suppose uninformative uniform priors are used for both parameters, with $f(p) = f(\psi) = 1$. The posterior distribution is

$$f(p, \psi|\mathbf{y}) = \frac{36}{5}p\psi(1 - p\psi).$$

The exact marginal posterior distribution for p can be found by integrating $f(p, \psi|\mathbf{y})$ with respect to ψ over the range of ψ of 0 to 1. The marginal posterior distribution for p is then

$$f(p|\mathbf{y}) = \frac{1}{5}p(18 - 12p).$$

The prior and marginal posterior distributions for the parameter p are plotted in Figure 6.2a. In this case we find the exact overlap between the posterior and prior distribution. The two lines cross at $\frac{1}{5}p(18 - 12p) = 1$, which has a

TABLE 6.5

Estimated percentage overlap of the prior and posterior
distributions for the mark-recovery model

Parameter	Uninformative	Informative	Disinformative
ϕ_1	85.0%	87.0%	41.8%
ϕ_a	33.0%	76.8%	24.9%
λ_1	58.0%	73.9%	28.2%
λ_a	68.7%	45.4%	42.1%

value of 0.3681 in the range 0 to 1. The overlap can then be calculated as

$$\tau = \int_0^{0.3681} \frac{1}{5}p(18 - 12p)dp + (1 - 0.3681) = 0.84.$$

The overlap is much larger than 0.35, therefore the parameter p is considered
weakly identifiable.

If the prior for both parameters followed a beta distribution with $f(p) = 6p(1 - p)$ and $f(\psi) = 6\psi(1 - \psi)$, then the posterior distribution is

$$f(p, \psi|\mathbf{y}) = 225(p\psi)^2(1 - p\psi)(1 - p)(1 - \psi).$$

Again the exact marginal posterior distribution for p can be found by integrating $f(p, \psi|\mathbf{y})$ with respect to ψ over the range of ψ of 0 to 1. The marginal
posterior distribution for p is then

$$f(p|\mathbf{y}) = \frac{15}{4}p^2(3p - 5)(p - 1).$$

The prior and marginal posterior distributions for the parameter p are plotted
in Figure 6.2b. Again we can find the exact overlap between the posterior and
prior distribution. This time the two curves cross when $\frac{15}{4}p^2(3p - 5)(p - 1) = 6p(1 - p)$, with one of the solutions being 0.4319. The overlap can be then
calculated as

$$\tau = \int_0^{0.4319} \frac{15}{4}p^2(3p - 5)(p - 1)dp + \int_{0.4319}^1 6p(1 - p)dp = 0.88.$$

Again the parameter p is classified as weakly identifiable as the overlap is
much larger than 0.35.

6.4.2   Mark-Recovery Model

We next consider the mark-recovery example from Section 6.3.2. From the
same analysis, presented in Table 6.2, the overlap between the posterior and
prior can be also estimated using a Monte Carlo approximation. The R code
markrecbay.R demonstrates how this overlap is calculated. Table 6.5 gives
the estimated percentage overlap for all three sets of priors. Figure 6.3 shows

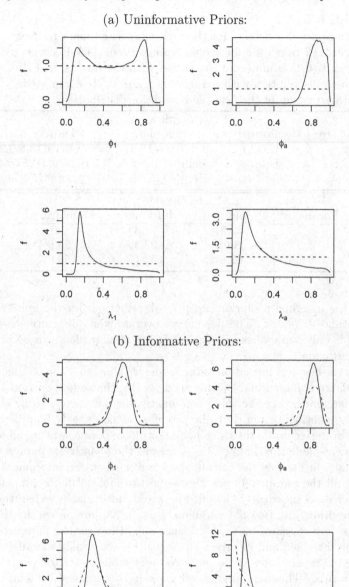

FIGURE 6.3
Prior and posterior overlap for the mark-recovery model example. Figures show the kernel density of a sample from the posterior distribution (solid line) with the prior density (dotted line) for each of the four parameters, using (a) uninformative priors and (b) informative priors.

TABLE 6.6

Prior and posterior overlap for the reparameterised mark-recovery example. (a) Priors used for the reparameterised Mark-Recovery Example. U(0,1) is uniform distribution with the range 0 to 1. Beta(a,b) is the beta distribution with parameters a and b. (b) Estimated percentage overlap of the prior and posterior distributions.

	(a) Priors		
Parameter	Uninformative	Informative	Disinformative
ϕ_a	U(0,1)	Beta(12,3)	Beta(13.8,9.2)
β_1	U(0,1)	Beta(6.216,45.584)	Beta(9.4125,53.375)
β_2	U(0,1)	Beta(1.632,25.568)	Beta(47.6,71.4)

	(b) Overlap		
Parameter	Uninformative	Informative	Disinformative
ϕ_a	32.5%	74.3%	0.0%
β_1	3.0%	24.3%	24.3%
β_2	41.8%	52.1%	56.0%

plots of the marginal posterior and prior distributions for the uninformative and informative priors. (The percentage overlap when disinformative priors are used is only summarised in Table 6.5, but similar plots can be produced from the R code.)

From the classical formation of the model we know that the parameters ϕ_1, λ_1 and λ_a are non-identifiable, whereas ϕ_a is identifiable (see Section 4.2). We would therefore expect the three non-identifiable parameters to be classified as weakly identifiable and ϕ_a to be classified as not weakly identifiable. For the uninformative priors the three non-identifiable parameters are all over the weakly identifiable threshold of 35%, whereas the identifiable parameter ϕ_a's overlap falls just below the threshold. However when the informative priors are used all the parameters are above the threshold of 35%. Therefore the parameter ϕ_a is incorrectly classified as weakly identifiable. When the disinformative priors are used the parameters ϕ_1 and λ_a are above the threshold of 35%, so are classified as weakly identifiable. However the parameter λ_1 is below the threshold and incorrectly classified as not weakly identifiable.

We could reparameterise the mark-recovery model, so that $\beta_1 = (1-\phi_1)\lambda_1$ and $\beta_2 = \phi_1\lambda_a$. This reparameterised mark-recovery model is identifiable. A similar study of prior and posterior overlap is considered for three types of priors: uninformative uniform priors, informative priors and disinformative priors. As previously the informative priors are based on the parameters values used to create the data set and the disinformative priors do not match the parameters used to create the data set. The prior distributions are given in Table 6.6a.

The estimated percentage overlap is given in Table 6.6b. A burn-in of 100 000 and 100 000 MCMC samples are used to create the kernel posterior density and estimate the percentage overlap.

TABLE 6.7

Prior and posterior overlap for the age dependent mark-recovery model.

Parameter	Overlap
ϕ_1	40.0%
ϕ_2	57.0%
ϕ_3	91.7%
λ	29.1%

When the parameters all have uniform priors, two of the parameters are below the 35% cut off, but the parameter β_2 is technically classified as weakly identifiable, when it is in fact identifiable. In the case where the informative priors are used the overlap is a lot greater than 35% for two of the parameters, classifying this model as weakly identifiable, when in fact it is identifiable. There is only one parameter with an overlap that classifies the parameter as weakly identifiable when the disinformative priors were used. This overlap stems from matching of prior and likelihood information, rather than non-identifiability. A more detailed study of the effect of different priors, known as a prior sensitivity analysis, would be able to pick up this problem.

Another parameterisation of a mark-recovery model has survival fully dependent on age, ϕ_i, and constant reporting probability, λ. For three years of marking and recapture the probabilities of marking in year i and recapture in year j, $P_{i,j}$, are summarised by the matrix

$$\mathbf{P} = \begin{bmatrix} (1-\phi_1)\lambda & \phi_1(1-\phi_2)\lambda & \phi_1\phi_2(1-\phi_3)\lambda \\ 0 & (1-\phi_1)\lambda & \phi_1(1-\phi_2)\lambda \\ 0 & 0 & (1-\phi_1)\lambda \end{bmatrix}.$$

Table 6.7 shows the percentage overlap between the prior and marginal posterior distributions when uniform priors are used. Examining the overlap leads us to conclude that λ is not weakly identifiable, but the other parameters are. However all the parameters in this model are non-identifiable (see Cole et al., 2012). Brooks et al. (2000) explain that there is a ridge in the likelihood in this example, and because of the uniform priors distributions there is also a ridge in the posterior. They discuss the effect of integration on the ridge, and show that narrow credible intervals are obtained for λ (and for more than three years of data for ϕ_1 as well). The same explanation extends to the overlap between prior and marginal posterior, demonstrating it is possible in some examples to obtain an overlap of less than 35% for non-identifiable parameters.

This example demonstrates that using the overlap of the posterior and prior is unreliable. Whilst a non-identifiable parameter will most likely have a large overlap in the prior and marginal posterior, an overlap in the prior and marginal posterior can also occur because an informative prior happens

TABLE 6.8

Estimated percentage overlap of the posterior and prior for the logistic growth example without constraints.

Parameter	Uninformative	Vague	Informative	Disinformative
μ	89.0%	85.9%	85.1%	35.1%
$\alpha_{D,1}$	77.2%	75.6%	82.1%	41.5%
$\alpha_{D,2}$	75.6%	68.4%	82.5%	8.8%
$\alpha_{T,1}$	83.3%	78.6%	82.2%	64.3%
$\alpha_{T,2}$	83.8%	73.3%	83.1 %	17.7%
β_1	77.2%	53.8%	78.0%	0.6%
β_2	78.0%	74.3%	80.9%	73.5%
β_3	76.7%	52.1%	83.2%	1.7%

TABLE 6.9

Percentage overlap of the posterior and prior for the logistic growth example with constraints.

Parameter	Uninformative	Vague	Informative	Disinformative
μ	10.8%	19.2%	64.5%	16.7%
$\alpha_{D,1}$	11.1%	22.7%	68.9%	32.9%
$\alpha_{T,1}$	11.4%	31.9%	68.1%	46.0%
β_1	12.0%	10.6%	67.4%	14.0%
β_2	14.5%	30.7%	75.6%	73.3%

to match with the posterior and it is possible for there to be little overlap for a non-identifiable parameter.

6.4.3   Logistic Growth Curve Model

We return to the logistic growth curve example from Section 6.3.3. For the same examples presented in Tables 6.3 and 6.4, the percentage overlap between the posterior and prior distributions have been calculated and are given in Tables 6.8 and 6.9.

For the original logistic growth curve example without constraints the uninformative, vague and informative priors all have overlaps much larger than the 35% cut off, so are all classified as weakly identifiable. However the disinformative prior has some parameters that are above the cut-off and some that are below the cut-off.

When constraints are added to the logistic growth curve example, when uniformative or vague priors are used the overlaps are all less than 35%, so the parameters are not weakly identifiable, as would be expected. However when informative priors are used all the parameters have overlaps over 35%. When disinformative priors are used several of the parameters have overlaps over 35%, incorrectly classifying the model as weakly identifiable.

This again demonstrates that examining the overlap between prior and posterior distribution can lead to the wrong conclusion.

6.5 Detecting Bayesian Identifiability using Data Cloning

Data cloning is a method for finding approximate maximum likelihood estimates and their standard errors using Bayesian methods (Lele et al., 2007). The method can also be used to explore identifiability in models (Lele et al., 2010) as discussed in Section 4.1.3 for the classical formation of a model.

Data cloning involves creating a replicate data set consisting of K cloned data sets, that is, the original data set repeated K times. If the likelihood of the model is identifiable then the variance of the posterior tends to zero as K tends to infinity. However, if the likelihood of the model is non-identifiable the variance of the posterior does not tend to zero. The proof of this method is based on whether or not the likelihood is identifiable (Appendix A of Lele et al., 2010), hence this method can be used to investigate Bayesian identifiability, but not the identifiability of the posterior distribution.

The same algorithm described in Section 4.1.3 can be used with different prior distributions at step 1, rather than an uninformative prior. This is demonstrated in examples 6.5.1 and 6.5.2 below.

6.5.1 Basic Occupancy Model

Data cloning for the classical formation of the basic occupancy model was discussed in Section 4.1.3.1. In the occupancy model it is possible to find the exact posterior distribution, as shown in Section 6.3.1. It is also possible to calculate the exact variance of the marginal posterior distribution. To simplify calculation we consider the same simple data set used in as Section 6.3.1. In Section 6.4.1 we showed the marginal posterior distribution is

$$f(p|\mathbf{y}) = \frac{18}{5}p - \frac{12}{5}p^2,$$

when uniform priors are used on both parameters. The variance of the marginal posterior distribution is then

$$\int_0^1 p^2 f(p|\mathbf{y})dp + \left\{ \int_0^1 pf(p|\mathbf{y})dp \right\}^2 = \frac{3}{50}.$$

This is the variance for $K = 1$. Values for $K = 10, 20, \ldots, 100$ are calculated in a similar way, as demonstrated in the Maple code, basicoccupancy.mw. We also consider the reparameterised model with $\beta = p\psi$, which has an identifiable likelihood. In both models, uniform priors are used on each parameter as

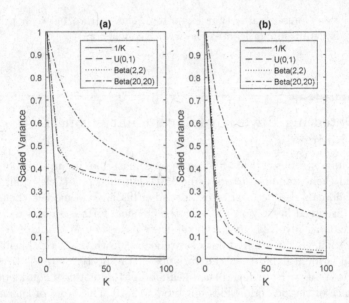

FIGURE 6.4
Data cloning occupancy example showing the scaled variance for a varying number of data clones, K. Figure (a) shows the results for the parameter p in the original model with two parameters. Figure (b) shows the results for the parameter β in the reparameterised model. The prior distribution of the parameter is given in the legend.

well as priors following a Beta(2,2) or a Beta(20,20) distribution. Results are presented in Figure 6.4. The figures show the number of clones, K, plotted against the scaled variance. The scaled variance is calculated as the variance of the marginal posterior for K clones divided by the variance when $K = 1$. The scaled variance is always 1 for $K = 1$.

For the basic occupancy model with a non-identifiable likelihood, the scaled variance tends to about 0.4, so therefore would be correctly classified as non-identifiable. There are slight differences depending on the prior used. This demonstrates Corollary A.2 from Lele et al. (2010), which states that as K increases the posterior distribution tends to a truncated prior distribution. The truncated prior distribution depends on the prior distribution chosen, therefore the limit of the scaled distribution will also vary with prior distribution.

For the reparameterised basic occupancy model, which has an identifiable likelihood, the variance tends to 0. The rate of convergence is approximately $1/K$ for uniform priors, but the rate is slower the more informative the prior is. For the Beta distribution with both parameters equal to 20, the rate of convergence to 0 is considerably slower, and more than $K = 100$ clones would be needed to reach the correct conclusion.

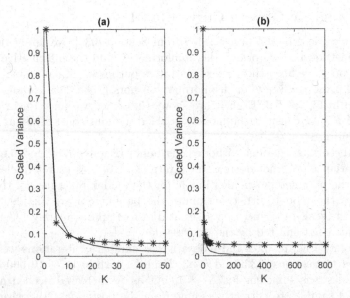

FIGURE 6.5
Data cloning occupancy example showing the scaled variance for varying number of data clones, K, for an example close to the boundary. Both figures show the scaled variance for p ($-*$) alongside $1/K$ ($-$). Figure (a) shows the results for K from 1 to 50. Figure (b) shows the results for K from 1 to 800.

Data cloning does not always give the correct result; see Section 4.1.3.2. Here, we demonstrate another issue with data cloning when parameters are close to the boundary. Suppose that 1000 sites were observed as occupied and only one site was observed as unoccupied. Whilst this is unlikely to occur in practice, this results in a posterior mean for p of 0.9985, close to the boundary of 1. In this case, as above, the exact posterior can be found, as well as its mean and variance. Figure 6.5 shows the scaled variance for different values of K. If only Figure 6.5 (a) is presented, with values of K from 1 to 50, then it appears that the scaled variance is close to $1/K$, so therefore p is identifiable. However when larger values of K are considered it can be seen that the scaled variance is tending to a fixed limit rather than to infinity, so the parameter is not identifiable.

Whilst trying increasingly larger values of K may identify whether a parameter really is identifiable, this is not practical in all cases, as K may need to be very large. This method should therefore be used with caution, particularly if there is a parameter close to a boundary.

6.5.2   Logistic Growth Curve Model

For the logistic growth curve model, from Section 6.3.3, we apply the data cloning method for two priors: the uninformative and the informative priors. (Two priors have been chosen for illustrative purposes, similar results can be obtained using the vague or disinformative priors.) The data cloning results are given in Figure 6.6. Each figure shows the scaled variance for increasing values of K. We consider the model with and without constraints, as well as the two priors.

For the original model without constraints (Figures 6.6a and 6.6b), the scaled variance does not decrease at a rate of $1/K$. It is therefore clear that none of the parameters are identifiable for the model, regardless of the prior used. For the model with constraints, the parameters are all identifiable, because the scaled variance decreases at a rate of approximately $1/K$. Again this result is not effected by the prior used.

To create Figures 6.6a and 6.6b a larger number of iterations were used compared to Sections 6.3.3 and 6.4.3. Using a burn-in in of 10 000 and 10 000 samples was sufficient for $K = 1$, but as K increased needed a larger number of samples to correctly sample from the posterior. Otherwise, if the number of samples was not increased Figures 6.6a and 6.6b looked similar to Figures 6.6c and 6.6d. This would have led to the wrong conclusion: that all the parameters were estimable. Instead we increased the burn-in and number of samples to 100 000 to produce Figure 6.6.

When using data cloning caution is needed when considering the number of samples, as non-identifiable models can preform badly in terms of poor mixing and convergence. The number of samples that is sufficient for $K = 1$ may not be sufficient for larger values of K.

6.6   Symbolic Method for Detecting Bayesian Identifiability

The symbolic method (Sections 3.2 and 4.2) can be used to detect identifiability in a Bayesian formation of a model. The only requirement is a suitable exhaustive summary. The exhaustive summary used depends on the definition of identifiability used.

Bayesian identifiability (Definition 13) can be considered to be equivalent to identifiability of the likelihood (Gelfand and Sahu, 1999), therefore any appropriate exhaustive summary derived from the log-likelihood exhaustive summary could be used. Results on Bayesian identifiability would be identical to the classical formation of the same model. As this is the same as earlier chapters we do not consider this definition further in this section.

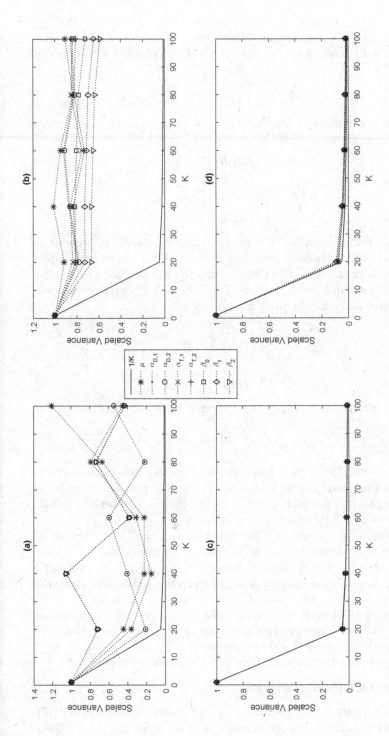

FIGURE 6.6

Data cloning for the logistic growth curve example, showing scaled variance for different numbers of data clones, K. Figures (a) and (b) show results for the original model. Figures (c) and (d) show results for the model with constraints. Figures (a) and (c) use uninformative priors. Figures (b) and (d) use informative priors. Note that the parameters $\alpha_{D,2}$, $\alpha_{T,2}$ and β_2 do not appear in Figures (c) and (d) as the constraint $\alpha_{D,2} = \alpha_{T,2} = \beta_2 = 0$ has been applied.

Instead here we consider posterior identifiability (Definition 12). An exhaustive summary for posterior identifiability consists of terms for the likelihood, $f(\mathbf{y}|\boldsymbol{\theta})$, and terms for the prior, $f(\boldsymbol{\theta})$. If the likelihood was formed as

$$f(\mathbf{y}|\boldsymbol{\theta}) = \prod_{i=1}^{n} f(y_i|\boldsymbol{\theta}),$$

for a sample of n independent observations, then an exhaustive summary for the likelihood component is

$$\kappa_L = \left[\begin{array}{c} f(y_1|\boldsymbol{\theta}) \\ f(y_2|\boldsymbol{\theta}) \\ \vdots \\ f(y_n|\boldsymbol{\theta}) \end{array} \right].$$

As taking natural logarithm is a one-to-one transformation it is possible to use the log-likelihood exhaustive summary from Section 3.1.4, or any simpler exhaustive summary that could be derived from the log-likelihood exhaustive summary. If there are independent priors, $f(\theta_i)$, on each of p parameters, then an exhaustive summary for the prior component is

$$\kappa_P = \left[\begin{array}{c} f(\theta_1) \\ f(\theta_2) \\ \vdots \\ f(\theta_p) \end{array} \right].$$

The posterior exhaustive summary, for checking posterior identifiability, is then

$$\kappa = \left[\begin{array}{c} \kappa_L \\ \kappa_P \end{array} \right].$$

Now we have established an exhaustive summary to check for identifiability of the posterior distribution, we can use the exhaustive summary based methods outlined in Chapters 3, 4 and 5. We demonstrate how the symbolic method can be used for Bayesian models in Sections 6.6.1 to 6.6.3.

If uninformative priors are used on all parameters then there will be no information on the parameters in the exhaustive summary κ_P. The information on identifiability will then be limited to the likelihood exhaustive summary, κ_L. A natural consequence of the posterior exhaustive summary is that the identifiability results will then be identical to the equivalent classical formation of the model, which also has the same results as Bayesian identifiability. This demonstrates that Bayesian identifiability is equivalent to considering identifiability without the influence of prior information.

6.6.1 Basic Occupancy Model

For the basic occupancy model the likelihood component exhaustive summary can be either of the exhaustive summaries discussed in Section 3.1.4.1. Here the simplest of those exhaustive summaries,

$$\kappa_L = \begin{bmatrix} \psi p \\ 1 - \psi p \end{bmatrix},$$

is used.

When uniform priors are used for both parameters, as $f(p) = 1$ and $f(\psi) = 1$, the prior component of the exhaustive summary is

$$\kappa_P = \begin{bmatrix} 1 \\ 1 \end{bmatrix}.$$

The joint exhaustive summary is then

$$\kappa = \begin{bmatrix} \psi p \\ 1 - \psi p \\ 1 \\ 1 \end{bmatrix},$$

with parameters $\boldsymbol{\theta} = [\psi, p]$. Obviously the posterior is non-identifiable, although formally this can be shown using the symbolic method by forming the derivative matrix

$$\mathbf{D} = \frac{\partial \kappa}{\partial \boldsymbol{\theta}} = \begin{bmatrix} p & -p & 0 & 0 \\ \psi & -\psi & 0 & 0 \end{bmatrix}.$$

The derivative matrix has rank 1, but there are two parameters, therefore the posterior is non-identifiable.

If both parameters have beta prior distributions with $f(p) \propto p(1 - p)$ and $f(\psi) \propto \psi(1 - \psi)$, the prior component of the exhaustive summary is

$$\kappa_P = \begin{bmatrix} p(1 - p) \\ \psi(1 - \psi) \end{bmatrix}.$$

The joint exhaustive summary is then

$$\kappa = \begin{bmatrix} \psi p \\ 1 - \psi p \\ p(1 - p) \\ \psi(1 - \psi) \end{bmatrix}.$$

The derivative matrix,

$$\mathbf{D} = \frac{\partial \kappa}{\partial \boldsymbol{\theta}} = \begin{bmatrix} p & -p & 0 & 1 - 2\psi \\ \psi & -\psi & 1 - 2p & 0 \end{bmatrix},$$

now has full rank 2, therefore the posterior is identifiable. This result is again obvious as the last two terms each contain a single parameter, so from the two priors alone the model is identifiable.

6.6.2   Logistic Growth Curve Model

An exhaustive summary for the likelihood of the logistic growth curve example, discussed in Sections 6.3.3, 6.4.3 and 6.5.2, consists of the design matrix multiplied by the vector of parameters,

$$
\kappa_L = \mathbf{X}\theta =
\begin{bmatrix}
1 & 1 & 0 & 1 & 0 & -1 & -1 & 0 \\
1 & 1 & 0 & 1 & 0 & 0 & 0 & 0 \\
1 & 1 & 0 & 1 & 0 & 1 & 1 & 0 \\
1 & 1 & 0 & 0 & 1 & -1 & 0 & -1 \\
1 & 1 & 0 & 0 & 1 & 0 & 0 & 0 \\
1 & 1 & 0 & 0 & 1 & 1 & 0 & 1 \\
1 & 0 & 1 & 1 & 0 & -1 & -1 & 0 \\
1 & 0 & 1 & 1 & 0 & 0 & 0 & 0 \\
1 & 0 & 1 & 1 & 0 & 1 & 1 & 0 \\
1 & 0 & 1 & 0 & 1 & -1 & 0 & -1 \\
1 & 0 & 1 & 0 & 1 & 0 & 0 & 0 \\
1 & 0 & 1 & 0 & 1 & 1 & 0 & 1
\end{bmatrix}
\begin{bmatrix}
\mu \\
\alpha_{D,1} \\
\alpha_{D,2} \\
\alpha_{T,1} \\
\alpha_{T,2} \\
\beta_0 \\
\beta_1 \\
\beta_2
\end{bmatrix}.
$$

The derivative matrix will be the design matrix, so that

$$
\mathbf{D} = \frac{\partial \kappa_L}{\partial \theta} =
\begin{bmatrix}
1 & 1 & 0 & 1 & 0 & -1 & -1 & 0 \\
1 & 1 & 0 & 1 & 0 & 0 & 0 & 0 \\
1 & 1 & 0 & 1 & 0 & 1 & 1 & 0 \\
1 & 1 & 0 & 0 & 1 & -1 & 0 & -1 \\
1 & 1 & 0 & 0 & 1 & 0 & 0 & 0 \\
1 & 1 & 0 & 0 & 1 & 1 & 0 & 1 \\
1 & 0 & 1 & 1 & 0 & -1 & -1 & 0 \\
1 & 0 & 1 & 1 & 0 & 0 & 0 & 0 \\
1 & 0 & 1 & 1 & 0 & 1 & 1 & 0 \\
1 & 0 & 1 & 0 & 1 & -1 & 0 & -1 \\
1 & 0 & 1 & 0 & 1 & 0 & 0 & 0 \\
1 & 0 & 1 & 0 & 1 & 1 & 0 & 1
\end{bmatrix}.
$$

As stated in Gelfand and Sahu (1999), and in Section 6.3.3, this is not identifiable because the design matrix has rank 5, but there are eight parameters. Using uninformative priors on all parameters will also result in a non-identifiable posterior distribution.

Now consider using informative priors. Suppose normal distribution priors are used on the parameters $\alpha_{D,2}$, $\alpha_{T,2}$ and β_2, and uninformative priors are used on the other parameters. The known means of the normal priors are $m_{\alpha_{D,2}}$, $m_{\alpha_{T,2}}$ and m_{β_2} respectively and the known variances are $s^2_{\alpha_{D,2}}$, $s^2_{\alpha_{T,2}}$ and $s^2_{\beta_2}$ respectively. A suitable exhaustive summary would consist of the prior distributions for each parameter, which are of the form

$$
f(\theta_i) = \frac{1}{\sqrt{2\pi s^2_{\theta_i}}} \exp\left\{ \frac{(\theta_i - m_{\theta_i})^2}{2 s^2_{\theta_i}} \right\}.
$$

As multiplying by a constant or taking logs are one-to-one transformations a simpler exhaustive summary is

$$
\kappa_P = \begin{bmatrix} \frac{(\alpha_{D,2} - m_{\alpha_{D,2}})^2}{s^2_{\alpha_{D,2}}} \\ \frac{(\alpha_{T,2} - m_{\alpha_{T,2}})^2}{s^2_{\alpha_{T,2}}} \\ \frac{(\beta_2 - m_{\beta_2})^2}{s^2_{\beta_2}} \end{bmatrix}.
$$

To check identifiability of the posterior we can use the exhaustive summary

$$
\kappa = \begin{bmatrix} \kappa_L \\ \kappa_P \end{bmatrix}.
$$

We show in the Maple code, `logisticgrowth.mw`, that the derivative matrix, $\mathbf{D} = \partial\kappa/\partial\theta$, has full rank of 8. By a trivial application of the extension theorem (Theorem 6 from Section 3.2.6.1), it does not matter which priors are used on the other parameters; the posterior will be identifiable.

Consider a modified PLUR decomposition of the derivative matrix, making use of Theorem 7 from Section 3.2.7. Writing $\mathbf{D} = \mathbf{PLUR}$, the determinant of the matrix $\mathbf{U}$ is

$$
\det(\mathbf{U}) = \frac{(\alpha_{D,2} - m_{\alpha_{D,2}})(\alpha_{T,2} - m_{\alpha_{T,2}})(\beta_2 - m_{\beta_2})}{s^2_{\alpha_{D,2}} s^2_{\alpha_{T,2}} s^2_{\beta_2}}.
$$

As either $s^2_{\alpha_{D,2}}$, $s^2_{\alpha_{T,2}}$ or $s^2_{\beta_2}$ tend to infinity the determinant of $\mathbf{U}$ tends to zero. In a classical formation of the model we would then state that the model will be near-redundant for very large values of $s^2_{\alpha_{D,2}}$, $s^2_{\alpha_{T,2}}$ or $s^2_{\beta_2}$ (see Section 3.5). Using the same principle, as the prior variance becomes large the model would be weakly identifiable. This is equivalent to using a vague prior on the parameters $\alpha_{D,2}$, $\alpha_{T,2}$ and β_2. So whilst using a vague prior on these parameters will result in a posterior that is technically identifiable, the posterior would be close to being non-identifiable, and there would be similar problems as those associated with a non-identifiable posterior.

6.6.3 Mark-Recovery Model

Consider again the mark-recovery model. The likelihood component of the exhaustive summary for the mark-recovery model would consist of the terms

$$
P_{i,j}^{N_{i,j}} \text{ for } i = 1, \ldots, n_1 \; j = i, \ldots, n_2
$$

and

$$
\left(1 - \sum_{j=i}^{n_2} P_{i,j} \right)^{F_i - \sum_{j=i}^{n_2} N_{i,j}} \quad \text{for } i = 1, \ldots, n_1
$$

where F_i is the number of animals marked in year i, $N_{i,j}$ is the number of animals marked in year i that are recovered dead in year j, $P_{i,j}$ is the corresponding probability that an animal marked in year i is recovered dead in year j, n_1 is the number of years of marking and n_2 is the number of years of recovery. For the mark-recovery model described in Section 6.3.2 with $n_1 = n_2 = 3$ and

$$P_{i,j} = \begin{cases} (1-\phi_1)\lambda_1 & j = i \\ \phi_1 \phi_a^{j-i-1}(1-\lambda_a) & j > i \end{cases}.$$

A simpler exhaustive summary consists of the probabilities $P_{i,j}$ (see Section 3.2.4), so here we use the exhaustive summary

$$\kappa_L = \begin{bmatrix} (1-\phi_1)\lambda_1 \\ \phi_1(1-\phi_a)\lambda_a \\ \phi_1\phi_a(1-\phi_a)\lambda_a \end{bmatrix}.$$

Firstly consider the case when there are uniform priors on all four parameters. The prior component of the exhaustive summary is

$$\kappa_P = \begin{bmatrix} 1 \\ 1 \\ 1 \\ 1 \end{bmatrix}.$$

The joint exhaustive summary is then

$$\kappa = \begin{bmatrix} (1-\phi_1)\lambda_1 \\ \phi_1(1-\phi_a)\lambda_a \\ \phi_1\phi_a(1-\phi_a)\lambda_a \\ 1 \\ 1 \\ 1 \\ 1 \end{bmatrix},$$

with parameters $\boldsymbol{\theta} = [\lambda_1, \lambda_a, \phi_1, \phi_a]$. Given that we have already stated in Section 1.2.4 that the classical model is parameter redundant, the posterior is obviously non-identifiable. This can be shown formally by forming the derivative matrix

$$\mathbf{D} = \frac{\partial \kappa}{\partial \theta} = \begin{bmatrix} 1-\phi_1 & 0 & 0 & 0 & 0 & 0 & 0 \\ 0 & \phi_1(1-\phi_a) & \phi_1\phi_a(1-\phi_a) & 0 & 0 & 0 & 0 \\ -\lambda_1 & (1-\phi_a)\lambda_a & \phi_a(1-\phi_a)\lambda_a & 0 & 0 & 0 & 0 \\ 0 & -\phi_1\lambda_a & \phi_1(1-\phi_a)\lambda_a - \phi_1\phi_a\lambda_a & 0 & 0 & 0 & 0 \end{bmatrix},$$

which has rank 3. As there are four parameters the posterior is non-identifiable.

If informative beta priors are used, for example the informative priors listed in Table 6.1, the prior component exhaustive summary will be of the form

$$\kappa_p = \begin{bmatrix} \lambda_1^{a_1}(1-\lambda_1)^{b_1} \\ \lambda_a^{a_2}(1-\lambda_a)^{b_2} \\ \phi_1^{a_3}(1-\phi_1)^{b_3} \\ \phi_a^{a_4}(1-\phi_a)^{b_4} \end{bmatrix},$$

where a_i and b_i are known constants. The parameters are $\boldsymbol{\theta} = [\lambda_1, \lambda_a, \phi_1, \phi_a]$.

Rather than forming $\boldsymbol{\kappa} = [\boldsymbol{\kappa}_L, \boldsymbol{\kappa}_p]^T$, we can apply the extension theorem (Theorem 6 from Section 3.2.6.1) starting with κ_p. The derivative matrix

$$\mathbf{D}_1 = \frac{\partial \kappa_p}{\partial \boldsymbol{\theta}} = \begin{bmatrix} A_1 & 0 & 0 & 0 \\ 0 & A_2 & 0 & 0 \\ 0 & 0 & A_3 & 0 \\ 0 & 0 & 0 & A_4 \end{bmatrix},$$

where

$$A_i = \theta_i^{a_i-1} a_i (1-\theta_i)^{b_i} - \theta_i^{a_i}(1-\theta_i)^{b_i-1} b_i,$$

has full rank 4. Adding additional exhaustive summary terms κ_L adds no additional parameters, therefore by a trivial application of the extension theorem (Remark 1 from Section 3.2.6.1) the posterior is identifiable. This demonstrates the fact that adding individual informative priors on all parameters will result in an identifiable posterior.

6.7   Conclusion

If a classical formation of a model is non-identifiable it could lead to potential identifiability issues in the posterior of the equivalent Bayesian model, depending on the priors used. Using uniformative priors will result in the identical identifiability results to the classical formation of the model. Using informative priors on unidentifiable parameters results in a posterior that is identifiable, though the posterior will be heavily reliant on the prior distribution used. This fits with the definition of Bayesian identifiability, which is equivalent to identifiability in the likelihood. Bayesian identifiability, in effect, does not take into account prior information, so that there will be no problem associated with the choice of prior in a model that is identifiable under this definition.

However in hierarchical models, such as the example from Swartz et al. (2004) in Section 6.2, considering identifiability of the likelihood results in a classification of non-identifiable for a model that there is no practical problem in using. Therefore it is recommended that both identifiability of the posterior and the likelihood are considered.

If a classical formation of a model is non-identifiable it is only recommended that the Bayesian version of the model is used in that form if there is reliable prior information for non-identifiable parameters. Otherwise any results will be heavily dependent on the prior used, noting that the prior affects any confounded parameters as well as the parameter itself.

7

Identifiability in Continuous State-Space Models

As stated in Section 1.3.2, compartmental models are deterministic models used to represent a system where materials or energies are transmitted between multiple compartments (see, for example, Godfrey, 1983, Walter and Contreras, 1999). They are a type of continuous state-space models that can be used to mathematically describe a specific biological process in areas such as pharmacokinetics, epidemiology and biomedicine (Anderson, 1983). The mathematical expressions to describe these models are given in Section 7.1, along with several simple illustrative examples.

There is a long history of studying identifiability in continuous state-space models; see, for example, Walter (1982), Walter (1987) and Chis et al. (2011a). In Section 7.2 we discuss methods for checking for identifiability in continuous state-space models. Section 7.3 discusses the software available for checking identifiability in compartmental models.

7.1 Continuous State-Space Models

A linear continuous state-space model with m states is defined by the equations

$$\mathbf{y}(t, \boldsymbol{\theta}) = \mathbf{C}(\boldsymbol{\theta})\mathbf{x}(t, \boldsymbol{\theta}), \tag{7.1}$$

and

$$\frac{\partial}{\partial t}\mathbf{x}(t, \boldsymbol{\theta}) = \mathbf{A}(\boldsymbol{\theta})\mathbf{x}(t, \boldsymbol{\theta}) + \mathbf{B}(\boldsymbol{\theta})\mathbf{u}(t), \tag{7.2}$$

where $\mathbf{y}(t, \boldsymbol{\theta})$ is the output of the system at time t, $\mathbf{u}(t)$ is the input into the system at time t and $x_i(t, \boldsymbol{\theta})$ is the ith state variable at time t. The system depends on a set of unknown parameters, $\boldsymbol{\theta}$, and known matrices $\mathbf{A}$, $\mathbf{B}$ and $\mathbf{C}$.

In a compartmental model, the states are compartments representing components of a system, so that $x_i(t)$ represents a measure of the ith compartment at time t. The parameters are then used to describe how the material flows between compartments within and out of the system. These parameters appear in the model through the matrices $\mathbf{A}$, $\mathbf{B}$ and $\mathbf{C}$ known as compartmental,

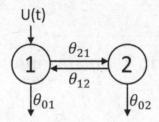

FIGURE 7.1
A representation of a two-compartment linear model from Bellman and Astrom (1970). The circles represent the compartments and the arrows represent flow from one compartment to the next. There is initial input into compartment 1 then output from the same compartment is observed.

input and output matrices respectively (see, for example, Godfrey and Distefano, 1987). Illustrative examples of linear compartmental models are given in Sections 7.1.1 and 7.1.2 below.

A non-linear continuous state-space model is defined by the equations

$$y(t, \boldsymbol{\theta}) = \mathbf{h}\{\mathbf{x}(t, \boldsymbol{\theta}), \boldsymbol{\theta}\}, \tag{7.3}$$

and

$$\frac{\partial}{\partial t}\mathbf{x}(t, \boldsymbol{\theta}) = \mathbf{f}\{\mathbf{x}(t, \boldsymbol{\theta}), \boldsymbol{\theta}\} + \mathbf{u}(t)\mathbf{g}\{\mathbf{x}(t, \boldsymbol{\theta}), \boldsymbol{\theta}\}, \tag{7.4}$$

where $y(t, \boldsymbol{\theta})$ is the output of the system, $\mathbf{u}(t)$ is the input into the system and $\mathbf{x}(t, \boldsymbol{\theta})$ is the state variable, depending on the time t and a set of unknown parameters, $\boldsymbol{\theta}$. Here $\mathbf{h}$, $\mathbf{f}$ and $\mathbf{g}$ are known functions (see, for example, Evans and Chappell, 2000). An example of a non-linear state-space model is given in Section 7.1.3 below.

7.1.1 Two-Compartment Linear Model

Bellman and Astrom (1970) present a compartmental model with two states represented in Figure 7.1. This linear compartmental model is defined by the equations

$$\begin{cases} \dfrac{\partial x_1(t)}{\partial t} = -(\theta_{01} + \theta_{21})x_1(t) + \theta_{12}x_2(t) + u(t) \\ \dfrac{\partial x_2(t)}{\partial t} = \theta_{21}x_1(t) - (\theta_{12} + \theta_{02})x_2(t) \end{cases}$$

and

$$y(t) = x_1(t)$$

where $x(0) = 0$, $u(0) = 1$ and $u(t) = 0$ for $t > 0$.

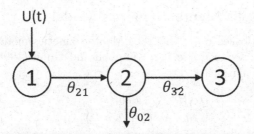

FIGURE 7.2
A representation of a three-compartment linear model. This is model K from Godfrey and Chapman (1990). The circles represent the compartments and the arrows represent flow from one compartment to the next. There is initial input into compartment 1 then only output from compartment 3 is observed.

The compartmental, input and output matrices are

$$\mathbf{A} = \left[\begin{array}{cc} -(\theta_{01} + \theta_{21}) & \theta_{12} \\ \theta_{21} & -(\theta_{12} + \theta_{02}) \end{array} \right], \mathbf{B} = \left[\begin{array}{c} 1 \\ 0 \end{array} \right], \mathbf{C} = \left[\begin{array}{cc} 1 & 0 \end{array} \right].$$

The model has four parameters $\boldsymbol{\theta} = [\theta_{01}, \theta_{02}, \theta_{12}, \theta_{21}]$.
The Maple code for this example is given in `twocompartmentmodel.mw`.

7.1.2 Three-Compartment Linear Model

Section 1.3.2 presented an example from Godfrey and Chapman (1990). An alternative adaptation of the model is given in Figure 7.2. This linear compartmental model, with three states, is defined by the equations

$$\left\{ \begin{array}{l} \dfrac{\partial x_1(t)}{\partial t} = -\theta_{21} x_1(t) + u(t) \\ \dfrac{\partial x_2(t)}{\partial t} = \theta_{21} x_1(t) - (\theta_{32} + \theta_{02}) x_2(t) \\ \dfrac{\partial x_3(t)}{\partial t} = \theta_{32} x_2(t) \end{array} \right.$$

and

$$y(t) = x_3(t),$$

where $x(0) = 0$, $u(0) = 1$ and $u(t) = 0$ for $t > 0$.
The matrices for this linear compartmental model are:

$$\mathbf{A} = \left[\begin{array}{ccc} -\theta_{21} & 0 & 0 \\ \theta_{21} & -\theta_{32} - \theta_{02} & 0 \\ 0 & \theta_{32} & 0 \end{array} \right], \mathbf{B} = \left[\begin{array}{c} 1 \\ 0 \\ 0 \end{array} \right], \mathbf{C} = \left[\begin{array}{ccc} 0 & 0 & 1 \end{array} \right].$$

The model has three parameters $\boldsymbol{\theta} = [\theta_{02}, \theta_{21}, \theta_{32}]$.
The Maple code for this example is given in `threecompartmentmodel.mw`.

7.1.3   Michaelis Menten Kinetics Model

This example is based on a Michaelis Menten kinetics model used to model enzyme kinetics, see example 1 of Chappell and Gunn (1998). The model is defined by the equations

$$\frac{\partial x_1(t)}{\partial t} = \frac{\theta_1 x_1(t)}{\theta_2 + \theta_3 x_1(t)} + u(t),$$

and

$$y(t) = c x_1(t),$$

where $x(0) = 0$, $u(0) = 1$ and $u(t) = 0$ for $t > 0$.

The model has four parameters $\boldsymbol{\theta} = [c, \theta_1, \theta_2, \theta_3]$.

The Maple code for this example is given in `nonlinearmodel.mw`.

7.2   Methods for Investigating Identifiability in Continuous State-Space Models

Walter and Pronzato (1996) and Chis et al. (2011a) discuss several different methods for investigating identifiability in continuous state-space models. In Sections 7.2.1 to 7.2.3, we explain three of these methods: the transfer function approach, similarity of transforms and the Taylor series approach. Other methods are referenced in Section 7.2.4.

7.2.1   Transfer Function Approach

The transfer function approach is a method for checking identifiability in linear state-space models. This method involves taking Laplace transforms of $\mathbf{y}$ in the system described by equations (7.1) and (7.2). The transfer function is then defined as

$$\mathbf{Q}(s) = \frac{\mathbf{y}(s)}{\mathbf{u}(s)} = \mathbf{C}(s\mathbf{I} - A)^{-1},$$

where $\mathbf{y}(s)$ is the Laplace transform of $\mathbf{y}(t, \boldsymbol{\theta})$ and $\mathbf{u}(s)$ is the Laplace transform of $\mathbf{u}(t)$. The ith term in the transfer function, $\mathbf{Q}(s)$, will be of the form

$$Q_i(s) = \frac{b_{m-1}s^{m-1} + b_{m-2}s^{m-2} + \ldots + b_1 s + b_0}{s^m + a_{m-1}s^{m-1} + a_{m-2}s^{m-2} + \ldots + a_1 s + a_0}.$$

An exhaustive summary is formed from these terms. For example, if only one compartment is observed and $\mathbf{Q}(s) = Q_1(s)$ then

$$\boldsymbol{\kappa}(\boldsymbol{\theta}) = \begin{bmatrix} a_0 \\ a_1 \\ \vdots \\ a_{m-2} \\ a_{m-1} \\ b_0 \\ b_1 \\ \vdots \\ b_{m-2} \\ b_{m-1} \end{bmatrix}.$$

If there is unique solution to the equations $\kappa_i(\boldsymbol{\theta}) = k_i$ for the parameters $\boldsymbol{\theta}$ then the model is globally identifiable. If there is a countable number of solutions the model is locally identifiable and if there are infinite solutions the model is non-identifiable (Bellman and Astrom, 1970). Godfrey and Distefano (1987) show that it is possible to check for local identifiability by checking the rank of an appropriate Jacobian matrix. This is the symbolic method, of Section 3.2. We demonstrate this method in Sections 7.2.1.1 and 7.2.1.2. This method is not applicable for non-linear models, so for example could not be applied to the non-linear example of Section 7.1.3.

7.2.1.1 Two-Compartment Linear Model

For the two-compartment linear model described in Section 7.1.1, the transfer function is

$$Q(s) = \frac{s + \theta_{12} + \theta_{02}}{s^2 + s(\theta_{01} + \theta_{02} + \theta_{12} + \theta_{21}) + \theta_{01}\theta_{02} + \theta_{01}\theta_{12} + \theta_{21}\theta_{02}}.$$

The non-constant coefficients of s is the numerator and denominator result in exhaustive summary

$$\boldsymbol{\kappa}(\boldsymbol{\theta}) = \begin{bmatrix} \theta_{12} + \theta_{02} \\ \theta_{01} + \theta_{02} + \theta_{12} + \theta_{21} \\ \theta_{01}\theta_{02} + \theta_{01}\theta_{12} + \theta_{21}\theta_{02} \end{bmatrix}.$$

The parameters are $\boldsymbol{\theta} = [\theta_{01}, \theta_{02}, \theta_{12}, \theta_{21}]$, giving a derivative matrix

$$\mathbf{D} = \frac{\partial \boldsymbol{\kappa}}{\partial \boldsymbol{\theta}} = \begin{bmatrix} 0 & 1 & \theta_{12} + \theta_{02} \\ 1 & 1 & \theta_{01} + \theta_{21} \\ 1 & 1 & \theta_{01} \\ 0 & 1 & \theta_{0,2} \end{bmatrix}.$$

We show in the Maple code that the rank is 3. As there are four parameters the model is non-identifiable.

Following Theorem 5, in Section 3.2.5, we find estimable parameter combinations by first solving $\boldsymbol{\alpha}(\boldsymbol{\theta})^T \mathbf{D}(\boldsymbol{\theta}) = 0$, which has solution

$$\boldsymbol{\alpha}(\boldsymbol{\theta})^T = \begin{bmatrix} -1 & \frac{\theta_{12}}{\theta_{21}} & -\frac{\theta_{12}}{\theta_{21}} & 1 \end{bmatrix}.$$

Then we solve the partial differential equation

$$-\frac{\partial f}{\partial \theta_{01}} + \frac{\partial f}{\partial \theta_{02}} \frac{\theta_{12}}{\theta_{21}} - \frac{\partial f}{\partial \theta_{21}} \frac{\theta_{12}}{\theta_{21}} + \frac{\partial f}{\partial \theta_{21}} = 0$$

and find a solution of $\theta_{0,1} + \theta_{2,1}$, $\theta_{1,2}\theta_{2,1}$ and $\theta_{1,2} + \theta_{0,2}$. This is a reparameterisation that results in a model that at least locally identifiable. To check whether it is globally or locally identifiable we need to reparameterise the exhaustive summary in terms of $\beta_1 = \theta_{0,1} + \theta_{2,1}$, $\beta_2 = \theta_{1,2}\theta_{2,1}$ and $\beta_3 = \theta_{1,2} + \theta_{0,2}$. This gives

$$\boldsymbol{\kappa} = \begin{bmatrix} \beta_3 \\ \beta_1 + \beta_3 \\ \beta_1\beta_3 - \beta_2 \end{bmatrix}.$$

Then we solve $\kappa_i(\boldsymbol{\beta}) = k_i$, which gives $\beta_1 = k_2 - k_1$, $\beta_2 = k_1 k_2 - k_1^2 - k_3$, $\beta_3 = k_1$. As this is a unique solution this reparameterisation is globally identifiable.

7.2.1.2   Three-Compartment Linear Model

The three-compartment linear model, described in Section 7.1.2, has transfer function

$$Q(s) = \frac{\theta_{21}\theta_{32}}{s^3 + s^2(\theta_{02} + \theta_{21} + \theta_{32}) + s(\theta_{02}\theta_{21} + \theta_{21}\theta_{32})}.$$

The exhaustive summary is therefore

$$\boldsymbol{\kappa} = \begin{bmatrix} \theta_{21}\theta_{32} \\ \theta_{02} + \theta_{21} + \theta_{32} \\ \theta_{21}(\theta_{02} + \theta_{32}) \end{bmatrix}.$$

Differentiating this exhaustive summary with respect to the parameters $\boldsymbol{\theta} = [\theta_{02}, \theta_{21}, \theta_{32}]$ gives derivative matrix

$$\mathbf{D} = \frac{\partial \boldsymbol{\kappa}}{\partial \boldsymbol{\theta}} = \begin{bmatrix} 0 & 1 & \theta_{21} \\ \theta_{32} & 1 & \theta_{02} + \theta_{32} \\ \theta_{21} & 1 & \theta_{21} \end{bmatrix}.$$

The derivative matrix is of full rank 3, therefore the model is at least locally identifiable. Solving the set of equations $\kappa_i = k_i$ for $i = 1, 2, 3$ results in the solutions

$$\theta_{02} = \frac{k_3 - k_1}{\frac{1}{2}k_2 \pm \frac{1}{2}\sqrt{k_2^2 - 4k_3}},$$

$$\theta_{21} = \frac{1}{2}k_2 \pm \frac{1}{2}\sqrt{k_2^2 - 4k_3},$$

$$\theta_{32} = \frac{k_1}{\frac{1}{2}k_2 \pm \frac{1}{2}\sqrt{k_2^2 - 4k_3}}.$$

As there are two solutions for each parameter this model is locally identifiable.

7.2.2 Similarity of Transform or Exhaustive Modelling Approach

The similarity of transformation approach, which is also known as the exhaustive modelling approach, involves generating the set of all possible models that have the same input and output behaviour as the original model (Walter and Lecourtier, 1981, Walter, 1982, Godfrey and Distefano, 1987).

Consider an m state linear model which depends on the matrices $\mathbf{A}(\boldsymbol{\theta})$, $\mathbf{B}(\boldsymbol{\theta})$ and $\mathbf{C}(\boldsymbol{\theta})$. We refer to this as the original model. First check that the following are satisfied:

1. $\text{rank}\left[\begin{array}{cccc} \mathbf{B}(\boldsymbol{\theta}) & \mathbf{A}(\boldsymbol{\theta})\mathbf{B}(\boldsymbol{\theta}) & \ldots & \mathbf{A}(\boldsymbol{\theta})^{m-1}\mathbf{B}(\boldsymbol{\theta}) \end{array}\right] = m$ (known as the controllability rank condition),

2. $\text{rank}\begin{bmatrix} \mathbf{C}(\boldsymbol{\theta}) \\ \mathbf{C}(\boldsymbol{\theta})\mathbf{A}(\boldsymbol{\theta}) \\ \vdots \\ \mathbf{C}(\boldsymbol{\theta})\mathbf{A}(\boldsymbol{\theta})^{m-1} \end{bmatrix} = m$ (known as the observability rank condition).

Equivalent models would have matrices $\mathbf{A}(\boldsymbol{\theta}')$, $\mathbf{B}(\boldsymbol{\theta}')$ and $\mathbf{C}(\boldsymbol{\theta}')$. Any equivalent models are related to the original matrices via a similarity of transformation satisfying the equations:

$$\mathbf{A}(\boldsymbol{\theta}')\mathbf{T} = \mathbf{T}\mathbf{A}(\boldsymbol{\theta}) \qquad (7.5)$$

$$\mathbf{B}(\boldsymbol{\theta}') = \mathbf{T}\mathbf{B}(\boldsymbol{\theta}) \qquad (7.6)$$

$$\mathbf{C}(\boldsymbol{\theta}')\mathbf{T} = \mathbf{C}(\boldsymbol{\theta}), \qquad (7.7)$$

where $\mathbf{T}$ is an $m \times m$ matrix with entries t_{ij}. If there is a unique $\mathbf{T}$ satisfying equations (7.5) to (7.7) then the model is globally identifiable. If there are a finite set of $\mathbf{T}$ satisfying equations (7.5) to (7.7) then the model is locally identifiable. Otherwise the model is non-identifiable (Walter and Lecourtier, 1981, Walter, 1982, Godfrey and Distefano, 1987).

This method is extended to non-linear models in Vajda et al. (1989). However the PDEs involved are usually difficult to solve and in some cases the controllability and observability conditions are difficult to prove (Chis et al., 2011a). Therefore here we only consider linear models. Two linear examples are given in Sections 7.2.2.1 and 7.2.2.2.

7.2.2.1   Two-Compartment Linear Model

Consider the two-compartment linear model in 7.1.1. For controllability, the matrix

$$\begin{bmatrix} \mathbf{B}(\theta) & \mathbf{A}(\theta)\mathbf{B}(\theta) \end{bmatrix} = \begin{bmatrix} 1 & -\theta_{01} - \theta_{21} \\ 0 & \theta_{21} \end{bmatrix}$$

has rank 2, as required. For observability, the matrix

$$\begin{bmatrix} \mathbf{C}(\theta) \\ \mathbf{C}(\theta)\mathbf{A}(\theta) \end{bmatrix} = \begin{bmatrix} 1 & 0 \\ -\theta_{01} - \theta_{21} & \theta_{12} \end{bmatrix}$$

again has rank 2 as required.

Then consider equations (7.5) to (7.7). Equation (7.6) gives

$$\mathbf{B}(\theta') = \begin{bmatrix} 1 \\ 0 \end{bmatrix} = \begin{bmatrix} t_{11} \\ t_{21} \end{bmatrix} = \begin{bmatrix} t_{11} & t_{12} \\ t_{21} & t_{22} \end{bmatrix} \begin{bmatrix} 1 \\ 0 \end{bmatrix} = \mathbf{TB}(\theta).$$

Therefore $t_{11} = 1$ and $t_{21} = 0$. Then equation (7.7) results in

$$\mathbf{C}(\theta')\mathbf{T} = \begin{bmatrix} 1 & t_{12} \end{bmatrix} = \begin{bmatrix} 1 & 0 \end{bmatrix} = \mathbf{C}(\theta),$$

which gives $t_{12} = 0$. Evaluating equation (7.5) gives

$$\mathbf{A}(\theta')\mathbf{T} = \mathbf{TA}(\theta)$$

$$\begin{bmatrix} -\theta'_{01} - \theta'_{21} & \theta'_{12}t_{22} \\ \theta'_{21} & -(\theta'_{12} + \theta'_{02})t_{22} \end{bmatrix} = \begin{bmatrix} -\theta_{01} - \theta_{21} & \theta_{12} \\ t_{22}\theta_{21} & -(\theta_{12} + \theta_{02})t_{22} \end{bmatrix}$$

so that $t_{22} = \theta_{12}/\theta'_{12}$ or $t_{22} = \theta'_{21}/\theta_{21}$. This results in

- $\theta_{12}\theta_{21} = \theta'_{12}\theta'_{21}$
- $\theta_{01} + \theta_{21} = \theta'_{01} + \theta'_{21}$
- $\theta_{12} + \theta_{02} = \theta'_{12} + \theta'_{02}$.

As there is not a unique or finite number of solutions the model is non-identifiable.

7.2.2.2   Three-Compartment Linear Model

For the three-compartment linear model in 7.1.1, we start by checking for controllability and observability as follows:

$$\begin{bmatrix} \mathbf{B}(\theta) & \mathbf{A}(\theta)\mathbf{B}(\theta) & \mathbf{A}(\theta)^2\mathbf{B}(\theta) \end{bmatrix} = \begin{bmatrix} 1 & -\theta_{21} & \theta_{21}^2 \\ 0 & \theta_{21} & -\theta_{21}^2 - (\theta_{32} + \theta_{02})\theta_{21} \\ 0 & 0 & \theta_{32}\theta_{21} \end{bmatrix},$$

$$\begin{bmatrix} \mathbf{C}(\boldsymbol{\theta}) \\ \mathbf{C}(\boldsymbol{\theta})\mathbf{A}(\boldsymbol{\theta}) \\ \mathbf{C}(\boldsymbol{\theta})\mathbf{A}(\boldsymbol{\theta})^2 \end{bmatrix} = \begin{bmatrix} 0 & 0 & 1 \\ 0 & \theta_{32} & 0 \\ \theta_{32}\theta_{21} & -\theta_{32}(\theta_{32} + \theta_{02}) & 0 \end{bmatrix}.$$

Both matrices have rank 3 as required.

Then we consider equations (7.5) to (7.7). Equation (7.6) gives

$$\mathbf{B}(\boldsymbol{\theta}') = \begin{bmatrix} 1 \\ 0 \\ 0 \end{bmatrix} = \begin{bmatrix} t_{11} \\ t_{21} \\ t_{31} \end{bmatrix} = \mathbf{TB}(\boldsymbol{\theta}).$$

Therefore $t_{11} = 1$, $t_{21} = 0$ and $t_{31} = 0$. Then equation (7.7) results in

$$\mathbf{C}(\boldsymbol{\theta}')\mathbf{T} = \begin{bmatrix} 0 & t_{32} & t_{33} \end{bmatrix} = \begin{bmatrix} 0 & 0 & 1 \end{bmatrix} = \mathbf{C}(\boldsymbol{\theta}).$$

Thus $t_{32} = 0$ and $t_{33} = 1$. We show in the Maple code, `threecompartment` `model.mw`, that solving equation (7.5) gives $t_{23} = 0$, $t_{13} = 0$ and $t_{12} = (\theta_{21} - \theta'_{21})/\theta_{21}$ and two options for t_{22}: $t_{22} = \theta_{32}/\theta'_{32}$ and $t_{22} = \theta'_{21}/\theta_{21}$. This gives two solutions:

- $\theta_{02} = \theta'_{02}$, $\theta_{21} = \theta'_{21}$, $\theta_{32} = \theta'_{32}$

- $\theta_{02} = \theta'_{02}\theta'_{21}/(\theta'_{02} + \theta'_{32})$, $\theta_{21} = \theta'_{02} + \theta'_{32}$, $\theta_{32} = \theta'_{21}\theta'_{32}/(\theta'_{02} + \theta'_{32})$.

As there are two solutions this model is locally identifiable.

7.2.3 Taylor Series Expansion Approach

The Taylor series expansion approach involves a Taylor series expansion of $\mathbf{y}$ to create a system of equations that give information on the identifiability of linear and non-linear compartmental models.

Let

$$\mathbf{a}_i(t) = \frac{\partial^i \mathbf{y}(t, \boldsymbol{\theta})}{\partial t^i}.$$

Pohjanpalo (1978) shows that if the equations

$$\mathbf{a}_i(0) = \mathbf{k}_i, \quad i = 0, 1, \ldots, \infty$$

have a unique solution the model is globally identifiable. They also show that a derivative matrix can be formed from these terms and the rank used to check for local identifiability. This is the same as symbolic method (Section 3.2) using an exhaustive summary of

$$\boldsymbol{\kappa} = \begin{bmatrix} \mathbf{a}_0(0) \\ \mathbf{a}_1(0) \\ \mathbf{a}_2(0) \\ \vdots \end{bmatrix}.$$

However, a symbolic algebraic packages such as Maple require a fixed dimension derivative matrix. Therefore it in not possible in practice to have an infinite number of terms forming a derivative matrix. For a linear system of compartmental models with m compartments it is sufficient to use

$$\kappa = \begin{bmatrix} \mathbf{a}_0(0) \\ \mathbf{a}_1(0) \\ \mathbf{a}_2(0) \\ \vdots \\ \mathbf{a}_{2m}(0) \end{bmatrix}$$

(Thowsen, 1978). Magaria et al. (2001) provide similar upper limits on the numbers of terms for certain non-linear models. The alternative is use a fixed number of terms with the extension theorem (Theorem 6 of Section 3.2.6.1).

The Taylor series expansion approach is demonstrated in examples in Sections 7.2.3.1 to Section 7.2.3.3 below.

7.2.3.1 Two-Compartment Linear Model

For the two-compartmental model described in Section 7.1.1 the first few Taylor series terms are

$$
\begin{aligned}
a_0(0) &= 0, \\
a_1(0) &= 1, \\
a_2(0) &= -\theta_{01} - \theta_{21}, \\
a_3(0) &= \theta_{01}^2 + 2\,\theta_{01}\theta_{21} + \theta_{12}\theta_{21} + \theta_{21}^2, \\
a_4(0) &= -\theta_{01}^3 - 3\theta_{01}^2\theta_{21} - 2\theta_{01}\theta_{12}\theta_{21} - 3\theta_{01}\theta_{21}^2 - \theta_{02}\theta_{12}\theta_{21} - \theta_{12}^2\theta_{21} - \\
&\quad\ 2\theta_{12}\theta_{21}^2 - \theta_{21}^3.
\end{aligned}
$$

To form an exhaustive summary we can exclude the constant terms $a_0(0)$ and $a_1(0)$. Using the rule from Thowsen (1978), as there are two compartments we only need terms up to term $a_4(0)$. Therefore an exhaustive summary consists of terms

$$\kappa = \begin{bmatrix} a_2(0) \\ a_3(0) \\ a_4(0) \end{bmatrix}.$$

Using this exhaustive summary we can also show this model is non-identifiable, with identical results to Section 7.2.1.1. The full details are given in the Maple code.

7.2.3.2 Three-Compartment Linear Model

The three-compartmental model, described in Section 7.1.2, would need Taylor series terms $a_0(0), a_1(0), \ldots, a_6(0)$ using the rule from Thowsen (1978). Instead here we use the extension theorem (Theorem 6 of Section 3.2.6.1) and use one less exhaustive summary term as a result. The terms $a_0(0) = a_1(0) = a_2(0) = 0$ are excluded from the exhaustive summary. Starting with the terms

$a_3(0)$, $a_4(0)$ and $a_5(0)$ we form the exhaustive summary

$$\boldsymbol{\kappa}_1 = \begin{bmatrix} \theta_{21}\theta_{32} \\ -\theta_{32}\theta_{21}\left(\theta_{21} + \theta_{32} + \theta_{02}\right) \\ \theta_{32}\theta_{21}\left(\theta_{02}^2 + \theta_{02}\theta_{21} + 2\theta_{32}\theta_{02} + \theta_{21}^2 + \theta_{21}\theta_{32} + \theta_{32}^2\right) \end{bmatrix},$$

which has parameters $\boldsymbol{\theta}_1 = [\theta_{02}, \theta_{21}, \theta_{32}]$. The derivative matrix

$$\mathbf{D}_1 = \frac{\partial \boldsymbol{\kappa}_1}{\partial \boldsymbol{\theta}_1}$$

is of full rank 3. The extension theorem (Theorem 6 Section 3.2.6.1) involves adding one extra exhaustive summary term $a_6(0)$ to give

$$\boldsymbol{\kappa}_2 = \begin{bmatrix} -\theta_{32}\theta_{21}(\theta_{02}^3 + \ldots + \theta_{32}^3) \end{bmatrix}.$$

However there are no additional parameters. Therefore by a trivial application of the extension theorem the model will always be full rank 3 for any number of additional exhaustive summary terms. As there are three parameters, this model is at least locally identifiable. In the Maple code we also show that the model is locally identifiable by solving the set of equations $a_i(0) = k_i$ for $i = 3, 4, 5$.

In this example, the extension theorem result is obvious, and using the extension theorem uses only one less exhaustive summary term than the rule from Thowsen (1978). However one problem with the Taylor series expansion approach is that each successive exhaustive summary term tends to be more complex than the previous. Reducing the number of exhaustive summary terms needed could be useful in more complex models.

7.2.3.3   Michaelis Menten Kinetics Model

For the non-linear model described in Section 7.1.3 the first four Taylor series terms are

$$
\begin{aligned}
a_0(0) &= 0, \\
a_1(0) &= c, \\
a_2(0) &= \frac{c\theta_1}{\theta_2}, \\
a_3(0) &= \frac{c\theta_1\left(\theta_1 - 2\theta_3\right)}{\theta_2^2}, \\
a_4(0) &= \frac{c\theta_1\left(\theta_1^2 - 8\theta_1\theta_3 + 6\theta_3^2\right)}{\theta_2^3}.
\end{aligned}
$$

We can ignore the zero term $a_0(0)$, then using the extension theorem, we construct an exhaustive summary with terms $a_1(0)$, $a_2(0)$ and $a_3(0)$ giving

$$\boldsymbol{\kappa} = \begin{bmatrix} c \\ \dfrac{c\theta_1}{\theta_2} \\ \dfrac{c\theta_1\left(\theta_1 - 2\theta_3\right)}{\theta_2^2} \end{bmatrix}.$$

As shown in the Maple code for this example, differentiating this exhaustive summary with respect to the four parameters, $\theta = [c, \theta_1, \theta_2, \theta_3]$ forms a derivative matrix, $\mathbf{D}$, with rank 3. As the rank is less than the number of parameters, the model is non-identifiable. A locally identifiable reparameterisation can be found by applying Theorem 5 from Section 3.2.5. Firstly solving $\boldsymbol{\alpha}^T \mathbf{D} = 0$ results in

$$\boldsymbol{\alpha} = [0, \theta_1/\theta_3, \theta_2/\theta_3, 1].$$

This results in the partial differential equation

$$\frac{\partial f}{\partial \theta_1} \frac{\theta_1}{\theta_3} + \frac{\partial f}{\partial \theta_2} \frac{\theta_2}{\theta_3} + \frac{\partial f}{\partial \theta_3} = 0,$$

which can be solved to give estimable parameter combinations c, θ_2/θ_1, θ_3/θ_1. The same method is used in Chappell and Gunn (1998) to obtain a similar reparameterisation. As we can reparameterise the original model in terms of this reparameterisation, the model must be non-identifiable. We can use a trival application the extension theorem to show this formally. We also show in the Maple code that this reparameterisation results in a globally identifiable model.

7.2.4   Other Methods

There are many methods used in compartmental models to check identifiability. In Sections 7.2.1 to 7.2.3 we discussed just three of them. In this section we we mention some of the other methods that can be used to check identifiability in compartmental models.

A direct test can be used, which is similar in concept to the examples in Section 3.4 (see, for example, Chapter 21 of Walter and Contreras, 1999, Denis-Vidal and Joly-Blanchard, 2000). However checking for identifiability using this method can be too complicated to be practical or an analytical solution may not exist (Chis et al., 2011a).

An alternative to the Taylor series approach is the generating series approach, which involves finding the lie derivatives of $y(t)$, rather than the standard derivatives used in the Taylor series approach. These coefficients can be used to form an alternative exhaustive summary (Walter and Lecourtier, 1982). This method can produce a structurally simpler exhaustive summary than the Taylor series approach (Walter and Pronzato, 1996).

Ljung and Glad (1994) used a differential algebra approach. Xia and Moog (2003) proposed a method based on an implicit function theorem. Craciun and Pantea (2008) and Davidescu and Jorgensen (2008) examine identifiability using reaction network theory. These methods are all discussed in Chis et al. (2011a).

An alternative method for finding identifiable parameter combinations, which involves Gröbner Bases, is given in Meshkat et al. (2009). This has the advantage of directly finding whether the identifiable parameter combinations are locally or globally identifiable. Theorem 5 from Section 3.2.5 only finds identifiable parameter combinations that are at least locally identifiable. As demonstrated in Section 7.2.1.1 an extra step is needed to verify global identifiability.

Janzen et al. (2018) and Janzen et al. (2019) develop methods for examining identifiability in the presence of random effects.

7.3 Software

Symbolic algebra packages such as Maple and Mathematica can be used to execute the algebraic methods used to determining identifiability. Alternatively software exists for checking identifiability in compartmental models.

The program DAISY (Bellu et al., 2007) uses a differential algebra algorithm to determine global identifiability. Meshkat et al. (2014) show that in some cases it is possible to find locally identifiable parameter combinations in non-identifiable models. However the program does not always work for complex models (Chis et al., 2011a).

The Matlab toolbox GenSSI (Chis et al., 2011b) uses the generating series approach combined with a method which involves the systematic computation of identifiability tableaus (Balsa-Canto et al., 2010). This program cannot find locally identifiable parameter combinations, although it can find groups of related parameters (Meshkat et al., 2014).

COMBOS is a web app for finding identifiable parameter combinations (Meshkat et al., 2014). For example, Figure 7.3 shows the input and output for the web app for the linear two-compartment model.

7.4 Discussion

Determining whether a compartmental model is identifiable is a challenging problem. This is especially true if there are a large number of compartments or the model is complex and non-linear (Chis et al., 2011a). For complex models, Chis et al. (2011a) compared using several methods and found the best methods to be the Taylor series or the generating series approach combined with identifiability tableaus. The latter is recommended by them as it is available in the software GenSSI. However we note that this program cannot find locally identifiable parameter combinations. An alternative is to use the simplest exhaustive summary and the extended symbolic method described in Section 5.2, as shown in the example in Section 5.2.2.2.

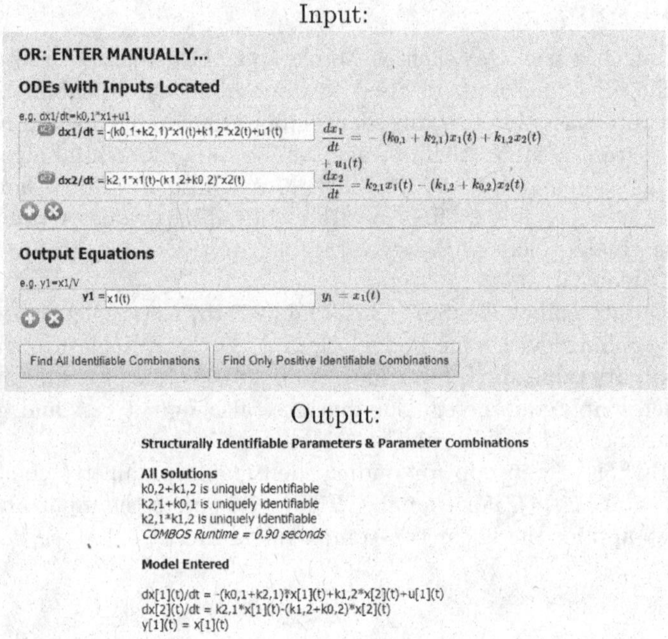
Input:
OR: ENTER MANUALLY...
ODEs with Inputs Located
e.g. dx1/dt=k0,1*x1+u1
dx1/dt = -(k0,1+k2,1)*x1(t)+k1,2*x2(t)+u1(t)
dx2/dt = k2,1*x1(t)-(k1,2+k0,2)*x2(t)
Output Equations
e.g. y1=x1/V
y1 = x1(t)
Find All Identifiable Combinations
Find Only Positive Identifiable Combinations
Output:
Structurally Identifiable Parameters & Parameter Combinations
All Solutions
k0,2+k1,2 is uniquely identifiable
k2,1+k0,1 is uniquely identifiable
k2,1*k1,2 is uniquely identifiable
COMBOS Runtime = 0.90 seconds
Model Entered
dx[1](t)/dt = -(k0,1+k2,1)*x[1](t)+k1,2*x[2](t)+u[1](t)
dx[2](t)/dt = k2,1*x[1](t)-(k1,2+k0,2)*x[2](t)
y[1](t) = x[1](t)

FIGURE 7.3
COMBOS web app input and output for the linear two-compartment example from Section 7.1.1. Note that k is used to represent the parameters rather than θ.

8

Identifiability in Discrete State-Space Models

In Chapter 7 we considered continuous state-space models, where the process is monitored in continuous time. In this chapter we consider discrete state-space models, where there are regular observations at discrete time intervals for a process. For example, one way of monitoring an ecological population is to count the number of animals seen over several time periods. When the population can be split into several states, such as juveniles and adults, this time-series of counts can be modelled using a discrete state-space model.

The chapter starts by introducing discrete state-space models in Section 8.1. We discuss how the Hessian and profile log-likelihood numerical methods can be used with discrete state-space models in Section 8.2. Next, we consider Bayesian methods to detect identifiability in discrete state-space models; Section 8.3 considers the prior and posterior overlap and Section 8.4 considers data cloning. Then in Section 8.5 we explain how to use the symbolic method with discrete-space models. Section 8.6 discusses prediction and identifiability. In Section 8.7 we very briefly discuss links between continuous state-space models and discrete state-space models. Finally in Section 8.8 we discuss a special case of state-space models, hidden Markov models.

8.1 Introduction to Discrete State-Space Models

A discrete state-space model consists of two stochastic processes, that is, processes that vary with time. The first is a state equation that describes the underlying process and how it evolves over time. The second is an observation equation that allows for the fact that not all states are observed or only a proportion of the state is observed.

For a discrete linear state-space model with n states, with m states or combination of states observed ($m \leq n$), the state equation is

$$\mathbf{x}_t = \mathbf{A}_t \mathbf{x}_{t-1} + \boldsymbol{\epsilon}_{t-1}, \, t = 1, 2, 3, \ldots, T$$

and the observation equation is

$$\mathbf{y}_t = \mathbf{C}_t \mathbf{x}_t + \boldsymbol{\eta}_t, \, t = 1, 2, 3, \ldots, T,$$

where $\mathbf{A}_t$ is an $n \times n$ transition matrix, $\mathbf{C}_t$ is an $m \times n$ measurement matrix, $\mathbf{x}_0$ is a vector of initial values, T is the number of observations and $\boldsymbol{\eta}_t$ and $\boldsymbol{\epsilon}_t$ are error processes. Observations are made a regular time points, for example, every year. The initial observation is at time $t = 1$ and a final observation at $t = T$. The error processes $\boldsymbol{\eta}_t$ and $\boldsymbol{\epsilon}_t$ are assumed to follow a suitable multivariate distribution with mean $\mathbf{0}$, for example a multivariate normal distribution (see, for example, McCrea et al., 2010, Cole and McCrea, 2016).

The discrete state-space model can be extended to allow non-linear relationships in the state and observation equations. The state equation is

$$\mathbf{x}_t = g(\mathbf{x}_{t-1}, \boldsymbol{\theta}) + \boldsymbol{\epsilon}_{t-1}, \, t = 1, 2, 3, \ldots$$

and the observation equation is

$$\mathbf{y}_t = h(\mathbf{x}_t, \boldsymbol{\theta}) + \boldsymbol{\eta}_t,$$

where $\boldsymbol{\theta}$ is a vector of parameters, $\boldsymbol{\eta}_t$ and $\boldsymbol{\epsilon}_t$ are error processes and $h(.)$ and $g(.)$ are known functions (see, for example, Newman et al., 2016, Cole and McCrea, 2016).

Below we provide several illustrative examples of discrete state-space models.

8.1.1 Abundance State-Space Model

Abundance data, consisting of counts of particular animals over time, can be modelled by a discrete state-space model to reflect the underlying population dynamics. For example Besbeas et al. (2002) use a discrete state-space model to describe yearly counts of Lapwings (*Vanellus vanellus*). This population of birds is split into two categories: birds in their first year of life (juveniles) and adult birds. Only the adult birds are observed in the count, which is taken yearly.

The state variable is $\mathbf{x}_t = [N_{1,t}, N_{a,t}]^T$, where $N_{1,t}$ is the number of juvenile birds and $N_{a,t}$ is the number of adult birds in year t. The underlying state equation, which describes the process of birth of new birds, survival from one year to the next and the process of moving from a juvenile bird to an adult bird, is

$$\mathbf{x}_t = \begin{bmatrix} 0 & \rho\phi_1 \\ \phi_a & \phi_a \end{bmatrix} \mathbf{x}_{t-1} + \begin{bmatrix} \epsilon_{1,t} \\ \epsilon_{a,t} \end{bmatrix},$$

where ρ denotes the annual productivity rate, ϕ_1 is the probability a bird survives its first year and ϕ_a is the annual survival probability for adult birds. The juvenile and adult error processes are $\epsilon_{1,t}$ and $\epsilon_{a,t}$ respectively.

As only adult birds are observed the observation equation is

$$y_t = \begin{bmatrix} 0 & 1 \end{bmatrix} \mathbf{x}_t + \eta_t.$$

The observation variable, y_t, represents the observed numbers of adult birds at time t.

This linear discrete state-space model has three parameters: ϕ_1, ϕ_a and ρ. Observe that the parameters ϕ_1 and ρ only appear in the state equation as the product $\rho\phi_1$. Therefore these two parameters are confounded and the model is parameter redundant. Parameter redundancy of this model is also discussed in Cole and McCrea (2016).

8.1.2 Multi-Site Abundance State-Space Model

This example extends the abundance state-space model of Section 8.1.1 above to multiple sites, and is based on a model from McCrea et al. (2010). This is also an example of a linear discrete state-space model.

Suppose that there are two sites where animals can be classified as newborn, immature and breeder. The number of newborns is denoted by $x_1(t)$ at site 1 and $x_2(t)$ at site 2. The number of immature animals are then represented by $x_3(t)$ and $x_4(t)$ for sites 1 and 2 respectively. The variables $x_5(t)$ and $x_6(t)$ refer to the number of animals who are breeders at sites 1 and 2 respectively. The transition matrix is

$$
\mathbf{A} = \begin{bmatrix}
0 & 0 & 0 & 0 & \rho_1\phi_{1,1} & 0 \\
0 & 0 & 0 & 0 & 0 & \rho_2\phi_{1,2} \\
\phi_{1,1} & 0 & \phi_{1,1}(1-\pi_1) & 0 & 0 & 0 \\
0 & \phi_{1,2} & 0 & \phi_{1,2}(1-\pi_2) & 0 & 0 \\
0 & 0 & \phi_{1,1}\pi_1 & 0 & \phi_{2,1}(1-\psi_{1,2}) & \phi_{2,2}\psi_{2,1} \\
0 & 0 & 0 & \phi_{1,2}\pi_2 & \phi_{2,1}\psi_{1,2} & \phi_{2,2}(1-\psi_{2,1})
\end{bmatrix},
$$

where $\phi_{1,k}$ is the survival probability of immature animals, $\phi_{2,k}$ is the survival probability of breeders, π_k is the probability an immature individuals move to a breeding state and ρ_k is the fecundity of a breeding animal. Each parameter is dependent on the site k. There is also a transition probability parameter, $\psi_{i,j}$, which is the probability of moving from site i to site j for breeders only. Assuming that animals are only observed when in the breeding state, the observation matrix is

$$
\mathbf{C} = \begin{bmatrix}
0 & 0 & 0 & 0 & 1 & 0 \\
0 & 0 & 0 & 0 & 0 & 1
\end{bmatrix}.
$$

The initial values of the state equation are known values $\mathbf{x}_0 = [x_{0,1}, x_{0,2}, x_{0,3}, x_{0,4}, x_{0,5}, x_{0,6}]^T$.

Cole and McCrea (2016) show that this discrete state-space model is not parameter redundant.

8.1.3 SIR State-Space Model

Epidemics can be described using an SIR state-space model (see, for example, Anderson and May, 1991 or Keeling and Rohani, 2008). The host, the person, animal or plant who may get the disease, can be in one of three stages: susceptible (S), infectious (I) or removed (R). The susceptible stage consists of all hosts who do not have the disease but could potentially be infected and move to the infectious stage. The infectious stage consists of all hosts who have the disease and can potentially infect hosts in the susceptible stage. Those in the infectious stage can move to the removed stage, where individuals have either recovered from the disease and are no longer infectious, or have been quarantined. The model also allows new hosts into the population in the susceptible stage, either through birth or immigration. Hosts can also leave the population at any stage, through death or emigration. Suppose data are collected on the number of hosts infected in discrete time, although not every host infected is reported. A discrete state-space model can then be used to represent this model. More traditionally a continuous state-space model is used to represent a SIR model, see, for example, King et al. (2016). In this chapter we consider the discrete version of the model, but we return to the continuous version in Section 8.7.

Let $x_{1,t}$ be the number of hosts in the susceptible stage, $x_{2,t}$ be the number of hosts in the infectious stage and $x_{3,t}$ be the number of hosts in the recovered stage. In this model the state equations are

$$x_{1,t} = x_{1,t-1} - \beta x_{1,t-1} - \mu_D x_{1,t-1} + \mu_B(x_{1,t-1} + x_{2,t-1} + x_{3,t-1}) + \epsilon_{1,t-1},$$

$$x_{2,t} = x_{2,t-1} + \beta x_{1,t-1} - \mu_D x_{2,t-1} - \gamma x_{2,t-1} + \epsilon_{2,t-1},$$

$$x_{3,t} = x_{3,t-1} + \gamma x_{2,t-1} - \mu_D x_{3,t-1} + \epsilon_{3,t-1},$$

where μ_B is the birth rate (including births and immigration), μ_D is the death rate (including death and emigration), β is the rate of infection and γ is the rate of transition from infectious to removed. A proportion, ρ, of the infections are reported as they occur, so that the observation equation is

$$y_t = \rho \beta x_{1,t} + \eta_t.$$

Initially the population sizes in each stage are $x_{1,0} = S_0$, $x_{2,0} = I_0$ and $x_{3,0} = R_0$

The parameters in this state-space model are

$$\boldsymbol{\theta} = [\beta, \mu_B, \mu_D, \rho, \gamma, S_0, I_0, R_0].$$

Here the model is simplified, in order to simplify symbolic algebra later. In a more realistic model the rate of infection β is likely to depend on the number of infected people in the population, for example β could be replaced by $\beta' x_{2,t}/(x_{1,t} + x_{2,t} + x_{3,t})$.

8.1.4 Stochastic Volatility State-Space Model

Financial data on asset returns can be described using a discrete state-space model to model stochastic volatility. Here we consider the canonical stochastic volatility model discussed in Kim et al. (1998). The state variable, x_t, represents the log volatility at time t. The state equation is

$$x_t = \mu + \phi(x_{t-1} - \mu) + \sigma_\epsilon \epsilon_{t-1}, \ t = 1, \ldots, T$$

where ϵ_t is an error process following a standard normal distribution (i.e., has mean 0 and variance 1). Here the initial state is at time $t = 1$, rather than $t = 0$. This initial state, x_1, is assumed to follow a normal distribution with mean μ and variance $\sigma^2/(1 - \phi^2)$. The parameter μ is a scaling factor, ϕ is the persistence in the volatility, σ_ϵ the volatility of the log-volatility and σ is a parameter specifying the variance of the initial value.

The observation variable, y_t, represents the mean-corrected return on holding the asset at time t. The observation equation is

$$y_t = \beta \exp(x_t/2)\eta_t, \; t = 1, \ldots, T,$$

where η_t is also an error process following a standard normal distribution and β is another scaling factor parameter. It is simpler to transform the observation equation by considering the logarithm of the square of the observations, which gives

$$\log(y_t^2) = 2\log(\beta) + x_t + \log(\eta_t^2)$$

(Kim et al., 1998). As η_t follows a standard normal distribution, $\log(\eta_t^2)$ follows a log chi-squared distribution on 1 degree of freedom, with mean -1.2704 and variance 4.9348.

In this non-linear discrete state-space model the parameters are $\boldsymbol{\theta} = [\beta, \mu, \phi, \sigma, \sigma_\epsilon]$. Kim et al. (1998) state that β must be set to 1 or $\mu = 0$ for identifiability reasons, i.e., this is a non-identifiable model without either of these constraints.

8.1.5 Stochastic Ricker Stock Recruitment SSM

A state-space model can be used to create stochastic version of a Ricker stock recruitment model (see Quinn and Deriso, 1999). For example, the number of juvenile Salmon alive in year t in a certain river can be modelled as

$$N_t|N_{t-1} \sim \text{Poisson}\{\alpha N_{t-1} \exp(-\beta N_{t-1})\}, \; \alpha > 0, \beta > 0,$$

where α represents survival and fecundity, and β represents density dependence. The population size is assumed to be a further parameter with $N_0 = n_0$. The observation process is

$$y_t|N_t \sim \text{lognormal}\{\log(N_t) - \sigma^2/2, \sigma^2\}$$

(Newman et al., 2016). Here σ^2 is assumed to be a fixed known value, however this could also be considered as a parameter.

This is another example of a non-linear discrete state-space model. The parameters in this model are $\boldsymbol{\theta} = [\alpha, \beta, n_0]$.

8.2   Hessian and Profile Log-Likelihood Methods

In discrete state-space models, classical statistical inference involves integration over the states to form the marginal distributions of the observations, and hence obtain the likelihood (see, for example, Newman et al., 2016, Section 4.3). However this integral is rarely tractable, so a numerical approximation or fitting algorithm is required.

As the Hessian method described in Section 4.1.1 requires an approximation of the Hessian matrix, this method only works if the method used to find maximum likelihood estimates also provides an approximation of the Hessian. Similarly the profile log-likelihood method described in Section 4.1.4 will only work if the method used to find maximum likelihood estimates can also be used to obtain the profile log-likelihood. One possible method that can be used with the Hessian and profile log-likelihood methods is the Kalman filter (see, for example, Newman et al., 2016, Section 4.4). This is illustrated in the example of Section 8.2.1 below.

Commonly a Bayesian approach is used as a alternative method to fit a discrete state-space model. In such cases, the Hessian and profile log-likelihood methods cannot be used to check for identifiability. The alternative would be to use the Bayesian methods described in Sections 8.3 and 8.4.

An alternative numerical method for checking identifiability is simulation; see Section 4.1.2. Auger-Méthé et al. (2016) use simulation to check identifiablity in state-space models.

8.2.1   Stochastic Ricker Stock Recruitment SSM

The Kalman filter is used to find maximum likelihood estimates for a set of simulated data in the R code, `ricker.R`. The Hessian method gives standardised eigenvalues of 1.000, 0.029 and 0.002, using a precision value of $\delta = 0.00001$. As the smallest eigenvalue is larger than the threshold of $\tau = q\delta = 3 \times 0.00001 = 0.00003$ the model is shown to not be parameter redundant.

The same R code also plots profile log-likelihoods for each of the parameters, which is shown in Figure 8.1. As all three profile log-likelihoods have a single maximum value, the model is again shown to not be parameter redundant. Note that in this example it was necessary to be careful when specifying upper and lower limits for the parameters in the profile log-likelihood, as the system becomes chaotic for certain parameter values.

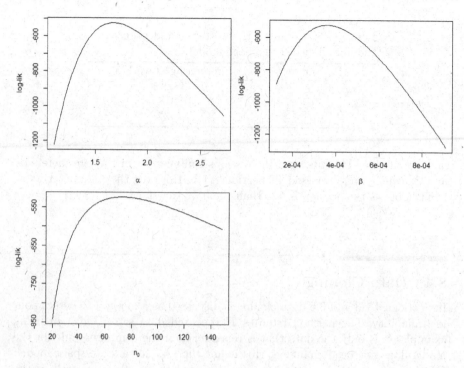

FIGURE 8.1

Profile log-likelihoods for stochastic Ricker stock recruitment state-space model.

8.3 Prior and Posterior Overlap

If a Bayesian method is used to fit a discrete state-space model, then the overlap of the prior and posterior can be examined to investigate Bayesian weak identifiability, as explained in Section 6.4. This method is demonstrated using the abundance state-space model in Section 8.3.1 below.

8.3.1 Abundance State-Space Model

The abundance state-space model, described in Section 8.1.1, is used to illustrate the Bayesian method of fitting a discrete state-space model, and to examine the overlap between the prior and posterior. This is executed in the R code abundance.R using the R package R2winbugs to call Winbugs to perform the MCMC (Sturtz et al., 2005). The percentage overlap for the three parameters is given in Table 8.1. The parameters ϕ_1 and ρ have a percentage overlap more than the threshold of 35%, suggesting these parameters are

TABLE 8.1

Percentage overlap of the posterior and prior for
the abundance state-space model.

Parameter	Percentage Overlap
ϕ_1	38.2%
ϕ_a	8.0%
ρ	49.2%

not identifiable. This fits with the obvious confounding of the parameters ϕ_1
and ρ, which was discussed in Section 8.1.1. However the parameter ϕ_a is
identifiable as the overlap is less than the 35% threshold.

8.4 Data Cloning

In Sections 4.1.3 and 6.5 data cloning is discussed as a method to detect non-
identifiability. The method requires K repeated data sets. Whilst in many
examples it is only the data that is repeated, and the computer code for the
model does not need changing, this is not the case for a state-space model.
In state-space models, the state-variables need to be cloned as well as the
observations. This adds an extra dimension, or extra iteration to the computer code. Not only does the computer code for the model need editing, the
extra dimension, which depends on the number of clones, K, will increase the
execution time of the method considerably. This is demonstrated in Section
8.4.1 below.

8.4.1 Abundance State-Space Model

The R code `abundance.R` also demonstrates data cloning for the abundance
state-space model, described in Section 8.1.1. In particular the Winbugs code
involves an extra iteration over each of K data clones.

Figure 8.4.1 shows a plot of the standardised variance for each of the three
parameters for $K = 1, 20, 40, 80$ and 100 clones of the data. For an identifiable
parameter we expect the variance to decrease at a rate of $1/K$. Although
this plot correctly shows the model is non-identifiable, it suggests that the
parameter ρ is identifiable, when it is confounded with ϕ_1. The plot correctly
suggests that ϕ_1 is non-identifiable and ϕ_a is identifiable.

This plot was creating using a burn-in of 10 000 samples and 10 000 further
samples. Increasing the number of samples considerably slows down computation, and still does not produce a figure where the standardised variance
behaves as expected, for a model where two parameters are confounded.

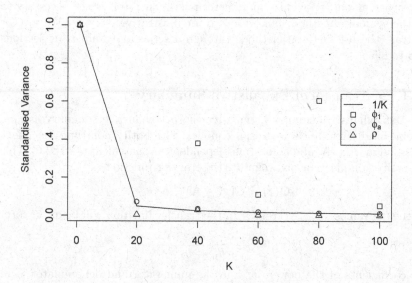

FIGURE 8.2

Plot showing data cloning for the abundance state-space model, showing the standardised variance as the number of clones, K, is increased. The solid line shows $1/K$, the line an identifiable parameter should follow.

8.5 Symbolic Method with Discrete State-Space Models

Parameter estimation is not straightforward in state-space models. The joint distribution of the states and observations can usually be expressed in an explicit form. However, maximum likelihood estimation requires the marginal distribution of the observations, which a high-dimensional integral, for which there is usually no explicit solution. Therefore the exhaustive summaries discussed in Section 3.1.4 do not apply. To use the symbolic method, exhaustive summaries for discrete state-space models are required.

Cole and McCrea (2016) provide two such exhaustive summaries that can be used for discrete state-space models. The first is the Z-transform exhaustive summary, which is described in Section 8.5.1. This is only suitable for linear discrete state-space models, with time invariant parameters, and no parameters in the variance component of the model. The second is the expansion exhaustive summary, which is described in Section 8.5.2. This method is more general, suitable for both linear and non-linear discrete state-space models and can include parameters in the variance component. Both methods are illustrated using the abundance model from Section 8.1.1. The abundance discrete state-space model is obviously parameter redundant, so it is not necessary to use the symbolic method, but we do so for illustrative purpose. We further illustrate the use of these methods through a series of examples in Sections 8.5.3 to 8.5.5.

8.5.1   Z-Transform Exhaustive Summary

The first exhaustive summary, the z-transform exhaustive summary, is only suitable for discrete linear state-space models with both a non-time dependent transition matrix, $\mathbf{A}$, and a non-time dependent measurement matrix, $\mathbf{C}$. This exhaustive summary involves finding the transfer function

$$\mathbf{Q}(z) = \mathbf{C}(z\mathbf{I} - \mathbf{A})^{-1}\mathbf{A}\mathbf{x}_0.$$

For the ith values in the vector $\mathbf{Q}(z)$, the transfer function will be of the form

$$Q(z)_i = \frac{a_0 + a_1 z + \dots}{b_0 + b_1 z + b_2 z^2 + \dots}.$$

The coefficients of the powers of z in the numerator and denominator form the exhaustive summary

$$\boldsymbol{\kappa}_i = \begin{bmatrix} a_0 \\ a_1 \\ \vdots \\ b_0 \\ b_1 \\ b_2 \\ \vdots \end{bmatrix}.$$

This is the discrete analogue of the Laplace transform approach used in continuous state-space models, which is discussed in Section 7.2.1.

For the abundance discrete state-space model described in Section 8.1.1 the matrices and vector needed to form the transfer function are

$$\mathbf{C} = \begin{bmatrix} 0 & 1 \end{bmatrix}, \ \mathbf{A} = \begin{bmatrix} 0 & \rho\phi_1 \\ \phi_a & \phi_a \end{bmatrix} \ \text{and} \ \mathbf{x}_0 = \begin{bmatrix} x_{0,1} \\ x_{0,2} \end{bmatrix}.$$

The transfer function is then

$$Q(z) = \mathbf{C}(z\mathbf{I} - \mathbf{A})^{-1}\mathbf{A}\mathbf{x}_0 = \frac{-\rho\phi_1\phi_a x_{0,2} - z\phi_a(x_{0,1} + x_{0,2})}{\rho\phi_1\phi_a + z\phi_a - z^2}.$$

The coefficients of z in the numerator and denominator form the exhaustive summary

$$\boldsymbol{\kappa} = \begin{bmatrix} -\rho\phi_1\phi_a x_{0,2} \\ -\phi_a(x_{0,1} + x_{0,2}) \\ \rho\phi_1\phi_a \\ \phi_a \end{bmatrix}. \tag{8.1}$$

The coefficient of z^2 in the denominator has been excluded from the exhaustive summary as it is a constant. The parameters are $\boldsymbol{\theta} = [\phi_1, \phi_a, \rho]$. The derivative matrix is then

$$\mathbf{D} = \frac{\partial \boldsymbol{\kappa}}{\partial \boldsymbol{\theta}} = \begin{bmatrix} -\rho\phi_a x_{0,2} & 0 & \rho\phi_a & 0 \\ -\rho\phi_1 x_{0,2} & -x_{0,1} - x_{0,2} & \rho\phi_1 & 1 \\ -\phi_1\phi_a x_{0,2} & 0 & \phi_1\phi_a & 0 \end{bmatrix}.$$

As the derivative matrix has rank 2, but there are three parameters, this model is parameter redundant with deficiency 1. Solving the equation $\boldsymbol{\alpha}^T \mathbf{D} = \mathbf{0}$ gives $\boldsymbol{\alpha}^T = [-\phi_1/\rho, 0, 1]$. The position of the zero indicates that the second parameter ϕ_a is individually identifiable. Solving the partial differential equation

$$-\frac{\partial f}{\partial \phi_1} \times \frac{\phi_1}{\rho} + \frac{\partial f}{\partial \rho} = 0$$

gives $\phi_1\rho$ as the other estimable parameter (Cole and McCrea, 2016). This is demonstrated in the Maple code `abundance.mw`.

8.5.2 Expansion Exhaustive Summary

The second exhaustive summary is the expansion exhaustive summary (Cole and McCrea 2016). For linear state-space models the exhaustive summary is

$$\boldsymbol{\kappa} = \begin{bmatrix} \mathbf{C}_1\mathbf{A}_1\mathbf{x}_0 \\ \mathbf{C}_2\mathbf{A}_2\mathbf{A}_1\mathbf{x}_0 \\ \mathbf{C}_3\mathbf{A}_3\mathbf{A}_2\mathbf{A}_1\mathbf{x}_0 \\ \vdots \end{bmatrix}$$

and for non-linear state-space models the exhaustive summary is

$$\boldsymbol{\kappa} = \begin{bmatrix} h\{g(\mathbf{x}_0)\} \\ h[g\{g(\mathbf{x}_0)\}] \\ h(g[g\{g(\mathbf{x}_0)\}]) \\ \vdots \end{bmatrix} = \begin{bmatrix} h\{g(\mathbf{x}_0)\} \\ h\{g^2(\mathbf{x}_0)\} \\ h\{g^3(\mathbf{x}_0)\} \\ \vdots \end{bmatrix}.$$

For the abundance discrete state-space model described in Section 8.1.1 the expansion exhaustive summary is

$$\boldsymbol{\kappa} = \begin{bmatrix} \mathbf{C}_1\mathbf{A}_1\mathbf{x}_0 \\ \mathbf{C}_2\mathbf{A}_2\mathbf{A}_1\mathbf{x}_0 \\ \mathbf{C}_3\mathbf{A}_3\mathbf{A}_2\mathbf{A}_1\mathbf{x}_0 \\ \vdots \end{bmatrix} = \begin{bmatrix} x_{0,1}\phi_a + x_{0,2}\phi_a \\ x_{0,2}\phi_1\phi_a\rho + x_{0,1}\phi_a^2 + x_{0,2}\phi_a^2 \\ x_{0,1}\phi_1\phi_a^2\rho + 2x_{0,2}\phi_1\phi_a^2\rho + x_{0,1}\phi_a^3 + x_{0,2}\phi_a^3 \\ \vdots \end{bmatrix}.$$

However this exhaustive summary does not necessarily have a fixed number of terms. Maple cannot form a derivative matrix and calculate the rank without a fixed number of terms. There are several possible options. If there is only a small number of time points then the exhaustive summary can stop with the term at the last time point. Cole and McCrea (2016) provide a theorem for a fixed number of terms needed in certain situations. Alternatively a fixed number of terms can be used and then the extension theorem applied (Theorem 6 of Section 3.2.6.1). Using the extension theorem is generally the best option, as it typically involves simpler derivative matrices and it is applicable in all cases.

Returning to the abundance discrete state-space model the first two terms of the expansion are

$$\kappa_1 = \left[\begin{array}{c} x_{0,1}\phi_a + x_{0,2}\phi_a \\ x_{0,2}\phi_1\phi_a\rho + x_{0,1}\phi_a^2 + x_{0,2}\phi_a^2 \end{array} \right].$$

The derivative matrix,

$$\mathbf{D} = \frac{\partial \kappa}{\partial \theta} = \left[\begin{array}{cc} 0 & \rho\,\phi_a x_{0,2} \\ x_{0,1} + x_{0,2} & \rho\,\phi_1 x_{0,2} + 2\,\phi_a x_{0,1} + 2\,\phi_a x_{0,2} \\ 0 & \phi_1\phi_a x_{0,2} \end{array} \right],$$

has rank 2. As shown previously this model is then parameter redundant and the estimable parameter combinations can be shown to be ϕ_a and $\phi_1\rho$. We can reparameterise the original model in terms of these estimable parameter combinations, so can then apply the extension theorem to the reparameterised model. As no additional parameters are added to the model when the next expansion term $(x_{0,2}\phi_1\phi_a\rho + x_{0,1}\phi_a^2 + x_{0,2}\phi_a^2 x_{0,1}\phi_1\phi_a^2\rho + 2x_{0,2}\phi_1\phi_a^2\rho + x_{0,1}\phi_a^3 + x_{0,2}\phi_a^3)$ is added, by a trivial application of the extension theorem (Remark 1, Section 3.2.6.1) this model is parameter redundant with deficiency 1. This is also demonstrated in the Maple code `abundance.mw`.

8.5.3   Stochastic Ricker Stock Recruitment SSM

The stochastic Ricker stock recruitment model is non-linear therefore the expansion method needs to be used to investigate identifiability of this model.

We can expand the expectation assuming that at each stage the value of x_{t-1} is fixed to the expectation at the previous stage. This provides an approximation to the actual expectation that is sufficient for examining identifiability. This gives

$$
\begin{aligned}
E(x_1|x_0) &= \alpha x_0 \exp(-\beta x_0) = \alpha n_0 \exp(-\beta n_0) \\
E\{x_2|x_1 = E(x_1|x_0)\} &= \alpha x_1 \exp(-\beta x_1) \\
&= \alpha^2 n_0 \exp(-\beta n_0) \exp\{-\beta\alpha n_0 \exp(-\beta n_0)\}
\end{aligned}
$$

If there were terms that only appeared in the variance component we would need to repeat the same expansion of the variance. The first three terms of the expansion gives the exhaustive summary,

$$\kappa = \begin{bmatrix} \alpha n_0 \exp(-\beta n_0) \\ \alpha^2 n_0 \exp(-\beta n_0) \exp\{-\beta \alpha n_0 \exp(-\beta n_0)\} \\ \alpha^3 n_0 \exp(-\beta n_0) \exp\{-\beta \alpha n_0 \exp(-\beta n_0)\} \exp[-\beta \alpha^2 n_0 \exp(-\beta n_0) \times \\ \exp\{-\beta \alpha n_0 \exp(-\beta n_0)\}] \end{bmatrix}.$$

In the Maple code, `ricker.mw`, the derivative matrix $\mathbf{D} = \partial\kappa/\partial\boldsymbol{\theta}$, is shown to have rank 3. As there are three parameters this demonstrates that this model is not parameter redundant. Note that as adding an extra year of expansion results in no extra parameters, a trivial application of the extension theorem (see Remark 1 in Section 3.2.6.1), shows that this result extends to any number of years of data.

8.5.4 Stochastic Volatility Model State-Space Model

The stochastic volatility model, described in Section 8.1.4 above, has parameters in the variance component of the model, therefore the expansion method needs to be used to investigate identifiability of this model, expanding both the expectation and variance.

Let $y'_t = \log(y_t^2)$. The expectation of y'_t is

$$E(y'_t) = 2\log(\beta) + E(x_t) - 1.2704,$$

where

$$E(x_t) = \mu + \phi\{E(x_{t-1}) - \mu\}.$$

As $x_1 = \mu$, it follows that $E(x_2) = \mu + \phi(\mu - \mu) = \mu$, $E(x_3) = \mu, \ldots$. Therefore $E(y'_t) = 2\log(\beta) + \mu - 1.2704$ for all t. The variance of y'_t is

$$\text{Var}(y'_t) = \text{Var}(x_t) + 4.9348,$$

where

$$\text{Var}(x_t) = \phi\text{Var}(x_{t-1}) + \sigma_\epsilon.$$

As

$$\text{Var}(x_1) = \frac{\sigma}{1 - \phi^2},$$

and

$$\text{Var}(y'_1) = \frac{\sigma}{1 - \phi^2} + 4.9348,$$

it follows that

$$\text{Var}(y'_2) = \frac{\phi\sigma}{1 - \phi^2} + \sigma_\epsilon^2 + 4.9348,$$

$$\text{Var}(y'_3) = \frac{\phi^2\sigma}{1 - \phi^2} + (\phi + 1)\sigma_\epsilon^2 + 4.9348,$$

$$\text{Var}(y'_4) = \frac{\phi^3\sigma}{1 - \phi^2} + (\phi^2 + \phi + 1)\sigma_\epsilon^2 + 4.9348.$$

The exhaustive summary combines the expectation and variance exhaustive summary terms for $t > 1$, ignoring repeated terms, giving

$$
\kappa = \begin{bmatrix}
2\log(\beta) + \mu - 1.2704 \\
\frac{\phi\sigma}{1-\phi^2} + \sigma_\epsilon^2 + 4.9348 \\
\frac{\phi^2\sigma}{1-\phi^2} + (\phi+1)\sigma_\epsilon^2 + 4.9348 \\
\frac{\phi^3\sigma}{1-\phi^2} + (\phi^2+\phi+1)\sigma_\epsilon^2 + 4.9348
\end{bmatrix}.
$$

From the exhaustive summary it is obvious that the model is non-identifiable without a constraint on β or μ, as the parameters only appear as the combination $\{2\log(\beta) + \mu\}$. This confirms what is stated in Kim et al. (1998) regarding identifiability of this model.

In the Maple code for this example, `volatility.mw`, we show that the appropriate derivative matrix has rank 4, but there are five parameters in the model, so the model is non-identifiable. The estimable parameter combinations are $\{2\log(\beta) + \mu\}$, ϕ, σ and σ_ϵ, which shows that the other parameters in the model are identifiable.

8.5.5   Multi-Site Abundance State-Space Model

This example, described in Section 8.1.2, is used to demonstrate how to deal with complex models, that is, when the exhaustive summary is too complex and Maple runs out of memory trying to calculate the rank of the appropriate derivative matrix. The Maple code for this example is `multistateabundance.mw`.

The expansion exhaustive summary has terms

$$
\kappa(\theta) = \begin{bmatrix}
\phi_{1,1}\pi_1 x_{0,3} - \phi_{2,1}\psi_{1,2}x_{0,5} + \phi_{2,2}\psi_{2,1}x_{0,6} + \phi_{2,1}x_{0,5} \\
\phi_{1,2}\pi_2 x_{0,4} + \phi_{2,1}\psi_{1,2}x_{0,5} - \phi_{2,2}\psi_{2,1}x_{0,6} + \phi_{2,2}x_{0,6} \\
-\pi_1^2\phi_{1,1}^2 x_{0,3} - \pi_1\phi_{1,1}\phi_{2,1}\psi_{1,2}x_{0,3} + \ldots + \phi_{2,1}^2 x_{0,5} \\
\vdots
\end{bmatrix},
$$

where the parameters are

$$
\theta = [\pi_1, \pi_2, \phi_{1,1}, \phi_{1,2}, \phi_{2,1}, \phi_{2,2}, \psi_{1,2}, \psi_{2,1}, \rho_1, \rho_2].
$$

Even with only five exhaustive summary terms Maple runs out of memory trying to calculate the rank of an appropriate derivative matrix.

The first option is to use the hybrid symbolic-numerical method (Section 4.3). Using Maple, the rank of the derivative matrix, $\partial\kappa(\theta)/\partial\theta$, is calculated to be 10 at each of five random points in the parameter space. As the original model has 10 parameters the model is not parameter redundant.

The second option is to use the reparameterisation Theorem (Theorem 13 in Section 5.2.2). Using the reparameterisation

$$
\mathbf{s} =
\begin{bmatrix}
s_1 \\
s_2 \\
s_3 \\
s_4 \\
s_5 \\
s_6 \\
s_7 \\
s_8 \\
s_9 \\
s_{10}
\end{bmatrix}
=
\begin{bmatrix}
\phi_{1,1}\pi_1 x_{0,3} + \phi_{2,1}\bar{\psi}_{1,2} x_{0,5} + \phi_{2,2}\psi_{2,1} x_{0,6} \\
\phi_{1,2}\phi_2 x_{0,4} + \phi_{2,1}\psi_{1,2} x_{0,5} + \phi_{2,2}\bar{\psi}_{2,1} x_{0,6} \\
\phi_{1,1}\pi_1(\phi_{1,1} x_{0,1} + \phi_{1,1}\bar{\pi}_1 x_{0,3}) \\
\phi_{1,2}\pi_2(\phi_{1,2} x_{0,2} + \phi_{1,2}\bar{\pi}_2 x_{0,4}) \\
\phi_{2,1}\bar{\psi}_{1,2} \\
\phi_{2,2}\psi_{2,1} \\
\phi_{2,1}\psi_{1,2} \\
\phi_{2,2}\bar{\psi}_{2,1} \\
\phi_{1,1}\pi_1\{\rho_1\phi_{1,1}^2 x_{0,5} + \phi_{1,1}\bar{\pi}_1(\phi_{1,1} x_{0,1} + \phi_{1,1}\bar{\pi}_1 x_{0,3})\} \\
\phi_{1,2}\pi_2\{\rho_2\phi_{1,2}^2 x_{0,6} + \phi_{1,2}\bar{\pi}_2(\phi_{1,2} x_{0,2} + \phi_{1,2}\bar{\pi}_2 x_{0,4})\}
\end{bmatrix},
$$

where $\bar{\theta} = 1 - \theta$, the exhaustive summary can be rewritten as

$$
\boldsymbol{\kappa}(\mathbf{s}) =
\begin{bmatrix}
s_1 \\
s_2 \\
s_1 s_5 + s_2 s_6 + s_3 \\
\vdots
\end{bmatrix}.
$$

The derivative matrix $\mathbf{D}_s = \partial \boldsymbol{\kappa}(\mathbf{s}) \, \partial \mathbf{s}$ is of full rank 10. As $\operatorname{rank}(\partial \mathbf{s}/\partial \boldsymbol{\theta}) = p_s = 10$ by the reparameterisation Theorem the original model has rank 10, again showing the model is not parameter redundant (Cole and McCrea, 2016).

In this example, both the hybrid symbolic-numerical method and the reparameterisation Theorem method give the same information: the model is not parameter redundant. The hybrid symbolic numeric method was much simpler to use. However if the model was parameter redundant, the reparameterisation Theorem method would have also been able to find the estimable parameter combinations and individually identifiable parameters, whereas the extended hybrid symbolic-numerical method (Section 5.3) would be needed to find the estimable parameter combinations.

8.6 Prediction

So far we have considered estimation of the model parameters as the main goal of statistical inference. However, in some state-space models the main goal may be the prediction of the observation process and/or the state process. It is possible that prediction does not always require the model to be identifiable.

In Section 2.1 we demonstrated that maximum likelihood inference for a non-identifiable model results in multiple maximum likelihood estimates. Then in Section 2.2 we showed that when fitted numerically the parameter estimates will converge to one of the maximum likelihood estimates. Suppose that a parameter redundant model with parameters $\boldsymbol{\theta}$ can be reparameterised in terms of a smaller set of parameters, $\mathbf{s}$, so that $\mathbf{s} = g(\boldsymbol{\theta})$, for some function g. Then regardless of which maximum likelihood value of $\hat{\boldsymbol{\theta}}$ is returned from a numerical algorithm, the values of $\hat{\mathbf{s}} = g(\hat{\boldsymbol{\theta}})$ will always be identical. For this reason, if the quantity which we wish to predict can be reparameterised in terms of $\mathbf{s}$, then prediction does not require a model to be identifiable. However, if the quantity which we wish to predict cannot be reparameterised in terms of $\mathbf{s}$, then prediction does require a model to be identifiable. By definition of the state-space model, identifiability is not a requirement for prediction of the observation process, but this will not necessarily be true for the state process, as we demonstrate in the example in Section 8.6.1 below.

8.6.1   SIR State-Space Model

Consider the SIR model described in Section 8.1.3. If we assume the variables follow Poisson distributions, there are no extra parameters in the variance terms, so an exhaustive summary can be formed from an expansion of the expectation alone.

Starting with $E(x_{1,0}) = S_0$, $E(x_{2,0}) = I_0$ and $E(x_{3,0}) = R_0$, we expand the expectation of the state-process to give,

$$
\begin{aligned}
E(x_{1,1}) &= S_0 - \beta S_0 - \mu_D S_0 + \mu_B(S_0 + I_0 + R_0) \\
E(x_{2,1}) &= I_0 + \beta S_0 - \mu_D I_0 - \gamma I_0 \\
E(x_{3,1}) &= R_0 + \gamma I_0 - \mu_D R_0 \\
E(y_1) &= \rho\beta\left\{S_0 - \beta S_0 - \mu_D S_0 + \mu_B(S_0 + I_0 + R_0)\right\} \\
E(x_{1,2}) &= \beta^2 S_0 - \beta I_0 \mu_B - \beta R_0 \mu_B - \beta S_0 \mu_B + \ldots - 2\mu_D S_0 + S_0 \\
E(x_{2,2}) &= -\beta^2 S_0 - \beta\gamma S_0 + \beta I_0 \mu_B + \beta R_0 \mu_B + \ldots - 2\mu_D I_0 + I_0 \\
E(x_{3,2}) &= \beta\gamma S_0 - \gamma^2 I_0 - 2\gamma I_0 \mu_D + R_0 \mu_D^2 + 2\gamma I_0 - 2R_0\mu_D + R_0 \\
E(y_2) &= \rho\beta\left(\beta^2 S_0 - \beta I_0 \mu_B - \beta R_0 \mu_B - \beta S_0 \mu_B + \ldots - 2\,\mu_D S_0 + S_0\right)
\end{aligned}
$$

$$\vdots$$

The values of $E(y_t)$ form the exhaustive summary, $\boldsymbol{\kappa}$. The derivative matrix, $\mathbf{D} = \partial\boldsymbol{\kappa}/\boldsymbol{\theta}$, has rank 4, but there are eight parameters so the model is non-identifiable. The estimable parameter combinations as a vector are

$$
\mathbf{s} = \begin{bmatrix} s_1 \\ s_2 \\ s_3 \\ s_4 \end{bmatrix} = \begin{bmatrix} \beta + \mu_B \\ \beta + \mu_D \\ \rho\beta S_0 \\ -\beta\rho\left\{(I_0 + R_0)\mu_B - \beta S_0\right\} \end{bmatrix}.
$$

Note that the parameter γ does not appear in the estimable parameter combinations, because it does not appear in $E(y_t)$. For this reason it will be impossible to write $E(x_{2,t})$ and $E(x_{3,t})$ in terms of $\mathbf{s}$, so it is not possible to predict these quantities. It is also not possible to write $E(x_{1,t})$ in terms of $\mathbf{s}$ alone, for example

$$E(x_{1,1}) = \frac{s_1 s_3 - s_2 s_3 + s_3 - s_4}{\beta \rho}.$$

Therefore it is not possible to predict $E(x_{1,t})$ either. However it is possible to write $E(y_t)$ in terms of the estimable parameter combinations. For example

$$E(y_1) = s_1 s_3 - s_2 s_3 + s_3 - s_4,$$

$$E(y_2) = s_1^2 s_3 - 2 s_1 s_2 s_3 + s_2^2 s_3 + 2 s_1 s_3 - s_1 s_4 - 2 s_2 s_3 + 2 s_2 s_4 + s_3 - 2 s_4.$$

This shows it is possible to predict $E(y_t)$.

8.7   Links to Continuous State-Space Models

The SIR state-space model, described in Section in 8.1.3, is traditionally formed as a continuous state-space model. This version would be formed from the equations defined by:

$$\begin{cases} \dfrac{\partial x_1(t)}{\partial t} = -x_1(t)(\mu_D + \beta) + \mu_B \{x_1(t) + x_2(t) + x_3(t)\} \\ \dfrac{\partial x_2(t)}{\partial t} = \beta x_1(t) - x_2(t)(\mu_D + \gamma) \\ \dfrac{\partial x_3(t)}{\partial t} = \gamma x_2(t) - \mu_D x_3(t) \end{cases}$$

and

$$y(t) = \rho \beta x_1(t)$$

with $x_1(0) = S_0$, $x_2(0) = I_0$ and $x_3(0) = R_0$. The parameters are $\boldsymbol{\theta} = [\beta, \mu_B, \mu_D, \rho, \gamma, S_0, I_0, R_0]$. We show in the Maple code (`SIRmodel.mw`) that the rank of an appropriate derivative matrix is 4, the same as the discrete version of the model. The estimable parameter combinations are also the same. Given analogous methods can be used to investigate identifiability it is expected (but not proven) that equivalent discrete and continuous state-space model will have the same rank and estimable parameter combinations.

8.8   Hidden Markov Models

A special case of discrete state-space models are hidden Markov models (HMM) (see, for example, Zucchini et al., 2016).

In a HMM we also have a time series of T observations which we represent by $\{X_t : t = 1, 2, \ldots, T\}$. Each X_t depends on a hidden Markov process represented by C_t. This hidden process can be in one of m states at time t. The HMM is defined by the probability of X_t, conditional on C_t, represented by

$$p_i(x) = \Pr(X_t = x | C_t = i)$$

and the transition probability between states represented by

$$\gamma_{ij} = \Pr(C_{t+1} = j | C_t = i),$$

with $\sum_{j=1}^{m} \gamma_{ij} = 1$. We let $\mathbf{P}(x)$ be a diagonal matrix with entries $p_i(x)$ and let $\boldsymbol{\Gamma}$ be the transition matrix whose (ij)th element is γ_{ij}. We let $\boldsymbol{\delta} = [\delta_1, \delta_2, \ldots, \delta_m]$, with $\sum_{j=1}^{m} \delta_j = 1$, represent the initial distribution for the states. The likelihood can then be written as a product of these matrices,

$$L_T = \boldsymbol{\delta}\mathbf{P}(x_1)\boldsymbol{\Gamma}\mathbf{P}(x_2)\boldsymbol{\Gamma}\mathbf{P}(x_3)\ldots\boldsymbol{\Gamma}\mathbf{P}(x_T)\mathbf{1}$$

where $\mathbf{1}$ is a column vector of 1s. If $\boldsymbol{\delta}$ is assumed to be the stationary distribution of the Markov chain then $\boldsymbol{\delta} = \boldsymbol{\delta}\boldsymbol{\Gamma}$, and this can replace $\boldsymbol{\delta}$ in the above (Zucchini et al., 2016).

Generic results on identifiability in HMM have been established by Petrie (1969) and Allman et al. (2009). Cole (2019) investigates identifiability in HMMs. The paper recommends not to use the Hessian method, as this can give the wrong results; this can be due to the threshold used, as well as an approximate Hessian matrix being a poor approximate to the true Hessian matrix (see, Visser et al., 2000). Cole (2019) also examines the log-likelihood profile method and provides an exhaustive summary to use the symbolic method for HMMs. We demonstrate the latter in an example in Section 8.8.1 below.

8.8.1   Poisson HMM

This example is a two-state Poisson HMM, used in Zucchini et al. (2016) for data on the number of of earthquakes. In this model, the transition matrix is

$$\boldsymbol{\Gamma} = \left[\begin{array}{cc} \gamma_{11} & 1 - \gamma_{11} \\ \gamma_{21} & 1 - \gamma_{21} \end{array} \right],$$

the initial state is

$$\boldsymbol{\delta} = [\delta_1, 1 - \delta_1]$$

and

$$p_i(x) = \frac{\exp(-\lambda_i)\lambda_i^x}{x!}.$$

There are four parameters in this model $\boldsymbol{\theta} = [\gamma_{11}, \gamma_{21}, \lambda_1, \lambda_2]$. We assume that this model is stationary. The Maple code for this example is given in `poissonHMM.mw`.

The exhaustive summary is similar to the expansion exhaustive summary in Section 8.5.2. For stationary HMM the exhaustive summary is

$$
\kappa = \begin{bmatrix}
\boldsymbol{\delta\Gamma}\mathbf{P}(x_1)\mathbf{1} \\
\boldsymbol{\delta\Gamma}\mathbf{P}(x_1)\boldsymbol{\Gamma}\mathbf{P}(x_2)\mathbf{1} \\
\boldsymbol{\delta\Gamma}\mathbf{P}(x_1)\boldsymbol{\Gamma}\mathbf{P}(x_2)\boldsymbol{\Gamma}\mathbf{P}(x_3)\mathbf{1} \\
\vdots \\
\boldsymbol{\delta\Gamma}\mathbf{P}(x_1)\boldsymbol{\Gamma}\mathbf{P}(x_2)\boldsymbol{\Gamma}\mathbf{P}(x_3)\dots\boldsymbol{\Gamma}\mathbf{P}(x_T)\mathbf{1}
\end{bmatrix}
$$

(Cole, 2019).

For the two-state Poisson HMM the exhaustive summary is

$$
\kappa = \begin{bmatrix}
\frac{(\delta_1\gamma_{11}+\bar\delta_1\gamma_{21})\exp(-\lambda_1)\lambda_1^{x_1}}{x_1!} + \frac{(\delta_1\bar\gamma_{11}+\bar\delta_1\bar\gamma_{21})\exp(-\lambda_2)\lambda_2^{x_1}}{x_1!} \\[2mm]
\frac{(\delta_1\gamma_{11}+\bar\delta_1\gamma_{21})\gamma_{11}\exp(-2\lambda_1)\lambda_1^{x_1}\lambda_1^{x_2}}{x_1!x_2!} + \frac{(\delta_1\bar\gamma_{11}+\bar\delta_1\bar\gamma_{21})\gamma_{21}\exp(-\lambda_1-\lambda_2)\lambda_2^{x_1}\lambda_1^{x_2}}{x_1!x_2!} + \\[2mm]
\frac{(\delta_1\gamma_{11}+\bar\delta_1\gamma_{21})\bar\gamma_{11}\exp(-\lambda_1-\lambda_2)\lambda_1^{x_1}\lambda_2^{x_2}}{x_1!x_2!} \quad \frac{(\delta_1\bar\gamma_{11}+\bar\delta_1\bar\gamma_{21})\bar\gamma_{21}\exp(-2\lambda_2)\lambda_1^{x_1}\lambda_2^{x_2}}{x_1!x_2!} \\[2mm]
\vdots
\end{bmatrix},
$$

where $\bar\theta = 1 - \theta$.

In the Maple code we show that when $T = 3$ the rank of the derivative matrix is 3, but there are are four parameters so the model is non-identifiable. For $T \geq 4$ the rank of the derivative matrix is 4, so the model is at least locally identifiable.

In this case, it is not possible to use the symbolic method to show that this model is in fact locally not globally identifiable. Cole (2019) use the log-likelihood profile method to show that this example is locally identifiable due to label switching (see Section 2.1.3). The model is globally identifiable if an order constraint is used, for example $\lambda_1 < \lambda_2$.

8.9 Conclusion

The symbolic method is the preferred method for analysing parameter redundancy in a state-space model, as it will give an accurate result, and the estimable parameter combinations can be obtained. Usually the expansion exhaustive summary is used (Section 8.5.2), as the z-transform exhaustive summary (Section 8.5.1) is only applicable for some linear discrete state-space models. However the exhaustive summaries for state-space models can be complex, as demonstrated in Section 8.5.5. In those cases the hybrid symbolic-numerical method (Section 4.3) or the methods discussed in Chapter 5 should be used. The other methods discussed in this chapter are applicable depending on the method used to fit the state-space model. If a classical approach is used, such as the Kalman filter, then the Hessian or profile log-likelihood could be used (see Section 8.2). If a Bayesian approach is followed then the prior and posterior overlap can be examined (Section 8.3) or data cloning can be used (Section 8.4), although all these methods can be inaccurate in their results.

State-space models can also be used as a general framework for other classes of models such as capture-recapture models and occupancy models (see McCrea and Morgan, 2014, Chapter 11). However it does not always follow that a state-space form for a model will be the simplest to examine identifiability. In fact for the symbolic method the exhaustive summaries can often be more complex (see Cole and McCrea, 2016).

9

Detecting Parameter Redundancy in Ecological Models

Due to climate change, changes in habitat and other such pressures on the environment, there is extensive interest in studying wildlife populations. In order to understand these populations, reliable representations of the underlying processes are required, which is achieved using various statistical models. As these models become more realistic they also increase in complexity. It is often not clear whether or not all the parameters can be estimated in these more complex models. This chapter considers parameter redundancy in a variety of different statistical models used in ecology, demonstrating the different techniques that can be used to investigate parameter redundancy in these models.

9.1 Marked Animal Populations

One method of monitoring animal populations is to mark the animals and then recapture live animals or recover dead animals at subsequent points in time. This section demonstrates how the symbolic method and extended symbolic methods can be used to obtain general results in mark-recovery models (Section 9.1.1), capture-recapture models (Section 9.1.2) and capture-recapture-recovery models (Section 9.1.3). Extensions to capture-recapture models include observing animals in different states, such as breeding and non-breeding, or observing animals at different sites. These more complex models are used to illustrate the extended symbolic method in Sections 9.1.4 and 9.1.5.

9.1.1 Mark-Recovery Model

The mark-recovery model was introduced in Section 1.2.4, and specific examples of mark-recovery models are considered in other chapters. Here, we consider a general mark-recovery model, with parameters either constant, time dependent, dependent on age structure or a mixture of both.

The general mark-recovery model has the probability of an animal being marked in year i and recovered in year j of

$$P_{i,j} = \left(\prod_{k=i}^{j-1} \phi_{k-i+1,k}\right)(1 - \phi_{j-i+1,j})\lambda_{j-i+1,j}, \text{ for } i = 1, \ldots, n_1, \ j = i, \ldots, n_2$$

(9.1)

where $\phi_{i,j}$ is the probability of an animal aged i surviving the jth year of the study, $\lambda_{i,j}$ is the probability of recovering a dead animal aged i in the jth year of the study, for n_1 years of marking and n_2 years of recovery. $N_{i,j}$ denotes the number of animals marked in year i, $i = 1, \ldots, n_1$ and recovered in year j, $j = i, \ldots, n_2$, and F_i denotes the number of animals marked in year i. The log-likelihood is

$$l = c + \sum_{i=1}^{n_1} \sum_{j=i}^{n_2} N_{i,j} \log(P_{i,j}) + \sum_{i=1}^{n_1} \left(F_i - \sum_{j=i}^{n_2} N_{i,j}\right) \log\left(1 - \sum_{j=i}^{n_2} P_{i,j}\right),$$

where c is a constant.

Parameter redundancy can be examined with the symbolic method using an exhaustive summary consisting of the $P_{i,j}$ terms. Cole et al. (2012) obtains general results for a wide variety of different mark-recovery models. One method to obtain general results is to use the symbolic method with the extension theorem (see Sections 3.2 and 3.2.6). Examples of using the symbolic methods with the extension theorem for mark-recovery models are given in Section 3.2.6.2.

It is also possible that the number of estimable parameters is limited by the number of exhaustive summary terms. The full model, with both parameters dependent on age and time, has $2n_1n_2 - n_1^2 + n_1$ parameters. For three years of marking and recovery, for example, this is 12 parameters. However the exhaustive summary, consisting of the probabilities $P_{i,j}$, has only $n_1 n_2 - \frac{1}{2}n_1^2 + \frac{1}{2}n_1$ terms. When there are three years of marking and recovery this is six terms. It is obvious when there are more parameters than there are exhaustive summary terms that the model will be parameter redundant. In general, the number of unique exhaustive summary terms can be the limiting factor in how many parameters can be estimated. The estimable parameter combinations in this case will be the exhaustive summary terms. If all distinct $P_{i,j}$ contain a parameter not in any other distinct $P_{i,j}$, then the rank is equal to the number of distinct $P_{i,j}$ (Cole et al. 2012).

For example, consider the model with age-dependent survival, ϕ_i, and age-dependent recovery, λ_i. When there are $n_1 = 3$ years of marking and $n_2 = 3$ years of recovery, the exhaustive summary is

$$\kappa = \begin{bmatrix} (1 - \phi_1)\lambda_1 \\ \phi_1(1 - \phi_2)\lambda_2 \\ \phi_1\phi_2(1 - \phi_3)\lambda_3 \end{bmatrix}.$$

Each exhaustive summary term contains a different parameter, λ_i. The number of distinct exhaustive summary terms limits the rank; therefore the rank of this model is 3. This can be extended to n_1 years of marking and n_2 years of recovery, where there are n_2 distinct terms. The model then has rank n_2. As there are $2n_2$ parameters, the model is always parameter redundant with deficiency $2n_2 - n_2 = n_2$ (Cole et al. 2012).

These methods are utilised in Cole et al. (2012) to obtain general tables of results for different mark-recovery models. Most commonly there will be the same number of years of marking and recovery, so that $n_1 = n_2 = n$. Table 9.1 provides general results for mark-recovery models with $n_1 = n_2 = n$. Note that all models in which one parameter is dependent on both age and time are parameter redundant, so are not included in the table, but are given in Cole et al. (2012).

The standard mark-recovery model can only be used to model age-dependence when the animals are of a known age when marked. McCrea et al. (2013) extends the mark-recovery model to allow age dependent parameters when the animals' ages are unknown. A general table of parameter redundancy results is given in the appendix of McCrea et al. (2013).

This section demonstrates how a table of general results can be generated for any number of years of marking and recovery. The advantage of creating a table of results is that anyone utilising the model can use the table as a reference to determine whether the model is parameter redundant without the need for further calculation.

9.1.2 Capture-Recapture Model

Rather than recovering dead animals, marked animals could be recaptured or resighted whilst they are still alive. Data collection involves marking animals, if they are not already uniquely distinguishable from one another, and then subsequently recapturing live animals. The Cormack Jolly Seber (CJS) model (Cormack 1964, Jolly 1965 and Seber 1965) is a capture-recapture model which can be used to estimate the probability of survival. It has been widely applied to a variety of contexts (for example, see Lebreton et al. 1992). The model assumes that not all animals are recaptured, therefore it also has parameters representing the probability of recapture as well as parameters for the probability of survival from one time period to the next. Parameters could be dependent on time or age of the animal. This model was discussed in Section 5.2.2.1.

Suppose that animals are marked over n_1 occasions and recaptured at n_2 subsequent occasions. Typically, marked and unmarked animals are caught on the same occasion, unmarked animals are marked, and all animals are released. The first marking occasion involves only capture and no recapture. For this standard set up there will be N capture and recapture occasions with $n_1 = n_2 = N - 1$. Data is normally aggregated so that an occasion represents one year, but other time periods are also possible. One way of representing the data is through capture histories for individual animals, which is typically represented as a string of 0s and 1s. An entry of 1 represents an occasion when

TABLE 9.1

General results for mark-recovery model for n years of marking and recovery. The first column gives the model parameters. Subscripts on parameters denote the parameter dependence with: no subscript = constant, t = time dependent, h = fully age dependent, 1 = 1st year, a = adult (> one year old). The second and third columns, r and d, denote the rank and deficiency of a model. The column labelled min n gives the minimum values of n for which the results are valid. The last column gives the estimable parameter combinations (Est. Par. Comb.). Estimable parameter combinations is blank if $d = 0$, as every parameter can be estimated. In some cases the estimable parameter combinations are the original exhaustive summary terms, indicated by κ. Full results are given in Cole et al. (2012).

Model	r	d	min n	Est. Par. Comb.
ϕ, λ	2	0	2	
ϕ, λ_t	$n+1$	0	2	
ϕ, λ_h	n	1	2	κ
ϕ_t, λ	$n+1$	0	2	
ϕ_t, λ_t	$2n-1$	1	2	$\lambda_1, \ldots, \lambda_{n-1}, \phi_1, \ldots, \phi_{n-1}, (1-\phi_n)\lambda_n$
ϕ_t, λ_h	$2n$	0	3	
ϕ_h, λ	n	1	2	κ
ϕ_h, λ_t	$2n-1$	1	2	κ
ϕ_h, λ_h	n	n	2	κ
ϕ_1, ϕ_a, λ	3	0	3	
$\phi_1, \phi_a, \lambda_t$	$n+2$	0	3	
$\phi_1, \phi_a, \lambda_h$	n	2	2	κ
$\phi_1, \phi_a \lambda_1, \lambda_a$	3	1	3	$\phi_a, \phi_1\lambda_a, \bar{\phi}_1\lambda_1$
$\phi_1, \phi_{a,t}, \lambda$	$n+1$	0	3	
$\phi_1, \phi_{a,t}, \lambda_t$	$2n$	0	3	
$\phi_1, \phi_{a,t}, \lambda_h$	$2n-1$	1	4	$\phi_{a,2}, \ldots, \phi_{a,n}, \bar{\phi}_1\lambda_1$ $\phi_1\lambda_i$ for $i = 2, \ldots, n$
$\phi_1, \phi_{a,t} \lambda_1, \lambda_a$	$n+1$	1	3	$\phi_{a,2}, \ldots, \phi_{a,n}, \phi_1\lambda_a, \bar{\phi}_1\lambda_1$
$\phi_{1,t}, \phi_a, \lambda$	$n+2$	0	3	
$\phi_{1,t}, \phi_a, \lambda_t$	$2n+1$	0	4	
$\phi_{1,t}, \phi_a, \lambda_h$	$2n$	1	3	$\phi_{1,1}, \ldots, \phi_{1,n}, \lambda_1,$ $\phi_a^{j-2}(1-\phi_a)\lambda_j$ for $j = 2, \ldots, n$
$\phi_{1,t}, \phi_a, \lambda_1, \lambda_a$	$n+3$	0	3	
$\phi_{1,t}, \phi_{a,t}, \lambda$	$2n$	0	3	
$\phi_{1,t}, \phi_{a,t}, \lambda_t$	$3n-3$	2	2	κ
$\phi_{1,t}, \phi_{a,t}, \lambda_h$	$3n-1$	0	5	
$\phi_{1,t}, \phi_{a,t}, \lambda_1, \lambda_a$	$2n+1$	0	4	
$\phi_{1,t}, \phi_{a,h}, \lambda$	$2n$	0	3	
$\phi_{1,t}, \phi_{a,h}, \lambda_t$	$3n-1$	0	5	
$\phi_{1,t}, \phi_{a,h}, \lambda_h$	$2n$	$n-1$	3	$\phi_{1,1}, \ldots, \phi_{1,n}, \lambda_1,$ $(1-\phi_j)\lambda_j \prod_{k=2}^{j-1} \phi_k$ for $j = 2, \ldots, n$
$\phi_{1,t}, \phi_{a,h}, \lambda_1, \lambda_a$	$2n$	1	3	$\phi_{1,1}, \ldots, \phi_{1,n}, \lambda_1,$ $(1-\phi_j)\lambda_a \prod_{k=2}^{j-1} \phi_k$ for $j = 2, \ldots, n$

an animal was captured and an entry of 0 represents an occasion where the animal was not recaptured. For example

$$h_1 = 01010$$

is a history for an individual first caught at occasion 2, then not recaptured at occasion 3, then recaptured at occasion 4 and not recaptured at occasion 5. At this last time point the animal could either have died or have not been recaptured, whereas at occasion 3 the animal was still alive, but not recaptured.

The probability that an animal of age i at time j survives until $j + 1$ is denoted by the parameter $\phi_{i,j}$. The probability that an animal of age i is recaptured at occasion j is $p_{i,j}$. If an animal was first recaptured, in its first year of life, at time a and was last recaptured at time b, with individual capture history entry δ_k at time k, then the probability a particular history, h, occurred is

$$\Pr(h) = \left\{ \prod_{k=a+1}^{b} \phi_{k-a,k-1} \left(\delta_k p_{k-a+1,k} + \bar{\delta}_k \bar{p}_{k-a+1,k} \right) \right\} \chi_{b-a+1,b}, \qquad (9.2)$$

where $\bar{\theta} = 1 - \theta$ and $\chi_{i,j} = \bar{\phi}_{i,j} + \phi_{i,j} \bar{p}_{i+1,j+1} \chi_{i+1,j+1}$ with $\chi_{i,n_2+1} = 1$ for all i (see, for example, Hubbard et al., 2014). In history h_1 above, $a = 2$, $b = 4$, $\delta_3 = 0$ and $\delta_4 = 1$. Therefore the probability of the history h_1 occurring is

$$\Pr(h_1) = \phi_{1,2} \bar{p}_{2,3} \phi_{2,3} p_{3,4} (\bar{\phi}_{3,4} + \phi_{3,4} \bar{p}_{4,5}).$$

If there are n individual histories the likelihood is

$$L = \prod_{m=1}^{n} \Pr(h_m).$$

Note this model is based on the assumption that animals are in their first year of life when first caught and marked. Frequently this is not the case; in which case the age dependence is dropped.

A possible exhaustive summary is the probability of each capture history, though this is not the simplest exhaustive summary. Hubbard et al. (2014) present an exhaustive summary for both time and age dependent models that consists of the terms:

- $s_{i,j} = \phi_{i,j} p_{i+1,j+1}$ for all $i = 1, \ldots, n_2$ and $j = i, \ldots, \min(n_1 + i - 1, n_2)$;

- $t_{i,j} = \phi_{i,j}(1 - p_{i+1,j+1})$ for all $i = 1, \ldots, n_2 - 1$ and $j = i, \ldots, \min(n_1 + i - 1, n_2 - 1)$.

When both parameters are time-dependent this exhaustive summary is the same as the reduced-form exhaustive summary found in Section 5.2.2.1.

For example, suppose that survival was time dependent, ϕ_t, and the recapture probability is constant, p. This example is given in the Maple code

CJS.mw. Consider the case with $n_1 = 2$ years of marking and $n_2 = 2$ years of recapture. The exhaustive summary has the terms

$$\kappa_1 = \left[\begin{array}{c} \phi_1 p \\ \phi_2 p \\ \phi_1 (1 - p) \end{array} \right].$$

Note that any repeated terms only appear once. The parameters are $\theta_1 = [\phi_1, \phi_2, p]$. The derivative matrix is then

$$\mathbf{D}_1 = \frac{\partial \kappa}{\partial \theta} = \left[\begin{array}{ccc} p & 0 & 1 - p \\ 0 & p & 0 \\ \phi_1 & \phi_2 & -\phi_1 \end{array} \right].$$

The rank of this derivative matrix is shown by the Maple code to be 3; the same as the number of parameters, therefore the model is not parameter redundant. Adding an extra year of recapture adds extra exhaustive summary terms

$$\kappa_2 = \left[\begin{array}{c} \phi_3 p \\ \phi_2 (1 - p) \end{array} \right],$$

and extra parameters $\theta_2 = [\phi_3]$. As we are adding only an extra parameter by a trivial application of the extension theorem (Theorem 6) the extended model will always be full rank (see Remark 1 of Section 3.2.6.1). (Adding an extra year of marking adds no new exhaustive summary terms and no extra parameters.) Therefore the general model with n_1 years of marking and n_2 years of recapture has rank $n_2 + 1$ for $n_1 \geq 2$ and $n_2 \geq 2$ (Hubbard et al., 2014).

The advantage of this simple exhaustive summary is that the derivative matrix is structurally very simple, so Maple can find its rank easily. It is also in a format that allows a straightforward application of the extension theorem (Theorem 6, Section 3.2.6.1) to extend results to any number of years of capture and recapture. Tables of general results for a wide range of capture-recapture models are given in Hubbard et al. (2014).

9.1.3   Capture-Recapture-Recovery Model

Some studies will record information on both live captures, recaptures and dead recoveries of marked animals (see for example Catchpole et al., 1998b or King and Brooks, 2003). The capture histories are extended to include dead recovery, which is represented by a 2. There can only be one dead recovery, and once an animal is dead it cannot be recaptured, therefore a 2 is always followed by a string of 0s for any remaining occasions. For example the history

$$h_2 = 101200$$

represents an individual first caught at occasion 1, then not recaptured at occasion 2, recaptured at occasion 3 and then recovered dead at occasion 4.

The parameter $\lambda_{i,j}$ denotes the probability that an animal of age i at time j who died between times j and $j+1$ is recovered. Other parameters are identical to the capture-recapture model in Section 9.1.2 above. If an animal, in its first year of life, is first recaptured at time a and is last recaptured, either dead or alive at time b, with individual capture history entry δ_k at time k, then the probability a particular history occurs is

$$
\Pr(h) = \begin{cases}
\displaystyle\prod_{k=a+1}^{b} \phi_{k-a,k-1}\big(\delta_k p_{k-a+1,k}+\bar{\delta}_k \bar{p}_{k-a+1,k}\big)\chi_{b-a+1,b} & \text{if } \delta_b=1 \\[2em]
\displaystyle\prod_{k=a+1}^{b-1} \phi_{k-a,k-1}\big(\delta_k p_{k-a+1,k}+\bar{\delta}_k \bar{p}_{k-a+1,k}\big)\bar{\phi}_{b-a,b-1}\lambda_{b-a,b-1} & \text{if } \delta_b=2,
\end{cases}
$$

where $\chi_{i,j} = \bar{\phi}_{i,j}\bar{\lambda}_{i,j} + \phi_{i,j}\bar{p}_{i+1,j+1}\chi_{i+1,j+1}$ represents the probability that an animal of age i at time j is not recaptured again, with $\chi_{i,n_2+1} = 1$ for all i (see, for example, Hubbard et al., 2014). From above, the history h_2 has $a = 1$, $b = 4$, $\delta_1 = 1$, $\delta_2 = 0$, $\delta_3 = 1$ and $\delta_4 = 2$. The probability is then

$$
\Pr(h_2) = \phi_{1,1}\bar{p}_{2,2}\phi_{2,2}p_{3,3}\bar{\phi}_{3,3}\lambda_{3,3}.
$$

As with the capture-recapture model the likelihood for n individual histories is

$$
L = \prod_{m=1}^{n} \Pr(h_m).
$$

Hubbard et al. (2014) present a simpler exhaustive summary for a general capture-recapture-recovery model that consists of terms

- $s_{i,j} = \phi_{i,j}p_{i+1,j+1}$ for all $i = 1,\ldots,n_2$ and $j = i,\ldots,\min(n_1+i-1,n_2)$,

- $t_{i,j} = \phi_{i,j}\bar{p}_{i+1,j+1}$ for all $i = 1,\ldots,n_2-1$ and $j = i,\ldots,\min(n_1+i-1,n_2-1)$,

- $r_{i,j} = (1 - \phi_{i,j})\lambda_{i,j}$ for all $i = 1,\ldots,n_2$ and $j = i,\ldots,\min(n_1+i-1,n_2)$.

As before n_1 is the number of years of marking and n_2 is the number of year of recapture and recovery.

For example, consider a capture-recapture-recovery model with survival probability that is dependent on age and time, $\phi_{h,t}$, recapture probability that is dependent on age, p_h, and recovery probability that is dependent on age and time, $\lambda_{h,t}$. The Maple code for this example is given in CRR.mw. When there are $n_1 = n_2 = 3$ years of marking and recapture/recovery the exhaustive summary is

$$\kappa = \begin{bmatrix} \phi_{1,1}p_2 \\ \phi_{1,2}p_2 \\ \phi_{1,3}p_2 \\ \phi_{2,2}p_3 \\ \phi_{2,3}p_3 \\ \phi_{3,3}p_4 \\ \phi_{1,1}(1-p_2) \\ \phi_{1,2}(1-p_2) \\ \phi_{2,2}(1-p_3) \\ (1-\phi_{1,1})\lambda_{1,1} \\ (1-\phi_{1,2})\lambda_{1,2} \\ (1-\phi_{1,3})\lambda_{1,3} \\ (1-\phi_{2,2})\lambda_{2,2} \\ (1-\phi_{2,3})\lambda_{2,3} \\ (1-\phi_{3,3})\lambda_{3,3} \end{bmatrix}.$$

The parameters are

$$\theta = [\phi_{1,1}, \phi_{1,2}, \phi_{1,3}, \phi_{2,2}, \phi_{2,3}, \phi_{3,3}, p_2, p_3, p_4, \lambda_{1,1}, \lambda_{1,2}, \lambda_{1,3}, \lambda_{2,2}, \lambda_{2,3}, \lambda_{3,3}].$$

Using Maple the derivative matrix $\mathbf{D} = \partial\kappa/\partial\theta$ can be shown to have rank 14, but there are 15 parameters, so the model is parameter redundant with deficiency 1.

To find the estimable parameter combinations we can apply Theorem 3.6 of Section 3.2.5. We first solve $\alpha^T \mathbf{D} = 0$ giving

$$\alpha^T = \begin{bmatrix} 0 & 0 & 0 & 0 & 0 & \frac{(1-\phi_{3,3})}{\lambda_{3,3}} & 0 & 0 & -\frac{(1-\phi_{3,3})p_4}{\phi_{3,3}\lambda_{3,3}} & 0 & 0 & 0 & 0 & 0 & 1 \end{bmatrix}.$$

The position of the zeros shows that the 12 parameters $\phi_{1,1}$, $\phi_{1,2}$, $\phi_{1,3}$, $\phi_{2,2}$, $\phi_{2,3}$, p_2, p_3, $\lambda_{1,1}$, $\lambda_{1,2}$, $\lambda_{1,3}$, $\lambda_{2,2}$ and $\lambda_{2,3}$ are individually identifiable. To find the combinations of the other parameters that can be estimated we solve the PDE

$$\frac{\partial f}{\partial \phi_{3,3}}\frac{(1-\phi_{3,3})}{\lambda_{3,3}} - \frac{\partial}{\partial p_4}\frac{(1-\phi_{3,3})p_4}{\phi_{3,3}\lambda_{3,3}} + \frac{\partial f}{\partial \lambda_{3,3}} = 0,$$

where f is an arbitrary function. A solution of this PDE results the estimable parameter combinations as $\phi_{3,3}p_4$ and $(1-\phi_{3,3})\lambda_{3,3}$.

As with the capture-recapture models, this is a much simpler exhaustive summary. For the capture-recapture-recovery model the exhaustive summary consisting of the probabilities of all possible capture histories has 26 terms for three years of recapture and three years of recovery, rather than the 15 terms of the simpler exhaustive summary. Similar to capture-recapture general results can be found, and are given in Hubbard et al. (2014).

9.1.4   Multi-State Model

Another extension of capture-recapture models is to have different states for animals, for example breeding and non-breeding. Multi-state models in some cases can have a simple enough form to apply the symbolic method directly (Gimenez et al., 2003), or can be more complex so that the symbolic method fails to find the rank (Hunter and Caswell, 2009). In this section we consider how the extended symbolic method can be used in structurally more complex multi-state capture-recapture models (Cole, 2012).

A general multi-state capture-recapture model has S different states. However it may not be possible to observe animals in every one of these S states. Animals are monitored over N different sampling occasions. Typically animals would be marked in years 1 to $(N-1)$ and recaptured in years 2 to N. The model can be represented by two different matrices: a transition matrix, $\mathbf{\Phi}_t$ and a recapture matrix, $\mathbf{P}_t$. The matrix $\mathbf{\Phi}_t$ has dimensions $S \times S$ and elements, $\phi_{i,j,t}$, which represent the probability of transition from state j at time t to state i at time $t+1$. The state transition includes both probabilities for survival and moving between different states. The matrix $\mathbf{P}_t$ is an $S \times S$ diagonal matrix, with diagonal entries, $p_{i,i}(t)$, representing the probability of recapturing an animal in state i at time t. In an unobservable state, $p_{i,i}(t) = 0$. Let $\Psi_{i,j,\tau,t}$ denote the probability that an individual released in state i at time τ is next recaptured in state j at time t. For an unobservable state i, $\Psi_{i,j,\tau,t} = 0$. In matrix form $\Psi_{i,j,\tau,t}$ can be represented as

$$\mathbf{\Psi}_{\tau,t} = \begin{cases} (\mathbf{P}_{\tau+1}\mathbf{\Phi}_\tau)^T & t = \tau+1 \\ \{\mathbf{P}_t\mathbf{\Phi}_{t-1}(\mathbf{I}-\mathbf{P}_{t-1})\mathbf{\Phi}_{t-2}\cdots & \\ (\mathbf{I}-\mathbf{P}_{\tau+1})\mathbf{\Phi}_\tau\}^T & t > \tau+1 \end{cases},$$

where $\mathbf{I}$ is the identity matrix (Hunter and Caswell, 2009).

The elements of $\mathbf{\Psi}_{\tau,t}$ form an exhaustive summary for this model. However Hunter and Caswell (2009) found that for many examples Maple runs out of memory trying to calculate the rank the derivative matrix, so the symbolic method cannot be applied directly. Cole (2012) developed a simpler exhaustive summary for a general multi-state model with S states. Out of these S states U are unobservable. We relabel the states, if necessary, so that the last U states are unobservable, i.e. states $1, \ldots, S - U$ are observable and states $(S - U + 1), \ldots, S$ are unobservable. We write the elements of the transition matrix as

$$\mathbf{\Phi}_t = \begin{bmatrix} a_{1,1,t} & a_{1,2,t} & \cdots & a_{1,S,t} \\ a_{2,1,t} & a_{2,2,t} & \cdots & a_{2,S,t} \\ \vdots & & & \vdots \\ a_{S,1,t} & a_{S,2,t} & \cdots & a_{S,S,t} \end{bmatrix}$$

and the elements of the recapture matrix are

$$\mathbf{P}_t = \begin{bmatrix} p_{1,t} & 0 & \cdots & 0 \\ 0 & p_{2,t} & \cdots & 0 \\ \vdots & & & \vdots \\ 0 & 0 & \cdots & p_{S,t} \end{bmatrix}.$$

Note that $p_{S-U+1,t} = ... = p_{S,t} = 0$, as these last states are unobservable. The simpler exhaustive summary consists of the terms,

- $p_{i,t+1}a_{i,j,t}$ for $t = 1, ..., N - 1$, $i = 1, .., S - U$ and $j = 1, .., S - U$,

- $p_{i,t}$ for $t = 2, ..., N - 1$ and $i = 1, .., S - U$,

- $p_{i,t+1}a_{i,j,t}a_{j,1,t}$ for $t = 2, ..., N - 1$, $i = 1, .., S - U$ and $j = S - U + 1, .., S$,

- $\dfrac{a_{j,i,t-1}}{a_{j,1,t-1}}$ for $t = 2, ..., N - 1$, $i = 2, .., S - U$ and $j = S - U + 1, .., S$,

- $\dfrac{a_{j,i,t-1}a_{i,1,t-2}}{a_{j,1,t-1}}$ for $t = 3, ..., N - 1$, $i = S - U + 1, .., S$ and $j = S - U + 1, .., S$.

The simpler exhaustive summary is valid if

$$(S^2 + S - 2U)(N - 1) - S - 2U(S - 1) < \frac{1}{2}(N^2 - N)(S - U)^2$$

and there is more than one observable state $(S - U > 1)$. Note that the last three terms of the exhaustive summary involves dividing through by $a_{j,1,t}$, so if any of the $a_{j,1,t}$ $(j = S - U + 1, .., S)$ are zero then $a_{j,1,t}$ can be replaced by $a_{j,2,t}$ in the last three types of exhaustive summary terms and the range changed appropriately. If $a_{j,2,t}$ is zero as well then use $a_{j,3,t}$ and so on. A Maple procedure for creating this exhaustive summary is given in Cole (2012) and the Maple worksheet `4stateexample.mw`.

For example, consider a 4-state breeding success model from Hunter and Caswell (2009). The model has the following four states:

- State 1, an observable state, for successful breeding.

- State 2, an observable state, for unsuccessful breeding.

- State 3, an unobservable state, for the year post-successful breeding.

- State 4, an unobservable state, for the year post-unsuccessful breeding.

The transition matrix is

$$\mathbf{\Phi}_t = \begin{bmatrix} \sigma_{1,t}\beta_{1,t}\gamma_{1,t} & \sigma_{2,t}\beta_{2,t}\gamma_{2,t} & \sigma_{3,t}\beta_{3,t}\gamma_{3,t} & \sigma_{4,t}\beta_{4,t}\gamma_{4,t} \\ \sigma_{1,t}\beta_{1,t}\bar{\gamma}_{1,t} & \sigma_{2,t}\beta_{2,t}\bar{\gamma}_{2,t} & \sigma_{3,t}\beta_{3,t}\bar{\gamma}_{3,t} & \sigma_{4,t}\beta_{4,t}\bar{\gamma}_{4,t} \\ \sigma_{1,t}\bar{\beta}_{1,t} & 0 & \sigma_{3,t}\bar{\beta}_{3,t} & 0 \\ 0 & \sigma_{2,t}\bar{\beta}_{2,t} & 0 & \sigma_{4,t}\bar{\beta}_{4,t} \end{bmatrix},$$

where $\sigma_{i,t}$ is the probability of survival at site i at time t, $\beta_{i,t}$ is probability of breeding given survival at site i at time t, $\gamma_{i,t}$ is probability of successful breeding given breeding and survival at site i at time t, and $\bar{\theta} = 1 - \theta$. The recapture matrix is

$$\mathbf{P} = \begin{bmatrix} p_{1,t} & 0 & 0 & 0 \\ 0 & p_{2,t} & 0 & 0 \\ 0 & 0 & 0 & 0 \\ 0 & 0 & 0 & 0 \end{bmatrix},$$

where $p_{i,t}$ is the probability of being recaptured in state i at time t. Note that $p_{3,t} = p_{4,t} = 0$ because states 3 and 4 are unobservable. When $N = 7$ we show in the Maple worksheet `4stateexample.mw` that the rank of the derivative matrix, derived from this simpler exhaustive summary, is $r = 62$. There are 78 parameters, therefore the model is parameter redundant with a deficiency of $d = 16$.

9.1.5 Memory Model

An alternative extension to capture-recapture models is to have multiple sites where animals are observed rather than multiple states. Animals will often return to the site they visited last rather than randomly choosing a new site, therefore a multi-site model with memory may be applicable. Rouan et al. (2009) compares two different memory models from Brownie et al. (1993) and Pradel (2005), with an equivalent model without memory known as the Arnason-Schwarz model (Arnason, 1973, Schwarz et al., 1993).

Here we consider the Brownie model of Brownie et al. (1993). Suppose there are N capture occasions, and S different sites. Using a similar notation as the previous sections the recapture history of an animal is represented by h, where 0 represents an animal was not encountered on that occasion and $i = 1, \ldots, S$ represents an animal was encountered at site i on that occasion. For example the history

$$h = 001102$$

corresponds to a study over $N = 6$ occasions. The animal was first encountered in site 1 at occasion $t = a$, in this case $a = 3$. It was then also encountered in site 1 at occasion $t = 4$. It was not encountered at occasion $t = 5$. Then it was encountered at site 2 at occasion $t = 6$. To determine the probability of history h, the general matrix notation of Rouan et al. (2009) is used. The matrices used are: the initial state matrix, $\mathbf{\Pi}_t$, the transition matrix, $\mathbf{\Phi}_t$ and the event matrix, $\mathbf{B}_t$. There is a separate event matrix, $\mathbf{B}_t^0$, for the first encounter. For each matrix there are $S + 1$ options: either an animal is in one of the S sites, or the animal dies or permanently emigrates and moves to the 'dead' state, denoted by $\dagger$. The matrices for $S = 2$ sites for the Brownie model are

$$\mathbf{\Pi}_t = [\pi_{1,t}, \pi_{2,t}, 0], \mathbf{\Phi}_{0,t} = \begin{bmatrix} \phi_{\star,1,1,t} & \phi_{\star,1,2,t} & 0 & 0 & \phi_{\star,1,\dagger,t} \\ 0 & 0 & \phi_{\star,2,1,t} & \phi_{\star,2,2,t} & \phi_{\star,2,\dagger,t} \\ 0 & 0 & 0 & 0 & 1 \end{bmatrix},$$

$$\mathbf{\Phi}_t = \begin{bmatrix} \phi_{1,1,1,t} & \phi_{1,1,2,t} & 0 & 0 & \phi_{1,1,\dagger,t} \\ 0 & 0 & \phi_{1,2,1,t} & \phi_{1,2,2,t} & \phi_{1,2,\dagger,t} \\ \phi_{2,1,1,t} & \phi_{2,1,2,t} & 0 & 0 & \phi_{2,1,\dagger,t} \\ 0 & 0 & \phi_{2,2,1,t} & \phi_{2,2,2,t} & \phi_{2,2,\dagger,t} \\ 0 & 0 & 0 & 0 & 1 \end{bmatrix},$$

$$\mathbf{B}_0 = \begin{bmatrix} 0 & 0 & 1 \\ 1 & 0 & 0 \\ 0 & 1 & 0 \end{bmatrix}, \mathbf{B}_t = \begin{bmatrix} \bar{p}_{1,t} & \bar{p}_{2,t} & \bar{p}_{1,t} & \bar{p}_{2,t} & 1 \\ p_{1,t} & 0 & p_{1,t} & 0 & 0 \\ 0 & p_{2,t} & 0 & p_{2,t} & 0 \end{bmatrix},$$

where $\phi_{i,j,k,t}$ is the transition probability for animals present at site k at occasion $t + 1$, at site j at occasion t and at site i at occasion $t - 1$, $\phi_{i,j,\dagger,t} = 1 - \phi_{i,j,1,t} - \phi_{i,j,2,t}$, $\phi_{\star,j,k,t}$ is the transition probability for animals at site k at occasion $t + 1$ and at site j when first captured, $\phi_{\star,j,\dagger,t} = 1 - \phi_{\star,j,1,t} - \phi_{\star,j,2,t}$, $\pi_{j,t}$ is the initial state probability for an animal at j site when first captured at occasion t and $p_{j,t}$ is the probability an animal is encountered at site j at occasion t. (The matrices for a general S are given in the supplementary material of Cole et al., 2014). The probability of any encounter history h is then

$$\Pr(h) = \mathbf{\Pi}_a \text{diag}\left\{\mathbf{B}_{0,a}(\nu_a,.)\right\} \left[\prod_{t=a+1}^{N} \mathbf{\Phi}_{t-1}\text{diag}\left\{\mathbf{B}_t(\nu_t,.)\right\}\right] \mathbf{1}_S,$$

where ν_t is the event observed at time t, $\mathbf{B}_t(\nu_t,.)$ is the row vector of $\mathbf{B}$ corresponding to event ν_t, $\mathbf{B}_0(\nu_t,.)$ is the row vector of $\mathbf{B}_0$ corresponding to event ν_t and $\mathbf{1}_S$ is a column vector consisting of S ones. The term diag $\{\mathbf{v}\}$, refers to creating a diagonal matrix from row vector $\mathbf{v}$. For n individual histories the likelihood is

$$L = \prod_{h=1}^{n} \Pr(h)$$

(Rouan et al., 2009).

Rouan et al. (2009) uses the hybrid symbolic-numeric method to investigate parameter redundancy in this model, and the two other models mentioned at the start of the Section. Cole et al. (2014) use the symbolic method to create general results on the parameter redundancy of these models. It involves developing simpler exhaustive summaries, similar to the exhaustive summary discussed in Section 9.1.4 above.

Cole et al. (2014) show that the Brownie model with time dependent initial state probabilities, time dependent transition probabilities and constant recapture probabilities is not parameter redundant, for $N \geq 3$ and $S \geq 2$. However this result is based on a perfect data set, that is, every possible history has been observed. In real data sets certain histories will not occur as they have a very small probability of occurring. For example for a model with all parameters constant if there were $N = 3$ years and $S = 2$ sites and parameter values $\pi_1 = 0.1$, $p_i = 0.2$, $\phi_{\star ij} = 0.3$ and $\phi_{ijk} = 0.3$ for $i,j,k = 1,2$, the probability of the history $h = 001$ is $P(001) = \pi_1 = 0.1$ and the probability of the history $h = 101$ is $P(101) = \pi_1\phi_{\star 11}(1 - p_1)\phi_{111}p_1 + \pi_1\phi_{\star 12}(1 - p_2)\phi_{121}p_1 = 0.00288$. If m animals were marked each year (across all sites) we would expect to see $E(h) = m \times P(h)$ animals with history h. If there were $m = 20$ animals marked per year we would expect to see the history $h = 001$ twice as $E(001) = 20 \times 0.1 = 2$, where as we would not expect to see the history $h = 101$ as $E(101) = 20 \times 0.00288 = 0.0576$.

The expected values can be used to create an 'expected data set'. To explore parameter redundancy caused by the data we only need to consider whether a history is present or absent from a data set. If $E(h)$ is greater than or equal to 1 then the history is assumed to be present in the 'expected data set'. However if $E(h)$ is less than 1 then the history is not in the 'expected data set'. In the above example the 'expected data set' is $h_1 = 100$, $h_2 = 200$, $h_3 = 010$, $h_4 = 020$, $h_5 = 001$, $h_6 = 021$, $h_7 = 002$ and $h_8 = 022$. This smaller data set may change the parameter redundancy result. Parameter redundancy of the 'expected data set' is checked using the hybrid symbolic-numeric method, using the probabilities of each history in the 'expected data set' as the exhaustive summary.

The parameter redundancy of a model with the 'expected data set' rather than perfect data can be checked using the hybrid symbolic-numerical method (see Section 4.3). When parameter redundancy is caused by the data rather than the model this is known as extrinsic parameter redundancy (see Sections 3.1.2 and 3.2.4). It is also possible to find how many animals need to be marked, m, to ensure a model is not parameter redundant. A Maple procedure to find the smallest value of m that ensures a model rank is the same as the model rank for perfect data is given in Cole et al. (2014) and the Maple worksheet `memory.mw`.

Consider now the Brownie model with $S = 2$ sites and for $N = 3$ years of data with time dependent initial state probabilities, time dependent transition probabilities and constant recapture probabilities. We use the following parameter values to generate an expected data set: $p_1 = 0.3$, $p_2 = 0.2$, $\phi_{1,1,1,t} = 0.5$, $\phi_{1,1,2,t} = 0.3$, $\phi_{2,2,2,t} = 0.5$, $\phi_{2,2,1,t} = 0.3$, $\phi_{2,1,1,t} = 0.4$, $\phi_{2,1,2,t} = 0.4$, $\phi_{1,2,2,t} = 0.4$, $\phi_{1,2,1,t} = 0.4$, $\phi_{\star,1,1} = 0.5$, $\phi_{\star,2,2} = 0.5$, $\phi_{\star,1,2} = 0.2$, $\phi_{\star,2,1} = 0.3$, $\pi_{1,t} = 0.5$. Whilst the model is not parameter redundant with perfect data, the Maple code `memory.mw` shows $m = 223$ different animals need to be marked each year for the 'expect data set' to remain full rank. That means that someone considering this experiment would need to consider a sample size of about $m = 223$ animals marked per year to be able to fit this model, if the parameter values were similar to those used to create the 'expected data set'.

By examining a wide range of possible parameter values it is possible to make general recommendations of sample sizes needed to have enough data to be able to fit a particular model, as shown in Cole et al. (2014).

9.2 Integrated Population Models

More realistic ecological models often describe a process that is parameter redundant. One way to overcome this parameter redundancy is to combine several sources of data into an integrated population model (see for example Besbeas et al., 2002, Schaub and Abadi, 2011). However, it does not necessarily follow that an integrated population model is identifiable. In this Section we discuss methods for checking parameter redundancy in integrated population models.

Suppose K independent data sets are collected. Whilst it is possible to fit separate models to each data set, if the data sets have parameters in common information can be shared using a joint log-likelihood of the form

$$l = \sum_{i=1}^{K} l_i,$$

where l_i is the log-likelihood for the ith data set, and each data set is assumed to be independent. Although the idea is simple, this method has lead to the advancement of modelling ecological data (see for example Chapter 9, Newman et al., 2016 or Chapters 11 and 12 of McCrea and Morgan, 2014). Integrated population modelling results in improved precision of parameter estimates and reduced correlation between parameters, as well as in some cases allowing estimation of parameters, which would be non-identifiable in a single model (Schaub and Abadi, 2011).

The simplest method to check for parameter redundancy in integrated models is to combine the exhaustive summaries of each model into one single exhaustive summary. If there are K data sets and the ith data set has exhaustive summary κ_i then the exhaustive summary for the integrated model is

$$\kappa = \begin{bmatrix} \kappa_1 \\ \kappa_2 \\ \kappa_3 \\ \vdots \\ \kappa_K \end{bmatrix}$$

(Cole and McCrea, 2016).

For example, consider an integrated population for both census and capture-recovery model from Besbeas et al. (2002). This combines a state-space model for census data with a mark-recovery model for mark-recovery data. The census model is the one discussed in Section 8.1, with two age categories for juveniles and adults. The mark-recovery model is the model first discussed in Section 1.2.4. There are two survival parameters, ϕ_1 and ϕ_a, for first year and adult survival respectively, which are common to both parts of the integrated model. The state-space model has the parameter ρ, which is the productivity rate. The mark-recovery model has two recovery parameters, λ_1 and λ_a, for first year and adults respectively.

This joint census and capture-recovery model is examined in the Maple file `basicIPM.mw`. The exhaustive summary for the state-space model comes from equation (8.1),

$$\kappa_1 = \begin{bmatrix} -\rho\phi_1\phi_a x_{0,2} \\ -\phi_a(x_{0,1} + x_{0,2}) \\ \rho\phi_1\phi_a \\ \phi_a \end{bmatrix}.$$

The exhaustive summary for the mark-recovery model consists of the probabilities of being marked in a specific year and recovered in a specific year (see Section 3.1.4.4). Ignoring repeated terms, the exhaustive summary is

$$\kappa_2 = \begin{bmatrix} (1 - \phi_1)\lambda_1 \\ \phi_1(1 - \phi_a)\lambda_a \\ \phi_1\phi_a(1 - \phi_a)\lambda_a \end{bmatrix}.$$

The joint exhaustive summary is then

$$\kappa = \begin{bmatrix} \kappa_1 \\ \kappa_2 \end{bmatrix} = \begin{bmatrix} -\rho\phi_1\phi_a x_{0,2} \\ -\phi_a(x_{0,1} + x_{0,2}) \\ \rho\phi_1\phi_a \\ \phi_a \\ (1 - \phi_1)\lambda_1 \\ \phi_1(1 - \phi_a)\lambda_a \\ \phi_1\phi_a(1 - \phi_a)\lambda_a \end{bmatrix}.$$

The parameters are $\theta = [\phi_1, \phi_a, \rho, \lambda_1, \lambda_a]$. Note that here $x_{0,1}$ and $x_{0,2}$ are treated as known constants, but could be included as additional parameters. We can then apply the symbolic method. The derivative matrix is

$$\mathbf{D} = \frac{\partial \kappa}{\partial \theta} =$$
$$\begin{bmatrix} -\phi_a\rho x_{0,2} & 0 & \rho\phi_a & 0 & -\lambda_1 & \bar{\phi}_a\lambda_a & \phi_a\bar{\phi}_a\lambda_a \\ -\rho\phi_1 x_{0,2} & -x_{0,1} - x_{0,2} & \rho\phi_1 & 1 & 0 & -\phi_1\lambda_a & \phi_1\lambda_a - 2\phi_1\phi_a\lambda_a \\ -\phi_1\phi_a x_{0,2} & 0 & \phi_1\phi_a & 0 & 0 & 0 & 0 \\ 0 & 0 & 0 & 0 & \bar{\phi}_1 & 0 & 0 \\ 0 & 0 & 0 & 0 & 0 & \phi_1\bar{\phi}_a & \phi_1\phi_a\bar{\phi}_a \end{bmatrix},$$

where $\bar{\theta} = 1 - \theta$. In the Maple code we show that the derivative matrix has rank 4. As there are five parameters this model is parameter redundant with deficiency 1. By solving an appropriate set of partial differential equations the estimable parameter combinations can be found to be ϕ_a, $\phi_1\rho$, $(1 - \phi_1)\lambda_1$ and $\phi_1\lambda_a$ (Cole and McCrea, 2016). Although this example is relatively simple, it demonstrates that combining two or more data sets does not necessarily mean the model will no longer be parameter redundant.

The potential problem with the simple combined exhaustive summary is one of structural complexity. Cole and McCrea (2016) provide an alternative method, which is similar to the extension theorem. This extension method is applicable if there are $K = 2$ data sets. Suppose the exhaustive summaries are κ_1 and κ_2 with parameters θ_1 and θ_2 of length p_1 and p_2 respectively. The extension method involves the following steps:

Step 1 Find the derivative matrix $\mathbf{D}_{1,1}(\theta_1) = \partial\kappa_1(\theta_1)/\partial\theta_1$ and calculate its rank, q_1.

Step 2　If $q_1 < p_1$ then reparameterise κ_1 in terms of its estimable parameters combinations, or another reparameterisation that results in a full rank matrix. Let $\mathbf{s}_1$ denote this reparameterisation. If $q_1 = p_1$ then set $\mathbf{s}_1 = \boldsymbol{\theta}$.

Step 3　Rewrite κ_2 in terms of $\mathbf{s}_1$ to give $\kappa_2(\boldsymbol{\theta}_2) = \kappa_2(\mathbf{s}_1, \boldsymbol{\theta}_{2,ex})$, where $\boldsymbol{\theta}_{2,ex}$ is a vector consisting of any additional parameters.

Step 4　Find the derivative matrix $\mathbf{D}_{2,2} = \partial\kappa_2(\mathbf{s}_1, \boldsymbol{\theta}_{2,ex})/\partial\boldsymbol{\theta}_{2,ex}$ and calcuate its rank r_{ex}.

Step 5　The integrated model will have rank $q_1 + r_{ex}$.

　　This method is useful because it requires the calculation of the rank of two simpler derivative matrices with smaller dimensions compared with the larger dimensions of the combined exhaustive summary method.

　　A consequence of this method is that if the two models are both individually full rank then the integrated model will be full rank. If $\boldsymbol{\theta}_{2,ex}$ consists of one additional parameter then the integrated model's rank will be $q_1 + 1$ without the need to find the rank of $\mathbf{D}_{2,2}$ (Cole and McCrea, 2016).

　　To demonstrate this extension method, consider the same integrated state-space model as above but with a constant recovery probability, so that $\lambda_1 = \lambda_a = \lambda$. The state-space exhaustive summary is

$$
\kappa_1 = \begin{bmatrix}
-\rho\phi_1\phi_a x_{0,2} \\
-\phi_a(x_{0,1} + x_{0,2}) \\
\rho\phi_1\phi_a \\
\phi_a
\end{bmatrix}.
$$

The parameters are $\boldsymbol{\theta}_1 = [\phi_1, \phi_a, \rho]$. In Section 8.5.1 and the Maple code `basicIPM.mw` we show that the derivative matrix $\partial\kappa_1/\partial\boldsymbol{\theta}_1$ has rank $q_1 = 2$, which is less than $p_1 = 3$. The estimable parameter combinations are ϕ_a and $\phi_1\rho$ giving the reparameterisation

$$
\mathbf{s}_1 = \begin{bmatrix} s_{1,1} \\ s_{1,2} \end{bmatrix} = \begin{bmatrix} \phi_a \\ \phi_1\rho \end{bmatrix}.
$$

The second exhaustive summary is the mark-recovery model with

$$
\kappa_2(\boldsymbol{\theta}_2) = \begin{bmatrix}
(1 - \phi_1)\lambda \\
\phi_1(1 - \phi_a)\lambda \\
\phi_1\phi_a(1 - \phi_a)\lambda
\end{bmatrix}.
$$

Rewriting this exhaustive summary in terms of the reparameterisation $\mathbf{s}_1$ gives

$$
\kappa_2(\mathbf{s}_1, \boldsymbol{\theta}_{2,ex}) = \begin{bmatrix}
\frac{\lambda(\rho - s_{1,2})}{\rho} \\
\frac{\lambda s_{1,2}(1 - s_{1,1})}{\rho} \\
\frac{\lambda s_{1,2} s_{1,1}(1 - s_{1,1})}{\rho}
\end{bmatrix},
$$

with $\boldsymbol{\theta}_{2,ex} = [\lambda, \rho]$. The derivative matrix,

$$\mathbf{D}_{2,2} = \frac{\partial \kappa_2(\mathbf{s}_1, \boldsymbol{\theta}_{2,ex})}{\partial \boldsymbol{\theta}_{2,ex}} = \begin{bmatrix} \frac{\rho - s_{1,2}}{\rho} & \frac{s_{1,2}(1 - s_{1,1})}{\rho} & \frac{s_{1,1}s_{1,2}(1 - s_{1,1})}{\rho} \\ \frac{\lambda s_{1,2}}{\rho^2} & -\frac{\lambda s_{1,2}(1 - s_{1,1})}{\rho^2} & -\frac{\lambda s_{1,1}s_{1,2}(1 - s_{1,1})}{\rho^2} \end{bmatrix},$$

has rank $r_{ex} = 2$. Therefore the rank of integrated model is $q_1 + r_{ex} = 2 + 2 = 4$. As there are four parameters in the model this model is not parameter redundant.

A more substantial example of using this method is given by the example in Section 9.2.1 below.

9.2.1 Four Data Set Integrated Model

On the Isle of May four different types of data are collected on colony of common guillemots. The four sets of data are:

- productivity data (productivity),

- capture-recapture data of adult birds (adult CR),

- capture-recapture-recovery data of birds ringed as chicks (chick CRR),

- count data (count).

The data sets have been collected for N years.

Reynolds et al. (2009) fit an integrated population model to the four data sets. Cole and McCrea (2016) show how the extension method described above can be used to investigate parameter redundancy in an integrated model that includes any combination of the four different data sets.

The productivity data is nest record data that monitors eggs laid and the number of resulting chicks. As a binomial model is fitted to each year of data an exhaustive summary consists of the binomial expectation. Assuming that n_t eggs are laid in year t with a productivity rate ρ_t then an exhaustive summary is

$$\kappa_1 = \begin{bmatrix} n_1 \rho_1 \\ n_2 \rho_2 \\ n_3 \rho_4 \\ \vdots \\ n_N \rho_N \end{bmatrix}$$

and the parameters are $\boldsymbol{\theta}_1 = [\rho_1, \rho_2, \ldots, \rho_N]$.

The adult CR data set is modelled using the capture-recapture model and exhaustive summary described in Section 9.1.2 above, with time-dependent survival probability ϕ_t and time-dependent recapture probability p_t.

The exhaustive summary is

$$
\kappa_2 = \begin{bmatrix}
\phi_{a,1}p_2 \\
\vdots \\
\phi_{a,N-1}p_N \\
\phi_{a,1}(1-p_2) \\
\vdots \\
\phi_{a,N-2}(1-p_{N-1})
\end{bmatrix}
$$

and the parameters are $\boldsymbol{\theta}_2 = [p_2, \ldots, p_N, \phi_{a,1}, \ldots, \phi_{a,N-1}]$.

The chick CRR data set follows the capture-recapture-recovery model discussed in Section 9.1.3 with the addition of permanent emigration and tag loss. The exhaustive summary, κ_3, consists of the terms

- $\psi_{i,t}\tau_{i,t}\phi_{i-1,t}q_{i,t}$ for $i = 1, \ldots, N-1$ and $t = i, \ldots, N-1$,

- $\psi_{i,j}\tau_{i,j}\phi_{i-1,j}(1-q_{i,j})$ for $i = 1, \ldots, N-2$, and $j = i, \ldots, N-2$,

- $\phi_{i-1,j}(1-\lambda_j)$ for $i = 1, \ldots, N-1$ and $j = i, \ldots, N-1$,

where λ_t is the recovery probability for year t, $\phi_{i,j}$ is the survival probability for an animal aged i at time t and $q_{i,t}$ is the resighting probability for an animal aged i at time t. (Note $q_{i,t}$ is different to p_t due to the method in which the data is collected). Emigration is assumed to only occur in the 4th year of life, with probability $\psi_{i,j} = \psi$, if $i = 4$ and 1 otherwise. Tag-loss is only assumed to occur for animals over four years old, with probability $\tau_{i,j} = \tau$, if $i > 4$ and 1 otherwise. The resighting probability $q_{i,t} = 0$ if $i = 1$ for all t, $q_{4,t} = q_4$ for all t and $q_{i,t} = q_a$ for all $i > 4$ and all t. Survival probability has $\phi_{i,t} = \phi_i$ for $0 < i < 5$ for all t and $\phi_{i,t} = \phi_{a,t}$ for $i \geq 5$ and all t. The parameters are

$$
\begin{aligned}
\boldsymbol{\theta}_3 = \; & [q_{2,2}, \ldots, q_{2,N-1}, q_{3,3}, \ldots, q_{3,N-1}, q_4, q_a, \phi_{0,1}, \ldots, \phi_{0,N-1}, \phi_1, \phi_2, \phi_3, \\
& \phi_{a,5}, \ldots, \phi_{a,N-1}, \lambda_1, \ldots, \lambda_N, \tau, \psi].
\end{aligned}
$$

The count data is modelled using a state-space model, and has exhaustive summary

$$
\kappa_4 = \begin{bmatrix}
\sigma^2 \\
x_{0,5}\phi_{a,5} + \frac{1}{2}x_{0,1}\rho_1\psi\phi_{0,1}\phi_1\phi_2\phi_3\phi_{a,5} \\
(x_{0,5}\phi_{a,5} + 1\frac{1}{2}x_{0,1}\rho_1\phi_{0,1}\psi\phi_1\phi_2\phi_3\phi_{a,5})\phi_{a,6} + \frac{1}{2}x_{0,2}\rho_2\psi\phi_{0,2}\phi_1\phi_2\phi_3 \\
\vdots
\end{bmatrix}.
$$

The parameters are

$$
\boldsymbol{\theta}_4 = [\phi_{0,1}, \ldots, \phi_{0,N-5}, \phi_1, \phi_2, \phi_3, \phi_{a,5}, \ldots, \phi_{a,N-1}, \psi, \rho_1, \ldots, \rho_{N-5}, \sigma].
$$

Table 9.2 gives the deficiency for each combination of model formed from one or more of the four data sets. These results are derived and given in

TABLE 9.2

Parameter redundancy results for integrated population model, for all different combinations of the four data sets. The x indicates the data set is included in the integrated model (or a single x indicates a model for just one data set). The last column, d, gives the deficiency of the model. A deficiency of greater than 1 indicates the integrated model (or single model) is parameter redundant. A deficiency of 0 indicates the integrated model (or single model) is not parameter redundant. (Results from Cole and McCrea, 2016.)

Productivity (κ_1)	Adult CR (κ_2)	Chick CRR (κ_3)	Count (κ_4)	d
x				0
	x			1
		x		0
			x	$2N - 6$
x	x			1
x		x		0
x			x	$N - 1$
	x	x		0
	x		x	N
		x	x	0
x	x	x		0
x	x		x	5
x		x	x	0
	x	x	x	0
x	x	x	x	0

Cole and McCrea (2016). Here we explain in more detail how these results are obtained. The results are valid for $N \geq 8$, although in some models the results are also applicable for smaller N.

We first consider all the models individually:

- For the productivity data alone the exhaustive summary, κ_1, contains one parameter per exhaustive summary term, so therefore it is obvious that the rank is the same as the length of the exhaustive summary and the number of parameters, N, and that the deficiency is 0.

- The adult CR data only is parameter redundant with rank $2N - 3$ and deficiency 1, as shown is Section 4.2.

- For the chick CRR alone parameter redundancy results are found using the standard symbolic method with the extension theorem (Theorem 6 of Section 3.2.6.1). When $N = 8$ the derivative matrix $\mathbf{D} = \partial\kappa_3/\partial\theta_3$ has full rank 35. Adding an extra year of data adds extra terms κ_{ex} and extra parameters θ_{ex}. The derivative matrix $\mathbf{D} = \partial\kappa_{ex}/\partial\theta_{ex}$ has full rank 5. Therefore by the extension theorem this model has deficiency 0.

- As the state-space model exhaustive summary for the count data, κ_4, contains at least one distinct parameter in each exhaustive summary terms, the rank of the derivative matrix will be the the number of exhaustive summary terms, $N - 4$. As there are the $3N - 10$ parameters the deficiency is $2N - 6$.

Next consider integrated data sets consisting of two data sets.

The productivity and adult CR combined do not have any parameters in common, therefore the rank is the sum of the rank for the two individual models $N + 2N - 3 = 3N - 3$, but there are $3N - 2$ parameters therefore the deficiency is 1.

The productivity and chick CRR are both individually full rank therefore the integrated model will also be full rank, so has deficiency 0.

The integrated model with productivity and count data has a joint exhaustive summary where every term contains a distinct parameter, therefore the rank is limited by the number exhaustive summary terms, $2N - 4$. There are $3N - 5$ parameters therefore the deficiency is $N - 1$.

For the adult CR and chick MRR data combined, we start with the chick MRR data exhaustive summary, κ_3, which is full rank with rank $q_1 = 5N - 5$, so $\mathbf{s}_1 = \boldsymbol{\theta}_3$. Then we examine the adult capture-recapture data set, with exhaustive summary κ_2. For $N = 7$ the extra parameters are $\boldsymbol{\theta}_{2,ex} = [p_{a,2}, \dots, p_{a,8}, \phi_{a,1}, \dots, \phi_{a,4}]$. The derivative matrix

$$\mathbf{D}_{2,2} = \frac{\partial \kappa_2}{\partial \boldsymbol{\theta}_{2,ex}}$$

is of full rank 10. As adding an extra year adds only one extra parameter, by a trivial application of the extension theorem $\mathbf{D}_{2,2}$ will always be full rank with rank $r_{ex} = N + 3$. The rank for this integrated model is then $q_1 + r_{ex} = 5N - 5 + N + 3 = 6N - 2$. This is identical to the number of parameters, therefore the deficiency is 0.

For the adult CR data and count data combined the first exhaustive summary is from the adult CR data set, κ_2. This is parameter redundant, with rank $q_1 = 2N - 3$ and all parameters are identifiable except for the last survival and recapture parameters, which are confounded as $\phi_{N-1} p_N$. The reparameterisation used is then

$$\mathbf{s}_1 = \begin{bmatrix} \phi_1 \\ \vdots \\ \phi_{N-1} \\ p_2 \\ \vdots \\ p_N \\ \phi_{N-1} p_N \end{bmatrix}.$$

When the second exhaustive summary, κ_4, is reparameterisated in terms of $\mathbf{s}_1$ the extra parameters are

$$\boldsymbol{\theta}_{2,ex} = [\psi, \rho_1, \dots, \rho_{N-5}, \phi_{0,1}, \dots, \phi_{0,N-5}, \phi_1, \phi_2, \phi_3, \sigma, p_N].$$

There is still at least one distinct parameter per exhaustive summary term, so the rank is again limited by the number of exhaustive summary terms, $r_{ex} = N - 4$. Therefore the integrated model has rank $q_1 + r_{ex} = 2N - 3 + N - 4 = 3N - 7$. There are $4N - 7$ parameters so the deficiency is N.

The chick CRR combined with the count data is similar, apart from the chick CRR data is full rank with rank $q_1 = 5N - 5$, and the extra parameters are $\theta_{2,ex} = [\rho_1, \ldots, \rho_{N-5}, \sigma]$. Now there is exactly one distinct parameter per exhaustive summary term in the second part which is also full rank $r_{ex} = N - 4$. The integrated model is full rank with rank $q_1 + r_{ex} = 5N - 5 + N - 4 = 6N - 9$ and deficiency 0.

When we consider three integrated data sets together we start with one of the two integrated data sets and add the third data set. For example consider the integrated model with productivity data, adult CR data and chick CRR data. We already know that the adult CR data and chick CRR data together is full rank. As the productivity data is also full rank, when we combine two individually full rank models the integrated model is also full rank so the deficiency is 0. Similarly when all four data sets are combined in one integrated model, this parameterisation of resulting model is full rank, as the productivity data set is full rank and the adult capture-recapture data, chick MRR data and census data combined is full rank.

9.2.2 Posterior and Prior Overlap in Integrated Population Model

A common method of fitting integrated population models involves a Bayesian method (see for example Brooks et al., 2004, Schaub et al., 2007, King et al., 2008). If a Bayesian method is used then it is a simple step to also consider weak identifiability within the model by examining the overlap between the prior and posterior distributions, as described in Section 6.4. For example this method is used in Abadi et al. (2010) and Korner-Nievergelt et al. (2014).

However, this method should be used with caution, as it can lead to incorrect conclusions. For example, Abadi et al. (2010) examine an integrated population model for a population of Little Owls, where the population is affected by immigration. The integrated population model consists of census data, capture-recapture data and productivity data. For the purpose of this example we do not consider the productivity data. The model for the census data is

$$y_t = \begin{bmatrix} 1 & 1 \end{bmatrix} \begin{bmatrix} x_{1,t} \\ x_{2,t} \end{bmatrix} + \eta_t,$$

$$\begin{bmatrix} x_{1,t} \\ x_{2,t} \end{bmatrix} = \begin{bmatrix} 0.5\rho\phi_1 & 0.5\rho\phi_1 \\ \phi_a + \gamma & \phi_a + \gamma \end{bmatrix} \begin{bmatrix} x_{1,t-1} \\ x_{2,t-1} \end{bmatrix} + \begin{bmatrix} \epsilon_{1,t} \\ \epsilon_{a,t} \end{bmatrix}$$

where $x_{1,t}$ is the number of one-year old females, $x_{2,t}$ is the number of females older than a year, y_t is the total number of females. The parameters are:

TABLE 9.3

Estimated percentage overlap of
the prior and posterior
distributions for the immigration
integrated model

Parameter	Percentage Overlap
ϕ_1	10.5%
ϕ_a	21.0%
ρ	36.4%
γ	7.5%
p	28.5%

ϕ_1 first year survival probability, ϕ_a adult survival, ρ productivity and γ the immigration rate. (η_t and $\epsilon_{i,t}$ are error processes.) The capture-recapture data consists of separate data for juveniles and adults. A CJS model (see Section 9.1.2) is used with parameters ϕ_1 , ϕ_a and p, where p is the recapture probability.

Abadi et al. (2010) examine the overlap between the posterior and prior distributions, which is replicated in the R code `immIPM.R` for a subset of the data. The results are given in Table 9.3. Using a cut-off of 35% only the parameter ρ is weakly identifiable. In Abadi et al. (2010) this leads to the conclusion that this integrated model, with census and capture-recapture data, can be used to estimate the immigration parameter, γ (as well as ϕ_1, ϕ_a and p). However in the Maple code `immIPM.mw` we use the symbolic method to show that the model is parameter redundant with estimable parameter combinations ϕ_1, ϕ_a, p and $\rho + 2\gamma/\phi_1$. That is, the immigration parameter is not identifiable. In fact the addition of the productivity data to the integrated population model is needed to separate ρ and γ.

9.3   Experimental Design

Experimental design can be key to whether or not a model is parameter redundant. A good experimental design can allow the estimation of the parameters of interest. Considering parameter identifiability in the model used before the experiment is conducted can help to inform the best experimental design to estimate key ecological parameters.

We illustrate this idea using two types of examples, the first is robust design (Section 9.3.1), and, for the second we return to the integrated population model involving four data sets (Section 9.3.2).

9.3.1   Robust Design

In ecology, robust design is used when referring to experiments with multiple secondary sampling occasions within primary sampling occasions. For example, the primary sampling occasions could be every year, the secondary sampling occasions could be several consecutive days. During the primary sampling occasions the population is assumed to be open to births, deaths, immigration and emigration. During the secondary sampling occasions, the population is assumed to be closed to births, deaths, immigration and emigration. The robust design model was first developed for capture-recapture models in Pollock (1982), and allows the estimation of temporary emigration (Kendall et al. 1995, 1997, Kendall and Bjorkland 2001). Introducing a robust design in other types of models can also remove parameter redundancy, as illustrated in Sections 9.3.1.1 and 9.3.1.2.

9.3.1.1   Robust Design Occupancy Model

The basic occupancy model, first introduced in Section 1.2.1, is obviously parameter redundant, because the parameters ψ and p only ever appear as a product. Therefore MacKenzie et al. (2002) developed the robust design occupancy model, where multiple sites are surveyed more than once (secondary sampling occasions) within a season (primary sampling occasion), assuming that ψ does not change between these surveys.

Hubbard (2014) shows, using the symbolic method, that as long as there are at least two surveys each season it is possible to estimate ψ and p. The parameter ψ can be season dependent, and the parameter p can be survey dependent and/or season dependent and this result still applies. Hubbard (2014) also explores identifiability in more complex occupancy models including multiple states (such as breeding and non-breeding), or multiple species with species interactions. Again these models also will theoretically not be parameter redundant as long as there are at least two surveys each season. However in practice specific data can result in parameter redundancy, as discussed in Hubbard (2014).

Identifiability in occupancy models is also considered in Gimenez et al. (2014) and Mosher et al. (2018).

9.3.1.2   Robust Design Removal Model

It can be a legal requirement to remove protected species from a site before development. Animals are captured and removed from location over several occasions. Usually the captured animals with be moved to a new location. A removal model can be used to model such a data set (see, for example, Pollock 1991). Some species live underground or shelter hidden from capture for part of the experiment, for example slow worms primarily live underground or under neither objects, but can be detected basking on the ground. Zhou et al. (2019) investigate removal models which include temporary emigration. Specifically a robust design removal model is developed, where there are at least two secondary samples for each primary sampling occasion. Temporary emigration is allowed to occur between each primary sampling, but not between secondary sampling occasions. For example, in a common lizard removal data set there were two removal samples each day for 47 days. Zhou et al. (2019) show that conducting two samples per day, rather than just one, can remove the problems of parameter redundancy and near redundancy.

9.3.2   Four Data Set Integrated Model

Returning to the four data set integrated model from Section 9.2.1 we consider which of the four data sets are necessary to estimate all of the parameters. To save money and effort needed, Lahoz-Monfort et al. (2014) investigated the effect of reducing the amount of data collected. This included the option of stopping collection of one of the data sets. As well as reducing precision, one possible side effect of stopping collection of a data set is introducing parameter redundancy.

From the results in Table 9.2 we consider which integrated models have deficiency 0, so that all the parameters are identifiable and could in theory be estimated. We also consider whether all the parameters appear in the model. For example, the chick CRR data set alone is not parameter redundant, but does not contain productivity parameters.

The results suggest that the chick CRR data set is key to being able to estimate all the parameters of interest, and should be still be collected. However it is possible to stop collecting the census data, or stop collecting both the productivity and the adult CR data, and still be able to estimate all the parameters of interest. Given that the census data is cheapest to collect, then in terms of being able to estimate parameters it is possible to stop collection of the productivity and the adult CR data. However the level of precision of the parameters will change, so other aspects of the modelling need to also be considered before stopping the collection of one or more data set, see Lahoz-Monfort et al. (2014).

9.4   Constraints

Suppose that an ideal model for a data set is parameter redundant. One way to use that model is to put some form of constraints on the model. One method of choosing those constraints is through basic knowledge of which parameters cannot be estimated and trial and error to choose the constraints. This is not necessarily the best method to give the optimal constraints for an example.

A more systematic method is possible, where every possible combination of constraints is considered, as demonstrated in the example in Section 9.4.1.

Alternatively once the estimable parameter combinations have been found, these can be used to determine the constraints, as demonstrated in the example in Section 9.4.2.

9.4.1   Multi-State Model

Hunter and Caswell (2009) consider multi-state mark recapture models with hidden states, which are described in Section 9.1.4 above. They consider all possible combinations of setting parameters of the same type with different states as being equal, using a version of the hybrid symbolic-numeric method. Below we illustrate one of their examples.

Consider the breeding success example described in Section 9.1.4 above, but with time invariant parameters so that the transition matrix is

$$
\Phi = \begin{bmatrix}
\sigma_1\beta_1\gamma_1 & \sigma_2\beta_2\gamma_2 & \sigma_3\beta_3\gamma_3 & \sigma_4\beta_4\gamma_4 \\
\sigma_1\beta_1\bar\gamma_1 & \sigma_2\beta_2\bar\gamma_2 & \sigma_3\beta_3\bar\gamma_3 & \sigma_4\beta_4\bar\gamma_4 \\
\sigma_1\bar\beta_1 & 0 & \sigma_3\bar\beta_3 & 0 \\
0 & \sigma_2\bar\beta_2 & 0 & \sigma_4\bar\beta_4
\end{bmatrix},
$$

where σ_i is the probability of survival at site i, β_i is the probability of breeding given survival at site i, γ_i is probability of successful breeding given breeding and survival at site i, and $\bar x = 1 - x$. The recapture matrix is

$$
\mathbf{P} = \begin{bmatrix}
p_1 & 0 & 0 & 0 \\
0 & p_2 & 0 & 0 \\
0 & 0 & 0 & 0 \\
0 & 0 & 0 & 0
\end{bmatrix},
$$

where p_i is the probability of being recaptured in state i, with $p_3 = p_4 = 0$ as states 3 and 4 are unobservable. The extended symbolic method was used in Cole et al., 2010 to show this model is parameter redundant with deficiency 2.

In Table 9.4 we show parameter redundancy results for various biologically feasible constraints on the non-identifiable parameters β and σ. This table was created using the simpler exhaustive summary given in Section 9.4.1; see 4stateexample.mw. Hunter and Caswell (2009) find the same results using the hybrid symbolic-numerical method.

9.4.2   Memory Model

In Section 9.1.5 we discussed the Brownie memory model (Brownie et al., 1993). Here we consider an alternative memory model, the Pradel model (Pradel, 2005); the identifiability of this model is also discussed in Rouan et al. (2009) and Cole et al. (2014).

The formation of the Pradel memory model is similar to the Brownie memory model, but the matrices used are slightly different. For $N = 2$ sites

TABLE 9.4

Parameter redundancy results for different combinations of
constraints in the time invariant multi-state breeding success
model. Constraints are considered on the non-identifiable
parameters β_i and σ_i. None indicated there is no constraint on
that parameter. The rank, r, and the deficiency d are given for
each model.

| | \multicolumn{8}{c}{σ constraint} | | | | | | | |
| β constraint | None | | $\sigma_1 = \sigma_2,$ $\sigma_3 = \sigma_4$ | | $\sigma_1 = \sigma_3,$ $\sigma_2 = \sigma_4$ | | $\sigma_1 = \sigma_2 =$ $\sigma_3 = \sigma_4$ | |
	r	d	r	d	r	d	r	d
None	12	2	12	0	12	0	11	0
$\beta_1 = \beta_2, \beta_3 = \beta_4$	11	1	9	1	10	0	9	0
$\beta_1 = \beta_3, \beta_2 = \beta_4$	10	2	10	0	10	0	9	0
$\beta_1 = \beta_2 = \beta_3 = \beta_4$	10	1	8	1	9	0	8	0

the initial state matrix is

$$\mathbf{\Pi}_t = \left[\pi_{1,1,t}, \pi_{1,2,t}, \pi_{2,1,t}, \pi_{2,2,t}, 0\right],$$

the transition matrix is

$$\mathbf{\Phi}_t = \begin{bmatrix} \phi_{1,1,1,t} & \phi_{1,1,2,t} & 0 & 0 & \phi_{1,1,\dagger,t} \\ 0 & 0 & \phi_{1,2,1,t} & \phi_{1,2,2,t} & \phi_{1,2,\dagger,t} \\ \phi_{2,1,1,t} & \phi_{2,1,2,t} & 0 & 0 & \phi_{2,1,\dagger,t} \\ 0 & 0 & \phi_{2,2,1,t} & \phi_{2,2,2,t} & \phi_{2,2,\dagger,t} \\ 0 & 0 & 0 & 0 & 1 \end{bmatrix},$$

and the encounter matrices are

$$\mathbf{B}_t^0 = \begin{bmatrix} 0 & 0 & 0 & 0 & 1 \\ 1 & 0 & 1 & 0 & 0 \\ 0 & 1 & 0 & 1 & 0 \end{bmatrix}, \mathbf{B}_t = \begin{bmatrix} \bar{p}_{1,t} & \bar{p}_{2,t} & \bar{p}_{1,t} & \bar{p}_{2,t} & 1 \\ p_{1,t} & 0 & p_{1,t} & 0 & 0 \\ 0 & p_{2,t} & 0 & p_{2,t} & 0 \end{bmatrix},$$

where $\phi_{i,j,k,t}$ is the transition probability for animals present at site k at
occasion $t+1$, at site j at occasion t and at site i at occasion $t-1$, $\pi_{i,j,t}$ is the
initial state probability for an animal at j site when first captured at occasion
t, which was at site i at occasion $t-1$ and $p_{j,t}$ is the probability an animal is
encountered at site j at occasion t. As with the Brownie model the matrices
for a general N are given in the supplementary material of Cole et al. (2014).
Note that $\pi_{N,N,t} = 1 - \sum_{i=1}^{N}\sum_{j=1}^{N-1} \pi_{i,j,t}$, to satisfy the fact that the initial
state probabilities must sum to 1.

For T occasions of marking and recovery and N sites the estimable parameter combinations are

- $p_{i,t}$, $i = 1, \ldots, N$, $t = 2, \ldots, T-1$,

- $\phi_{i,j,k,t}$, $i, j, k = 1, \ldots, N$, $t = 2, \ldots, T-2$,

- $\pi_{i,j,t}$, $i = 1, \ldots, N$, $j = 1, \ldots, N-1$, $t = 2, \ldots, T-1$,

- $\phi_{i,j,k,t} p_{k,T}$, $i, j, k = 1, \ldots, N$,

- $\sum_{i=1}^{N} \pi_{i,j,1} \phi_{i,j,k,1}$, , $k = 1, \ldots, N$,

- $\sum_{i=1}^{N} \pi_{i,j,1}$, , $j = 1, \ldots, N-1$,

- $\sum_{i=1}^{N} \pi_{i,j,T}$, , $j = 1, \ldots, N-1$.

From these estimable parameter combinations Cole et al. (2014) suggest the constraints

- $\phi_{i,j,k,1} = \phi_{\star,j,k,1}$, $i, j, k = 1, \ldots, N$,

- $p_{i,T} = 1$, $i = 1, \ldots, N$,

- $\pi_{i,j,1} = \pi_{\star,j,1}$, $i = 1, \ldots, N$, $j = 1, \ldots, N-1$,

- $\pi_{i,j,T} = \pi_{\star,j,T}$, $i = 1, \ldots, N$, $j = 1, \ldots, N-1$.

Rouan et al. (2009) used trial and error and knowledge of the model to obtain alternative constraints. Their constraints included constraints on all of the $\pi_{i,j,t}$ (as well as similar constraints on $\phi_{i,j,k,t}$ and $p_{i,j,t}$). This set of constraints is more restrictive than the estimable parameter combination restraints, as in fact some of the $\pi_{i,j,t}$ can be estimated.

9.5 Discussion

The statistical models discussed in this chapter represent some of the many models used in ecology and how to investigate parameter redundancy in those models. Other work on parameter redundancy and identifiability in ecological models includes:

- Models for fisheries data in Jiang et al. (2007), Cole and Morgan (2010b) (discussed Section 5.2.2.3), Allen et al. (2017) and Nater et al. (2019);

- Jolly-Seber and stop-over models in Pledger et al. (2009) and Matechou (2010);

- Tag loss Jolly-Seber models in Cai et al. (2019);

- Mixture model in capture-recapture in Pledger et al. (2010) and Yu (2015);

- Movement Patterns in Johnson et al. (2017);

- N-mixture model in Kery (2018) (and also discussed in Section 2.2.4).

Ecological models can be complex, and issues of identifiability and parameter redundancy may not be obvious. Utilising a parameter redundant model without knowledge of this can lead to bias and incorrect estimates of precision (as explained in Chapter 2). It is essential that anyone using ecological models has an understanding of when models are parameter redundant. However some of the methods described in this chapter, such as the symbolic method, require some mathematical background to use. One possible solution is to create general tables of parameter redundancy results when a new model is created, as demonstrated for the mark-recovery model in Section 9.1.1. A list of general tables of parameter redundancy results that have been created is given below:

- Mark-recovery models, in Cole et al. (2012) (some of which are summarised in Section 9.1.1);

- Age-dependent mixture mark-recovery model in McCrea et al. (2013);

- Mark-recovery models for historic data with missing total, in Jimenez-Munoz et al. (2019);

- Capture-recapture CJS models and capture-recapture-recovery model in Hubbard et al. (2014);

- Jolly-Seber and Stop-Over Models in Matechou (2010);

- Tag Loss Jolly Seber Models in Cai et al. (2019);

- Removal models in Zhou et al. (2019);

- Mixture model in capture-recapture, where general tables of results are found using simulation in Pledger et al. (2010) and confirmed using the symbolic method in Yu (2015).

Another solution is to include parameter redundancy checks as part of statistical packages. For example the program Mark (Cooch and White, 2017) uses the Hessian method described in Section 4.1.1, and the programs M-Surge (Choquet et al., 2004) and E-Surge (Choquet et al., 2009) includes the hybrid symbolic-numerical method described in Section 4.3.

10

Concluding Remarks

Hopefully this book has convinced you that checking for parameter redundancy or investigating identifiability is an important part of inference.

When should this checking take place? Ideally identifiability should first be considered at the experimental design stage. By collecting more information, or designing an experiment in a specific way problems with identifiability can be avoided before the experiment is conducted (see, for example, Section 9.3). However, it is not usually possible to influence the design stage of the experiment, in which case it is recommended that identifiability of the model is checked before a model is fitted or used, to detect identifiability issues caused by the inherent structure of the model. If a new model is developed, then checking identifiability and presenting identifiability results within the journal article on the model can be very useful to anyone who will use that paper. For example, McCrea et al. (2013) developed an age-dependent mixture model for mark-recovery data and Zhou et al. (2019) developed a robust design model for removal models (discussed in Section 9.3.1.2); both include a study on parameter redundancy of the models. As parameter redundancy or near redundancy can also be caused by a specific data set it is also important to include identifiability checks as part of the model fitting process.

As we have established the importance of checking identifiability, Section 10.1 below summarises the recommended methods for checking identifiability for different situations. The ideal model for a particular data set or problem may be non-identifiable, Section 10.2 discusses how to use a such a model.

10.1 Which Method to Use and When

When checking for identifiability caused by the inherent model structure it is recommended that the symbolic method is used (Sections 3.2 and 4.2), as this method is accurate, can be used to find estimable parameter combinations in parameter redundant models and can be used to generalise results. If the model is structurally too complex to obtain the rank of the appropriate derivative matrix, one option is the extended symbolic method (Section 5.2). However, as finding a simpler exhaustive summary is example dependent, this method is not always easy to use. An alternative is the hybrid symbolic-numerical method (Section 4.3), which is much simpler to use. Using the extended symbolic method (Section 5.3) can also obtain the estimable parameter combinations in parameter redundant models and can be used to generalise results.

When checking for identifiability for a specific data set (with a particular model) the hybrid symbolic-numeric method is ideal. It is more accurate than than the Hessian method (Section 4.1.1), or other numerical methods, and can be added to a computer package to provide automatic identifiability results. For example the hybrid symbolic-numeric method has been added the computer package M-Surge (Choquet et al. 2004) and E-Surge (Choquet et al. 2009), used for multi-state and multi-event models used in ecology.

10.2   Inference with a Non-Identifiable Model

Suppose that we have established that a model is parameter redundant or some of the the parameters are non-identifiable, but we wish to use that particular model. How should we proceed?

- **Do not use in current format.** If you are interested in the parameter estimates it is not recommended that the model is used in the current format, as this will result in biased parameter estimates for the non-identifiable parameters and misleading standard errors (see Sections 2.3.2 and 2.2).

- **Reparameterise model.** It is possible to reparameterise the model in terms of the estimable parameter combinations (see Sections 3.2.5 and 4.2). However it is possible that the reparameterised parameters are not useful, for example, they might not be biologically meaningful.

- **Add constraints on the parameters.** Parameters can be constrained, for example one confounded parameter can be set to zero. The estimable parameter combinations can be used to obtain suitable constraints (see Section 9.4). This is only recommended if the constraints are considered when making conclusions, and if the constraints are sensible.

- **Use informative priors.** A Bayesian analysis with informative priors on non-identifiable parameters can be used to over-come problems of non-identifiability. However, as demonstrated in Section 6.3, results will be heavily reliant on the priors used. This is only recommended if reliable prior information is available.

- **Add covariates or random effects.** Adding covariates and/or random effects to the certain parameters can separate out confounded parameters, see Section 5.1. This method is only recommend if the data has a strong dependence on the covariate or random effect. The model will be near redundant if the covariate or random effect is close to zero.

- **Collect more data:** It may be possible to collect more data. If the non-identifiability is caused by the model structure, then this would need to be a different type of data, with a different model that has information on some of the parameters. For example in ecology, integration population models are created, involving two or more different types of data. This was discussed in Section 9.2. However it does not follow that all integrated models will be identifiable, and identifiability still needs to be checked in the integrated models.

A

Maple Code

In this Appendix we explain how to use Maple code to execute the symbolic method and the hybrid symbolic-numerical method.

A.1 Checking Parameter Redundancy / Non-Identifiability

This Section explains how to use Maple to check for parameter redundancy or non-identifiability.

In Maple, to construct and manipulate matrices and vectors the `LinearAlgebra` package is used. This package is activated using the Maple code:

```
> with(LinearAlgebra):
```

The Maple procedure `Dmat`, given in Figure A.1, can be used to find the derivative matrix. If the exhaustive summary is stored in vector `kappa` and the parameters are stored in vector `pars`, then the Maple code for finding the derivative matrix is:

```
> D1 := Dmat(kappa,pars);
```

The rank of a matrix can be found using the intrinsic Maple procedure `Rank`. The following code is then used to find the rank of the derivative matrix and calculate the deficiency.

```
> r := Rank(D1); d := Dimension(pars)-r;
```

Note that the intrinsic Maple procedure `Dimension` finds the length of the vector of parameters, so calculates the number of parameters automatically.

Example Maple code for the basic occupancy model is given in Section A.1.1.

A.1.1 Basic Occupancy Model

In Section 3.2.1 we showed that the basic occupancy model, first introduced in Section 1.2.1, was parameter redundant using the derivative matrix method. The Maple code for finding the derivative matrix and its rank is

```
Dmat := proc (se, pars)
local DD1, i, j;
description "This procedure finds the derivative matrix ...";
with(LinearAlgebra);
DD1 := Matrix(1..Dimension(pars),1..Dimension(se));
for i to Dimension(pars) do
   for j to Dimension(se) do
      DD1[i, j] := diff(se[j],pars[i])
   end do
end do;
DD1
end proc
```

FIGURE A.1
The Maple procedure Dmat returns the derivative matrix given inputs se and pars. The vector se is the exhaustive summary. The vector pars is a vector of parameters.

```
> with(LinearAlgebra):
> kappa1 := <psi*p,-p*psi+1>:
> pars := <psi,p>:
> D1 := Dmat(kappa,pars);
> r := Rank(D1); d := Dimension(pars)-r;
```

The code returns a rank of $r = 1$ and a deficiency of $d = 1$, showing the model is parameter redundant.

A.2   Finding Estimable Parameter Combinations

The Maple procedure Estpar, given in Figure A.2, can be used to find the estimable parameter combinations or a locally identifiable reparameterisation.
   The code can be executed using

```
> Estpar(D1,pars,0)
```

to just obtain the estimable parameter combinations, where D1 is the derivative matrix and pars is a vector of parameters. The 0 indicates that just the estimable parameter combinations are displayed. Alternatively the code

```
> Estpar(D1,pars,1)
```

can be used obtain the null-space and PDEs as well.

```
Estpar := proc (DD1, pars, ret)
local r, d, alphapre, alpha, PDE, FF, i, ans;
description "Finds the estimable parameter combinations ...";
with(LinearAlgebra);
r := Rank(DD1);
d := Dimension(pars)-r;
alphapre := NullSpace(Transpose(DD1));
alpha := Matrix(d, Dimension(pars));
PDE := Vector(d);
FF := f(seq(pars[i],i=1..Dimension(pars)));
for i to d do
  alpha[i,1..Dimension(pars)]:=alphapre[i];
  PDE[i]:=add((diff(FF,pars[j]))*alpha[i,j],j=1..Dimension(pars))
end do;
if ret = 1 then
  ans:=<(pdsolve({seq(PDE[i]=0,i=1..d)}),{alpha},{PDE})>:
else
  ans:=pdsolve({seq(PDE[i]=0,i=1..d)}):
end if:
ans:
end proc
```

FIGURE A.2

The Maple procedure `Estpar` returns the estimable parameter combinations. The input `DD1` is a derivative matrix. The input `par` is a vector of parameters. The input `ret` specifies the output. If `ret` has a value of 1 then the PDEs and nulls-space is returned as well as the estimable parameter combinations. Otherwise only the estimable parameter combinations are returned.

A.2.1 Basic Occupancy Model

For the basic occupancy model the Maple code

```
> Estpar(D1, pars, 0)
```

returns

$$\{f(\psi,p) = \_F1(p\psi)\}$$

The estimable parameter combination is given in the bracket of the function $F1$, which in this case is $p\psi$.

A.3   Distinguishing Essentially and Conditionally Full Rank Models

To distinguish between essentially and conditionally full rank models we apply a PLUR decomposition and then find the determinate of the upper triangular matrix, see Theorem 7 of Section 3.2.7. Maple has an intrinsic function for finding a PLUR decomposition and a determinant. To perform a PLUR decomposition on derivative matrix D1 and then find the determinant of **U**, we use the following Maple code:

```
(pp,ll,u1,r1):=LUDecomposition(D1,output = ['P','L','U1','R']);
DetU := Determinant(u1);
```

The matrices **P**, **L**, **U** and **R** are stored in matrices pp, ll, u1 and r1 respectively. The determinant of **U** is stored in the variable DetU.

A.4   Hybrid Symbolic-Numerical Method

The Maple procedure Hybrid, given in Figure A.3, can be used to execute the hybrid symbolic-numerical method. The Maple code for the hybrid method is:

```
> results := Hybrid(D1,pars,minpars,maxpars,1);
```

The derivative matrix is stored in vector D1 and the parameters are stored in vector pars. The method involves evaluating the parameters at random points in the parameter space. The minimum and maximum values for the parameters are stored in minpars and maxpars. A value of 1 indicates the full results are returned.

A.4.1   Basic Occupancy Model

The Maple code for executing the hybrid symbolic-numerical method for the basic occupancy model is

```
> Hybrid(D1,pars,0.0,1.0,1)
```

```
Hybrid := proc(D1,pars,minpars,maxpars,ret)
local results, j, numpars, D1rand, ans, roll;
description "Hybrid symbolic-numerical method.";
results := Matrix(5, 2);
for j to 5 do
    roll:=rand(minpars..maxpars);
    numpars:=seq(pars[i]=evalf(roll()),i=1..Dimension(pars));
    D1rand:=eval(D1,{numpars});
    results[j,1]:=Rank(D1rand);
    results[j,2]:=NullSpace(Transpose(D1rand))
end do;
if ret = 1 then
    ans := results
else
    ans := max(results[1..5,1])
end if;
ans
end proc
```

FIGURE A.3

The Maple procedure Hybrid executes the hybrid symbolic-numerical method. The input D1 is a derivative matrix. The input par is a vector of parameters. The input ret specifies the output. If ret has a value of 1 then the full results are returned. Otherwise only the model rank is returned. The inputs minpars and maxpars are minimum and maximum values for the parameters, which should be given to at least one decimal place, e.g., 0.0 or 1.0.

One execution of this code returned:

$$
\begin{bmatrix}
1 & \left\{ \begin{bmatrix} -0.870187995940668202 \\ 0.492719851153537058 \end{bmatrix} \right\} \\
1 & \left\{ \begin{bmatrix} 0.707408188903686396 \\ -0.706805244938098931 \end{bmatrix} \right\} \\
1 & \left\{ \begin{bmatrix} 0.889311764810177042 \\ -0.457301415884762252 \end{bmatrix} \right\} \\
1 & \left\{ \begin{bmatrix} 0.836979437164332318 \\ -0.547234338984751045 \end{bmatrix} \right\} \\
1 & \left\{ \begin{bmatrix} -0.802310478949957440 \\ 0.596906940290603538 \end{bmatrix} \right\}
\end{bmatrix}
$$

Each row refers to a different set of random parameter values. The first column gives the rank, and the second column gives the left null space. In this case none of the entries in the null space are close to zero, therefore none of the parameters are individually identifiable. Note that running the code again will return slightly different values, as these depended on the random parameter values generated.

B

Winbugs and R Code

This appendix provides the R code for the Hessian method in Section B.1 and the Winbugs code for data cloning in Section B.2.

B.1 Hessian Method

In this section we explain how to execute the Hessian method, from Section 4.1.1, in R.

For log-likelihood $l = l(\boldsymbol{\theta}|\mathbf{y})$ the Hessian matrix is

$$\mathbf{H} = \begin{bmatrix} \frac{\partial^2 l}{\partial\theta_1\partial\theta_1} & \frac{\partial^2 l}{\partial\theta_1\partial\theta_2} & \cdots & \frac{\partial^2 l}{\partial\theta_1\partial\theta_p} \\ \frac{\partial^2 l}{\partial\theta_2\partial\theta_1} & \frac{\partial^2 l}{\partial\theta_2\partial\theta_2} & \cdots & \frac{\partial^2 l}{\partial\theta_2\partial\theta_p} \\ \vdots & & & \vdots \\ \frac{\partial^2 l}{\partial\theta_p\partial\theta_1} & \frac{\partial^2 l}{\partial\theta_p\partial\theta_2} & \cdots & \frac{\partial^2 l}{\partial\theta_p\partial\theta_p} \end{bmatrix}.$$

It is frequently not possible to find the exact Hessian matrix, instead a numerical method is used to approximate the partial derivatives. Several different methods exist to approximate the derivatives; see, for example, Dennis and Schnabel (1996) and Ridout (2009). The method used here approximates the partial derivative as

$$\frac{\partial^2 l(\boldsymbol{\theta})}{\partial\theta_i\partial\theta_j} = \frac{l(\boldsymbol{\theta}+\mathbf{e}_i\delta+\mathbf{e}_j\delta)-l(\boldsymbol{\theta}+\mathbf{e}_i\delta-\mathbf{e}_j\delta)-l(\boldsymbol{\theta}-\mathbf{e}_i\delta+\mathbf{e}_j\delta)+l(\boldsymbol{\theta}-\mathbf{e}_i\delta-\mathbf{e}_j\delta)}{4\delta^2}$$

where $\mathbf{e}_i$ is a vector with 1 in the ith position and 0 elsewhere and where δ is the step value that takes on a suitable small value. R code for finding the Hessian matrix numerically using this method is given in Figure B.1.

Theoretically δ should be chosen to be as small as possible to minimise the approximation error. However there is also potential to introduce round-off error for small delta. Ridout (2009) discuss formal methods for choosing the best choice of δ. Here we adopt an ad hoc approach. For the basic occupancy model in Section 3.2.1 we find the exact Hessian matrix and in Section 4.1.1.1 we use the approximate Hessian matrix. Comparing the exact

```
hessian <- function(funfcn,x,y,delta){
  # function to calculate the Hessian of funfcn at x
  t <- length(x)
  h <- matrix(0, t, t)
  Dx <- delta*diag(t)
  for (i in 1:t) {
    for (j in 1:i) {
      h[i,j] <- ( do.call("funfcn",list(x+Dx[i,]+Dx[j,],y)) -
                  do.call("funfcn",list(x+Dx[i,]-Dx[j,],y)) -
                  do.call("funfcn",list(x-Dx[i,]+Dx[j,],y)) +
                  do.call("funfcn",list(x-Dx[i,]-Dx[j,],y)) )
                / (4*delta^2)
      h[j,i] <- h[i,j]
    }
  }
  return(h)
}
```

FIGURE B.1

R code for finding an approximate Hessian matrix of function funfcn at x with data y for step length delta. Adapted from the Matlab program hessian.m in Figure 3.3 of Morgan (2009).

and approximate Hessian matrix we found that if $\delta = 0.0001$ or 0.00001 the approximation was correct to 4 decimal places, but if $\delta = 0.001$, 0.000001 or 0.0000001 the approximation was not correct to 4 decimal places. Experience with other examples suggest that $\delta = 0.000001$ is a good choice. However it is recommended that different values of δ are examined in every example.

For parameter redundant models the estimable parameters, that is parameters that are individually at least locally identifiable, can be found by examining the eigenvectors associated with the small eigenvalues. Any entries close to zero for all of these eigenvalues correspond to parameters that are at least locally identifiable. The cut-off for what counts as small is ϵ. The value of ϵ may vary from example to example. We use a value of $\epsilon = 0.01$, but also recommend examining the eigenvectors to decide what might be a suitable value for ϵ.

The Hessian method can be executed using the R function, hessianmethod, given in Figure B.2. For example, with the basic occupancy model suppose the log-likelihood is found using the function occlik, the maximum likelihood estimates are stored in the vector maxlikpar and the data is stored in the vector y. Then the R code for applying the Hessian method is:

```
results<-hessianmethod(occlik,maxlikpar,y,0.00001,0.01,print=TRUE)
```

```
hessianmethod<-function(funfcn,pars,y,delta,epsilon,print=TRUE){
  cutoff<-delta*length(pars)
  h<-do.call("hessian",list(funfcn,pars,y,delta))
  E<-eigen(h)
  standeigenvalues<-abs(E$values)/max(abs(E$values))
  noestpars<-0
  smalleig<-c( )
  for (i in 1:length(pars)) {
    if (standeigenvalues[i] >= cutoff) {
      noestpars<-noestpars + 1          }
    else {smalleig<-c(smalleig,i)}
  }
  identpars<-c( )
  if (min(standeigenvalues) < cutoff) {
    for (i in 1:length(pars)) {
        indent<-1
        for (j in 1:length(smalleig)) {
            if (abs(E$vectors[i,smalleig[j]]) > epsilon) {
              indent<-0 }
        }
        if (indent==1) {identpars <- c(identpars,i)}
    }
  }
  if (print) {
    if (min(standeigenvalues) < cutoff) {
      cat("model is non-identifiable or parameter redundant")
      cat("\n")
      if (is.null(identpars)) {
        cat('none of the original parameters are estimable') }
      else {cat('estimable parameters',identpars)}
    }
    else {cat("model is identifiable or not parameter redundant")}
    cat("\n")
    cat('smallest standardised eigenvalue',min(standeigenvalues))
    cat("\n")
    cat('number of estimable parameters',noestpars)
  }
  result<-list(standeigenvalues=standeigenvalues,
          noestpars=noestpars,identpars=identpars,E=E)
  return(result)
}
```

FIGURE B.2

R code for applying the Hessian method for log-likelihood function funfcn at
x, with data y, for step length delta, using a cut off of espilon to classify a
parameter as estimable / individually identifiable.

Here values of $\delta = 0.00001$ and $\epsilon = 0.01$ are used. These are the recommended values, but can be changed. (See `bassicocc.R` for full code).

As `print=TRUE` the output is

```
model is non-identifiable or parameter redundant
none of the original parameters are estimable
smallest standardised eigenvalue 1.735457e-06
number of estimable parameters 1
```

The output will specify whether the model is parameter redundant. If the model is parameter redundant it also specifies whether any of the original parameters are estimable. In this case none are; if any are, a list of the positions of the identifiable parameter is given. The smallest standardised eigenvalue and the number of estimable parameters are displayed.

Further output is stored in results: `results$standeigenvalues` stores the standardised eigenvalues, `results$noestpars` stores the number of estimable parameters, `results$identpars` the identifiable parameter positions (if none of the original parameters are identifiable will return null), `results$E$values` gives the eigenvalues and `results$E$vectors` gives the eigenvectors.

B.2   Data Cloning

Data cloning involves using an MCMC method to generate samples from the posterior distribution. Here we use the program Winbugs (Lunn et al., 2000).

The Winbugs code used in Section 4.1.3.1 for the basic occupancy model with parameterisation ψ, p is

```
> with(LinearAlgebra):
model {
# Specifying model which follows a binomial distribution
for(j in 1 :n) {
   beta[j]<-p*psi
   y[j]~ dbin(beta[j],1)
}
# Specifying uniform priors
p ~ dunif(0,1)
psi ~ dunif(0,1)
}
```

The Winbugs code for the parameterisation β is

```
model {
# Specifying model which follows a binomial distribution
for(j in 1 :n) {
```

```
    y[j]~ dbin(beta,1)
}
# Specifying uniform prior
beta ~ dunif(0,1)
}
```

Rather than using Winbugs directly, Winbugs can also be called from from R using the package R2WinBUGS (Sturtz et al., 2005). For example, this method is used in the R code pop.R.

Bibliography

Abadi, F., O. Gimenez, B. Ullrich, R. Arlettaz and M. Schaub (2010), Estimation of immigration rate using integrated population models. *Journal of Applied Ecology* **47**, 393–400 (Cited on pages 136, 209, and 210.)

Agresti, A. (1990), *Categorical data analysis*. Wiley (Cited on pages 132 and 134.)

Akaike, H. (1992), Information theory and an extension of the maximum likelihood principle. In S. Kotz and N. L. Johnson (eds.), *Breakthroughs in Statistics, Vol. I, Foundations and Basic Theory*, pp. 610–624, Springer (Cited on page 31.)

Albert, A. and J. A. Anderson (1984), On the existence of maximum likelihood estimates in logistic regression models. *Biometrika* **71**, 1–10 (Cited on page 22.)

Allen, S. D., W. H. Satterthwaite, D. G. Hankin, D. J. Cole and M. S. Mohr (2017), Temporally varying natural mortality: Sensitivity of a virtual population analysis and an exploration of alternatives. *Fisheries Research* **185**, 185–197 (Cited on page 215.)

Allman, E. S., C. Matias and J. A. Rhodes (2009), Identifiability of parameters in latent structure models with many observed variables. *The Annuals of Statistics* **37**, 3099–3132 (Cited on page 186.)

Anderson, D. H. (1983), *Compartment modelling and tracer kinetics*. Springer-Verlag (Cited on page 155.)

Anderson, R. M. and R. M. May (1991), *Infectious diseases of humans: dynamics and control*. Oxford University Press (Cited on page 171.)

Arnason, A. N. (1973), The estimation of population size, migration rates and survival in a stratified population. *Researches on Population Ecology* **15**, 1–8 (Cited on page 199.)

Audoly, S., L. D'Angio, M. P. Saccomani and C. Cobelli (1998), Global identifiability of linear compartmental models - a computer algebra algorithm. *IEEE Transactions on Biomedical Engineering* **45**, 36–47 (Cited on pages 92, 93, 110, and 111.)

Auger-Méthé, M., C. Field, C. M. Albertsen, A. E. Derocher, M. A. Lewis, I. D. Jonsen and J. Mills Flemming (2016), State-space models' dirty little secrets: even simple linear Gaussian models can have estimation problems. *Scientific Reports* **6**, 26677 (Cited on pages 77 and 174.)

Bailey, L. L., S. J. Converse and W. L. Kendall (2010), Bias, precision, and parameter redundancy in complex multistate models with unobservable states. *Ecology* **91**, 1598–1604 (Cited on pages 12, 70, and 77.)

Balsa-Canto, E., A. Alonso and J. R. Banga (2010), An iterative identification procedure for dynamic modeling of biochemical networks. *BMC Systems Biology* **4**, 11 (Cited on page 167.)

Bekker, P. A., A. Merckens and T. J. Wansbeek (1994), *Identification, equivalent models and computer algebra*. Academic Press (Cited on pages 36, 44, and 62.)

Bellman, R. and K. J. Astrom (1970), On structural identifiability. *Mathematical Biosciences* **7**, 329–339 (Cited on pages 40, 63, 156, and 159.)

Bellu, G., M. Saccomani, S. Audoly and L. D'Angio (2007), DAISY: A new software tool to test global identifiability of biological and physiological systems. *Computer Methods and Programs in Biomedicine* **88**, 52–61 (Cited on page 167.)

Bernardo, J. M. and A. F. M. Smith (1994), *Bayesian theory*. Wiley (Cited on page 125.)

Berry, J. and K. Houston (1995), *Mathematical modelling*. Butterworth-Heinemann (Cited on pages 4 and 5.)

Besbeas, P., S. N. Freeman, B. J. T. Morgan and E. A. Catchpole (2002), Integrating mark-recapture-recovery and census data to estimate animal abundance and demographic parameters. *Biometrics* **58**, 540–547 (Cited on pages 170, 201, and 202.)

Brooks, S. P., E. A. Catchpole, B. J. T. Morgan and S. C. Barry (2000), On the Bayesian analysis of ring-recovery data. *Biometrics* **56**, 951–956 (Cited on page 141.)

Brooks, S. P., R. King and B. J. T. Morgan (2004), A Bayesian approach to combining animal abundance and demographic data. *Animal Biodiversity and Conservation* **27**, 515–529 (Cited on page 209.)

Brownie, C., J. E. Hines, J. D. Nichols, K. H. Pollock and J. Hestbeck (1993), Capture-recapture studies for multiple strata including non-Markovian transitions. *Biometrics* **49**, 1173–1187 (Cited on pages 199 and 213.)

Burnham, K. P. and D. R. Anderson (2004), Multimodel inference: understanding AIC and BIC in Model Selection. *Sociological Methods and Research* **33**, 261–304 (Cited on page 31.)

Burnham, K. P., D. R. Anderson, G. C. White, C. Brownie and K. P. Pollock (1987), *Design and analysis of methods for fish survival experiments based on release–recapture.* American Fisheries Society (Cited on page 77.)

Cai, W., S. Yurchak, L. L. E. Cowen and D. J. Cole (2019), Parameter redundancy in Jolly-Seber tag loss models. *Submited for publication* (Cited on pages 215 and 216.)

Campbell, D. and S. Lele (2014), An ANOVA test for parameter estimability using data cloning with application to statistical inference for dynamic systems. *Computational Statistics and Data Analysis* **70**, 257–267 (Cited on pages 81 and 99.)

Carlin, B. P. and T. A. Louis (2000), *Bayes and empirical Bayes methods for data analysis.* Chapman and Hall/CRC (Cited on page 128.)

Catchpole, E. A., S. N. Freeman and B. J. T. Morgan (1998a), Estimation in parameter-redundant models. *Biometrika* **85**, 462–468 (Cited on pages 43 and 49.)

Catchpole, E. A., S. N. Freeman, B. J. T. Morgan and M. P. Harris (1998b), Integrated recovery/recapture data analysis. *Biometrics* **54**, 33–46 (Cited on page 194.)

Catchpole, E. A., P. M. Kgosi and B. J. T. Morgan (2001), On the near-singularity of models for animal recovery data. *Biometrics* **57**, 720–726 (Cited on pages 37, 67, 69, 85, and 94.)

Catchpole, E. A. and B. J. T. Morgan (1996), Model Selection in Ring-Recovery Models Using Score Tests. *Biometrics* **52**, 664–672 (Cited on page 32.)

Catchpole, E. A. and B. J. T. Morgan (1997), Detecting parameter redundancy. *Biometrika* **84**, 187–196 (Cited on pages 14, 36, 37, 41, 42, 44, 52, 62, and 83.)

Catchpole, E. A. and B. J. T. Morgan (2001), Deficiency of parameter-redundant models. *Biometrika* **88**, 593–598 (Cited on pages 52 and 53.)

Catchpole, E. A., B. J. T. Morgan, S. N. Freeman and W. J. Peach (1999), Modelling the survival of British lapwings Vanellus vanellus using weather covariates. *Bird Study* **46**, 5–13 (Cited on page 6.)

Chappell, M. J. and R. N. Gunn (1998), A procedure for generating locally identifiable reparameterisations of unidentifiable non-linear systems by the similarity transformation approach. *Mathematical Biosciences* **148**, 21–41 (Cited on pages 42, 49, 158, and 166.)

Chis, O.-T., J. R. Banga and E. Balsa-Canto (2011a), Structural identifiability of systems biology models: a critical comparison of methods. *PLoS ONE* **6**, e27755 (Cited on pages 101, 155, 158, 162, 166, and 167.)

Chis, O.-T., J. R. Banga and E. Balsa-Canto (2011b), GenSSI: a software toolbox for structural identifiability analysis of biological models. *Bioinformatics* **27**, 2610–2611 (Cited on page 167.)

Chis, O.-T., A. F. Villaverde, J. R. Banga and E. Balsa-Canto (2016), On the relationship between sloppiness and identifiability. *Mathematical Biosciences* **282**, 147–161 (Cited on pages 37, 68, and 94.)

Choquet, R. and D. J. Cole (2012), A hybrid symbolic-numerical method for determinig model structure. *Mathematical Biosciences* **236**, 117–125 (Cited on pages 12, 60, 90, and 91.)

Choquet, R., A.-M. Reboulet, R. Pradel, O. Gimenez and J.-D. Lebreton (2004), M-SURGE: New software specifically designed for multistate recapture models. *Animal Biodiversity and Conservation* **27**, 207–215 (Cited on pages 91, 99, 216, and 218.)

Choquet, R., L. Rouan and R. Pradel (2009), Program E-SURGE: A software application for fitting multievent models. In D. L. Thomson, E. G. Cooch and M. J. Conroy (eds.), *Modeling demographic processes in marked populations.*, pp. 845–865, Springer (Cited on pages 91, 99, 216, and 218.)

Claeskens, G. and N. L. Hjort (2008), *Model selection and model averaging.* Cambridge University Press (Cited on page 31.)

Cole, D. J. (2012), Determining parameter redundancy of multi-state mark-recapture models for sea birds. *Journal of Ornithology* **152**, 305–315 (Cited on pages 116, 197, and 198.)

Cole, D. J. (2019), Parameter redundancy and identifiability in hidden Markov models. *Metron* **77**, 105–118 (Cited on pages 68, 94, 186, and 187.)

Cole, D. J. and R. S. McCrea (2016), Parameter Redundancy in Discrete State-Space and Integrated Models. *Biometrical Journal* **58**, 1071–1090 (Cited on pages 170, 171, 177, 179, 180, 183, 188, 202, 203, 204, 205, and 207.)

Cole, D. J. and B. J. T. Morgan (2010a), Parameter redundancy with covariates. *Biometrika* **97**, 1002–1005 (Cited on pages 60, 61, 101, 102, 103, 104, and 105.)

Cole, D. J. and B. J. T. Morgan (2010b), A note on determining parameter redundancy in age-dependent tag return models for estimating fishing mortality, natural mortality and selectivity. *Journal of Agricultural Biological and Environmental Statistics* **15**, 431–434 (Cited on pages 113 and 215.)

Cole, D. J., B. J. T. Morgan, E. A. Catchpole and B. A. Hubbard (2012), Parameter redundancy in mark-reovery models. *Biometrical Journal* **54**, 507–523 (Cited on pages 7, 41, 48, 54, 85, 141, 190, 191, 192, and 216.)

Cole, D. J., B. J. T. Morgan, R. S. McCrea, R. Pradel, O. Gimenez and R. Choquet (2014), Does your (study) species have memory? Analysing capture-recapture data with memory models. *Ecology and Evolution* **4**, 2124–2133 (Cited on pages 116, 200, 201, 213, 214, and 215.)

Cole, D. J., B. J. T. Morgan and D. M. Titterington (2010), Determining the parametric structure of models. *Mathematical Biosciences* **228**, 16–30 (Cited on pages 12, 36, 37, 38, 42, 43, 49, 52, 53, 57, 58, 62, 63, 67, 68, 86, 93, 105, 108, 109, 110, 112, 117, 128, and 213.)

Cooch, E. and G. White (2017), *Program Mark. A gentle introduction.* http://www.phidot.org/software/mark/docs/book/ (Cited on pages 73, 81, 99, and 216.)

Corless, R. M. and D. J. Jeffrey (1997), The Turing factorization of a rectangular matrix. *SIGSAM Bulletin* **31**, 20–30 (Cited on pages 57 and 58.)

Cormack, R. M. (1964), Estimates of survival from the sighting of marked animals. *Biometrika* **51**, 429–438 (Cited on pages 109 and 191.)

Craciun, G. and C. Pantea (2008), Identifiability of chemical reaction networks. *Journal of Mathematical Chemistry.* **44**, 244–259 (Cited on page 166.)

Dasgupta, A., S. G. Self and S. Das Gupta (2007), Non-identifiable parametric probability models and reparametrization. *Journal of Statistical Planning and Inference* **137**, 3380–3393 (Cited on pages 14, 49, and 50.)

Davidescu, F. P. and S. B. Jorgensen (2008), Structural parameter identifiability analysis for dynamic reaction networks. *Chemical Engineering Science* **63**, 4754–4762 (Cited on page 166.)

Davison, A. C. (2003), *Statistical models.* Cambridge University Press (Cited on pages 4 and 6.)

Dawid, A. P. (1979), Conditional independence in statistical theory. *Journal of the Royal Statistical Society, Series B* **41**, 1–31 (Cited on page 127.)

Delforge, J. (1989), Relations between the main approaches to linear system identifiability: application to calculation of jacobian matrices determinants. *International Journal of Systems Science* **20**, 1079–1097 (Cited on page 42.)

Denis-Vidal, L. and G. Joly-Blanchard (2000), An easy to check criterion for (un)identifiability of uncontrolled systems and its applications. *IEEE Transactions on Automatic Control* **45**, 768–771 (Cited on page 166.)

Dennis, E. B., B. J. T. Morgan and M. S. Ridout (2015), Computational Aspects of N-Mixture Models. *Biometrics* **71**, 237–246 (Cited on pages 26, 27, and 28.)

Dennis, J. E. and R. J. Schnabel (1996), *Numerical methods for unconstrained optimization and nonlinear equations*. Society for Industrial and Applied Mathematics (Cited on page 227.)

Dufresne, E., H. A. Harrington and D. V. Raman (2018), The geometry of sloppiness. *Journal of Algebraic Statistics* **9**, 30–68 (Cited on pages 37 and 68.)

Eisenberg, M. C. and M. A. L. Hayashi (2014), Determining identifiable parameter combinations using subset profiling. *Mathematical Biosciences* **256**, 116–126 (Cited on pages 18 and 118.)

Evans, N. D. and M. J. Chappell (2000), Extensions to a procedure for generating locally identifiable reparameterisations of unidentifiable systems. *Mathematical Biosciences* **168**, 137–159 (Cited on pages 43, 49, and 156.)

Fisher, F. M. (1959), Generalization of the Rank and Order Conditions for Identifiability. *Econometrica* **27**, 431–447 (Cited on page 42.)

Fisher, F. M. (1961), Identifiability Criteria in Nonlinear Systems. *Econometrica* **29**, 574–590 (Cited on page 42.)

Fisher, F. M. (1966), *The identification problem in economics*. New York, McGraw-Hill Book Company (Cited on page 8.)

Fiske, I. J. and R. B. Chandler (2011), Unmarked: An R package for fitting hierarchical models of wildlife occurence and abundance. *Journal of Statistical Software* **43**, 1–23 (Cited on page 26.)

Forcina, A. (2008), Identifiability of extended latent class models with individual covariates. *Computational Statistics and Data Analysis* **52**, 5263–5268 (Cited on pages 12, 23, 91, 101, and 104.)

Freeman, S. N. and B. J. T. Morgan (1990), Studies in the analysis of ring-recovery data. *The Ring* **13**, 271–287 (Cited on page 6.)

Freeman, S. N. and B. J. T. Morgan (1992), A modelling strategy for recovery data from birds ringed as nestlings. *Biometrics* **48**, 217–235 (Cited on page 6.)

Freeman, S. N., B. J. T. Morgan and E. A. Catchpole (1992), On the augmentation of ring–recovery data with field information. *Journal of Animal Ecology* **61**, 649–657 (Cited on page 83.)

Gamerman, D. and H. F. Lopes (2006), *Markov chain Monte Carlo stochastic simulation for Bayesian inference*. Chapman and Hall/CRC (Cited on page 126.)

Garrett, E. S. and S. L. Zeger (2000), Latent class model diagnosis. *Biometrics* **56**, 1055–1067 (Cited on pages 128, 135, and 136.)

Gelfand, A. E. and S. K. Sahu (1999), Identifiability, improper priors, and Gibbs sampling for generalized linear models. *Journal of the American Statistical Association* **94**, 247–253 (Cited on pages 125, 127, 128, 132, 133, 134, 136, 146, and 150.)

Gimenez, O., L. Blanc, A. Besnard, R. Pradel, P. F. Doherty, E. Marboutin and R. Choquet (2014), Fitting occupancy models with E-SURGE: hidden Markov modelling of presence–absence data. *Methods in Ecology and Evolution* **5**, 592–597 (Cited on page 211.)

Gimenez, O., R. Choquet and J.-D. Lebreton (2003), Parameter redundancy in multistate capture-recapture models. *Biometrical Journal* **45**, 704–722 (Cited on pages 57, 61, and 197.)

Gimenez, O., B. J. T. Morgan and S. P. Brooks (2009), Weak identifiability in models for mark-recapture-recovery data. In D. L. Thomson, E. G. Cooch and M. J. Conroy (eds.), *Modeling demographic processes in marked populations.*, pp. 1055–1067, Springer (Cited on pages 125, 128, 129, and 136.)

Gimenez, O., A. Viallefont, E. A. Catchpole, R. Choquet and B. J. T. Morgan (2004), Methods for investigating parameter redundancy. *Animal Biodiversity and Conservation* **27**, 561–572 (Cited on pages 37, 71, 72, 73, 77, 78, and 83.)

Godfrey, K. (1983), *Compartmental models and their application.* Academic Press (Cited on pages 9 and 155.)

Godfrey, K. R. and M. J. Chapman (1990), Identifiability and indistinguishability of linear compartment models. *Mathematics and Computer Simulations* **32**, 273–295 (Cited on pages 10, 40, 63, and 157.)

Godfrey, K. R. and J. J. Distefano (1987), Identifiability of model parameters. In E. Walter (ed.), *Identifiability of parametric models.*, pp. 1–20, Pergamon Press (Cited on pages 63, 156, 159, and 161.)

Goodman, L. A. (1974), Exploratory latent structure analysis using both identifiable and unidentifiable models. *Biometrika* **61**, 215–231 (Cited on page 42.)

Graybill, F. A. (2001), *Matrices with applications in statistics, 2nd edition.* Cengage Learning (Cited on page 57.)

Hines, J. E. (2011), *Program PRESENCE 4.1 - Software to estimate patch occupancy and related parameters.* U.S. Geological Survey Patuxent Wildlife Research Center, Maryland. (Cited on page 26.)

Hogg, R. V., J. W. McKean and A. T. Craig (2014), *Introduction to mathematical statistics*. Pearson Education (Cited on page 4.)

Holzmann, H., A. Munk and W. Zucchini (2006), On identifiability in capture-recapture models. *Biometrics* **62**, 934–936 (Cited on pages 64 and 66.)

Hubbard, B. A. (2014), *Parameter redundancy with applications in statistical ecology*. Ph.D. thesis, University of Kent (Cited on page 211.)

Hubbard, B. A., D. J. Cole and B. J. T. Morgan (2014), Parameter redundancy in capture-recapture-recovery models. *Statistical Methodology* **17**, 17–29 (Cited on pages 193, 194, 195, 196, and 216.)

Hunter, C. M. and H. Caswell (2009), Rank and redundancy of multistate mark-recapture models for seabird populations with unobservable states. In D. L. Thomson, E. G. Cooch and M. J. Conroy (eds.), *Modeling demographic processes in marked populations.*, pp. 797–826, Springer (Cited on pages 12, 91, 101, 105, 197, 198, and 213.)

Janzen, D. L. I., M. Jirstrand, M. J. Chappell and N. D. Evans (2018), Extending existing structural identifiability analysis methods to mixed-effects models. *Mathematical Biosciences* **295**, 1–10 (Cited on page 167.)

Janzen, D. L. I., M. Jirstrand, M. J. Chappell and N. D. Evans (2019), Three novel approaches to structural identifiability analysis in mixed-effects models. *Computer Methods and Programs in Biomedicine* **171**, 141–152 (Cited on page 167.)

Jiang, H., K. H. Pollock, C. Brownie, J. E. Hightower, J. E. Hoenig and W. S. Hearn (2007), Age-dependent tag return models for estimating fishing mortality, natural mortality and selectivity. *Journal of Agricultural, Biological, and Environmental Statistics* **12**, 177–194 (Cited on pages 12, 91, 101, 105, 112, and 215.)

Jimenez-Munoz, M., D. J. Cole, S. N. Freeman, R. A. Robinson, S. R. Baillie and E. Matechou (2019), Estimating age-dependent survival from age-aggregated ringing data - extending the use of historical records. *Ecology and Evolution* **9**, 769–779 (Cited on page 216.)

Johnson, L. R., P. H. Boersch-Supan, R. A. Phillips and S. J. Ryan (2017), Changing measurements or changing movements? Sampling scale and movement model identifiability across generations of biologging technology. *Ecology and Evolution* **9**, 9257–9266 (Cited on page 215.)

Jolly, G. M. (1965), Explicit Estimates from Capture-Recapture data with both Death and Immigration-Stochastic Model. *Biometrika* **52**, 225–247 (Cited on pages 109 and 191.)

Kadane, J. B. (1975), The role of identification in bayesian theory. In S. E. Fienberg and A. Zellner (eds.), *Studies in Bayesian Econometrics and Statistics.*, pp. 175–191, North-Holland (Cited on pages 125 and 127.)

Keeling, M. J. and P. R. Rohani (2008), *Modeling infectious diseases in humans and animals.* Princeton University Press, Princeton, NJ (Cited on page 171.)

Kendall, W. L. and R. Bjorkland (2001), Using open robust design models to estimate temporary emigration from capture-recapture data. *Biometrics* **57**, 1113–1122 (Cited on page 211.)

Kendall, W. L. and J. D. Nichols (2002), Estimating state–transition probabilities for unobservable states using capture–recapture/resighting data. *Ecology* **83**, 3276–3284 (Cited on pages 77 and 78.)

Kendall, W. L., J. D. Nichols and J. E. Hines (1997), Estimating temporary emigration using capture-recapture data with Pollock's robust design. *Ecology* **78**, 563–578 (Cited on page 211.)

Kendall, W. L., K. H. Pollock and C. Brownie (1995), A likelihood-based approach to capture-recapture estimation of demographic parameters under the robust design. *Biometrics* **51**, 293–308 (Cited on page 211.)

Kery, M. (2018), Identifiability in N-mixture models: a large-scale screening test with bird data. *Ecology* **99**, 281–288 (Cited on pages 28 and 215.)

Kim, D. and B. G. Lindsay (2015), Empirical identifiability in finite mixture models. *Annals of the Institute of Statistical Mathematics* **67**, 745–772 (Cited on pages 19 and 22.)

Kim, S., N. Shephard and S. Chid (1998), Stochastic Volatility: Likelihood Inference and Comparison with ARCH Models. *Review of Economic Studies* **65**, 361–393 (Cited on pages 172, 173, and 182.)

King, A. A., D. Nguyen and E. L. Ionides (2016), Statistical inference for partially observed Markov processes via the R package pomp. *Journal of Statistical Software* **69**, 1–43 (Cited on page 172.)

King, R. and S. P. Brooks (2003), Closed-form likelihoods for Arnason-Schwarz models. *Biometrika* **90**, 435–444 (Cited on page 194.)

King, R., S. P. Brooks, C. Mazzetta, S. N. Freeman and B. J. T. Morgan (2008), Identifying and diagnosing population declines: a Bayesian assessment of lapwings in the UK. *Applied Statistics* **57**, 609–632 (Cited on page 209.)

Konstantinides, K. and K. Yao (1988), Statistical analysis of effective singular values in matrix rank determination. *IEEE Transactions on Acoutstics, Speech and Signal Processing* **36**, 757–763 (Cited on page 73.)

Koop, G., M. H. Pesaran and R. P. Smith (2013), On identification of Bayesian DSGE models. *Journal of Business and Economic Statistics* **31**, 300–314 (Cited on page 136.)

Koopmans, T. C., H. Rubin and R. B. Leipnik (1950), Measuring the equation systems of dynamic economics. In T. C. Koopmans (ed.), *Statistical inference in dynamic economic models.*, pp. 53–238, John Wiley (Cited on page 42.)

Korner-Nievergelt, F., F. Liechti and K. Thorup (2014), A bird distribution model for ring recovery data: where do the European robins go? *Ecology and Evolution* **4**, 720–731 (Cited on page 209.)

Lahoz-Monfort, J. J., M. P. Harris, B. J. T. Morgan, S. N. Freeman and S. Wanless (2014), Exploring the consequences of reducing survey effort for detecting individual and temporal variability in survival. *Journal of Applied Ecology* **51**, 534–543 (Cited on page 212.)

Lavielle, M. and L. Aarons (2016), What do we mean by identifiability in mixed effects models? *Journal of Pharmacokinetics and Pharmacodynamics* **43**, 111–122 (Cited on page 77.)

Lebreton, J.-D., K. P. Burnham, J. Clobert and D. J. Anderson (1992), Modeling survival and testing biological hypotheses using marked animals: A unified approach with case studies. *Ecological Monographs* **62**, 67–118 (Cited on pages 109 and 191.)

Lebreton, J.-D. and R. Pradel (2002), Multi–state recapture models: modelling incomplete individual histories. *Journal of Applied Statistics* **29**, 353–369 (Cited on page 83.)

Lee, P. M. (2012), *Bayesian statistics : an introduction*. Wiley (Cited on page 125.)

Lele, S. R., B. Dennis and F. Lutscher (2007), Data cloning: easy maximum likelihood rstimation for complex ecological models using Bayesian Markov chain Monte Carlo methods. *Ecology Letters* **10**, 551–563 (Cited on pages 79, 80, and 143.)

Lele, S. R., K. Nadeem and B. Schmuland (2010), Estimability and likelihood inference for generalized linear mixed models using data cloning. *Journal of the American Statistical Association* **105**, 1617–1625 (Cited on pages 71, 79, 80, 143, and 144.)

Lindley, D. V. (1971), *Bayesian statistics: a review*. SIAM, Philadelphia, PA (Cited on pages 125 and 127.)

Link, W. A. (2003), Nonidentifiability of population size from capture-recapture data with heterogeneous detection probabilities. *Biometrics* **59**(4), 1123–1130 (Cited on pages 65 and 66.)

Link, W. A. and R. J. Barker (2010), *Bayesian inference: with ecological applications*. Academic Press (Cited on page 126.)

Little, M., W. F. Heidenreich and G. Li (2009), Parameter Identifiability and Redundancy in a General Class of Stochastic Carcinogenesis Models. *PLoS ONE* **4**, e8520 (Cited on pages 44 and 72.)

Little, M., W. F. Heidenreich and G. Li (2010), Parameter Identifiability and Redundancy: Theoretical Considerations. *PLoS ONE* **5**, e8915 (Cited on pages 72 and 73.)

Ljung, L. and T. Glad (1994), On global identifiability of arbitrary model parameterizations. *Automatica* **30**, 265–276 (Cited on page 166.)

Lunn, D. J., A. Thomas, N. Best and D. Spiegelhalter (2000), WinBUGS - A Bayesian modelling framework: concepts, structure, and extensibility. *Statistics and Computing* **10**, 325–347 (Cited on pages 80, 83, 126, and 230.)

Lystig, T. C. and J. P. Hughes (2002), Exact computation of the observed information matrix for hidden Markov models. *Journal of Computational and Graphical Statistics* **11**, 678–689 (Cited on page 91.)

MacKenzie, D. I., J. D. Nichols, G. B. Lachman, S. Droege, J. A. Royle, and C. A. Langtimm (2002), Estimating site occupancy rates when detection probabilities are less than one. *Ecology* **83**, 2248–2255 (Cited on pages 5 and 211.)

MacKenzie, D. I., J. D. Nichols, J. A. Royle, K. H. Pollock, L. L. Bailey and J. E. Hines (2005), *Occupancy estimation and modeling: inferring patterns and dynamics of species occurrence*. Academic Press (Cited on page 5.)

Magaria, G., E. Riccomagno, M. J. Chappell and H. P. Wynn (2001), Differential algebra methods for the study of the structural identifiability of rational function state-space models in biosciences. *Mathematical Biosciences* **174**, 1–26 (Cited on page 164.)

Matechou, E. (2010), *Demographic models for wild animal survival*. Ph.D. thesis, University of Kent (Cited on pages 215 and 216.)

McCrea, R. S. and B. J. T. Morgan (2014), *Analysis of capture-recapture Data*. Chapman and Hall (Cited on pages 188 and 202.)

McCrea, R. S., B. J. T. Morgan, D. I. Brown and R. A. Robinson (2012), Conditional modelling of ring-recovery data. *Methods in Ecology and Evolution* **3**, 823–831 (Cited on page 102.)

McCrea, R. S., B. J. T. Morgan and D. J. Cole (2013), Age-dependent mixture models for recovery data on animals marked at unknown age. *Journal of the Royal Statistical Society, Series C, Applied Statistics* **62**, 101–113 (Cited on pages 191, 216, and 217.)

McCrea, R. S., B. J. T. Morgan, O. Gimenez, P. Besbeas, J.-D. Lebreton and T. Bregnballe (2010), Multi-site integrated population modelling. *Journal of Agricultural, Biological, and Environmental Statistics* **15**, 539–561 (Cited on pages 170 and 171.)

McCullagh, P. and J. A. Nelder (1989), *Generalized linear models*. Chapman and Hall (Cited on page 44.)

Meshkat, N., M. Eisenberg and J. I. DiStefano (2009), An algorithm for finding globally identifiable parameter combinations of nonlinear ODE models using Gröbner bases. *Mathematical Biosciences* **222**, 61–72 (Cited on page 166.)

Meshkat, N., C. E. Kuo and J. I. DiStefano (2014), On finding and using identifiable parameter combinations in nonlinear dynamic systems biology models and COMBOS: A novel web implementation. *PLoS ONE* **9**, e110261 (Cited on page 167.)

Morgan, B. J. T. (2009), *Applied stochastic modelling*. Chapman and Hall (Cited on pages 3, 4, 14, 22, 29, 30, 32, 136, and 228.)

Mosher, B. A., L. L. Bailey, B. A. Hubbard and K. Huyvaert (2018), Inferential biases linked to unobservable states in complex occupancy models. *Ecography* **41**, 32–39 (Cited on pages 77, 80, and 211.)

Nater, C. R., Y. Vindenesa, P. Aassb, D. J. Cole, O. Langangena, S. Jannicke Moed, A. Rustadbakkene, D. Turekf, L. A. Vollestada and T. Ergona (2019), Size- and stage-dependence in cause-specific mortality of migratory brown trout. *Submitted* (Cited on pages 101 and 215.)

Neath, A. A. and F. J. Samaniego (1997), On the efficacy of Bayesian inference for nonidentifiable models. *American Statistician* **51**, 225–232 (Cited on page 128.)

Newman, K. B., S. T. Buckland, B. J. T. Morgan, R. King, D. L. Borchers, D. J. Cole, P. Besbeas, O. Gimenez and L. Thomas (2016), *Modelling Population Dynamics: model formulation, fitting and assessment using state-space methods*. Springer (Cited on pages 170, 173, 174, and 202.)

Petrie, T. (1969), Probabilitic functions of finite state Markov chains. *Annals of Mathematical Statistics* **40**, 97–115 (Cited on page 186.)

Pledger, S., M. Efford, K. H. Pollock, J. A. Collazo and J. E. Lyons (2009), Stopover duration analysis with departure probability dependent on unknown time since arrival. In D. Thomson, E. Cooch and M. Conroy (eds.), *Modeling demographic processes in marked populations.*, pp. 349–363, Springer series (Cited on page 215.)

Pledger, S. A., K. H. Pollock and J. L. Norris (2010), Open capture-recapture models with heterogeneity: II Jolly-Seber model. *Biometrics* **66**, 883–890 (Cited on pages 215 and 216.)

Pohjanpalo, H. (1978), System identifiability based on the power series of the solution. *Mathematical Biosciences* **41**, 21–33 (Cited on page 163.)

Pohjanpalo, H. (1982), Identifiability of deterministic differential models in state space. Technical report, Research Centre of Finland Report No. 56 (Cited on page 42.)

Pollock, K. H. (1982), A capture-recapture design robust to unequal probability of capture. *Journal of Wildlife Management* **46**, 752–757 (Cited on page 211.)

Pollock, K. H. (1991), Modelling capture, recapture, and removal statistics for estimation of demographic parameters for fish and wildlife populations: past, present, and future. *Journal of the American Statistical Association* **86**, 225–238 (Cited on page 211.)

Ponciano, J. M., J. G. Burleigh, E. L. Braun and M. L. Taper (2012), Assessing parameter identifiability in phylogenetic models using data cloning. *Systematic Biology* **61**, 955–972 (Cited on page 80.)

Pradel, R. (1993), Flexibility in survival analysis from recapture data: handling trap dependence. In P. Lebreton, J.-D. North (ed.), *Marked Individuals in the Study of Bird Population*, pp. 29–37, Birkhauser (Cited on page 60.)

Pradel, R. (2005), Multi-event: an extension of multi-state capture-recapture models to uncertain states. *Biometrics* **61**, 442–447 (Cited on pages 199 and 213.)

Prakasa Rao, B. L. S. (1992), *Identifiability in stochastic models*. Academic Press (Cited on pages 62 and 64.)

Quinn, T. J. I. and R. B. Deriso (1999), *Quantitative fish dynamics*. Oxford University Press (Cited on page 173.)

Ran, Z.-Y. and B.-G. Hu (2014), Determining structural identifiability of parameter learning machines. *Neurocomputing* **127**, 88–97 (Cited on pages 11 and 38.)

Rannala, B. (2002), Identifiability of parameters in MCMC Bayesian inference of phylogeny. *Systematic Biology* **51**, 754–760 (Cited on page 128.)

Raue, A., C. Kreutz, T. Maiwald, J. Bachmann, M. Schilling, U. Klingmuller and J. Timmer (2009), Structural and practical identifiability analysis of partially observed dynamical models by exploiting the profile likelihood. *Bioinformatics* **25**, 1923–1929 (Cited on pages 38, 68, 69, 83, and 96.)

Reboulet, A.-M., A. Viallefont, R. Pradel and J.-D. Lebreton (1999), Selection of survival and recruitment models with SURGE 5.0. *Bird Study* **46**, S148–S156 (Cited on page 72.)

Redner, R. A. and H. F. Walker (1984), Mixture densities, maximum likelihood and the EM algorithm. *SIAM Review* **26**, 195–239 (Cited on page 19.)

Reynolds, T. J., R. King, J. Harwood, M. Frederikesen, M. P. Harris and S. Wanless (2009), Integrated data analyses in the presence of emigration and tag-loss. *Journal of Agricultural, Biological, and Environmental Statistics* **14**, 411–431 (Cited on page 205.)

Ridout, M. S. (2009), Statistical applications of the complex-step method of numerical differentiation. *The American Statistician* **63**, 66–74 (Cited on page 227.)

Rothenberg, T. J. (1971), Identification in parametric models. *Econometrica* **39**, 577–591 (Cited on pages 29, 36, 42, 44, and 62.)

Rouan, L., R. Choquet and R. Pradel (2009), A general framework for modeling memory in capture-recapture data. *Journal of Agricultural, Biological, and Environmental Statistics* **14**, 338–355 (Cited on pages 199, 200, 213, and 215.)

Royle, J. A. (2004), N-mixture models for estimating population size from spatially replicated counts. *Biometrics* **60**, 108–115 (Cited on pages 26 and 28.)

Royle, J. A., R. B. Chandler, C. Yackulic and J. D. Nichols (2012), Likelihood analysis of species occurrence probability from presence-only data for modelling species distributions. *Methods in Ecology and Evolution* **3**, 545–554 (Cited on page 101.)

San Martin, E. and J. Gonzalez (2010), Bayesian identifiability: Contributions to an inconclusive debate. *Chilean Journal of Statistics* **1**, 69–91 (Cited on page 125.)

Schaub, M. and F. Abadi (2011), Integrated population models: a novel analysis framework for deeper insights into population dynamics. *Journal of Ornithology* **152**, 227–237 (Cited on pages 201 and 202.)

Schaub, M., O. Gimenez, A. Sierro and R. Arlettaz (2007), Use of integrated modeling to enhance estimates of population dynamics obtained from limited data. *Conservation Biology* **21**, 945–955 (Cited on page 209.)

Schwarz, C., J. Schweigert and A. Arnason (1993), Estimating migration rates using tag-recovery data. *Biometrics* **49**, 177–193 (Cited on page 199.)

Schwarz, G. (1978), Estimating the dimension of a model. *Annals of Statistics* **6**, 461–464 (Cited on page 31.)

Seber, G. A. F. (1965), A note on the multiple-recapture census. *Biometrika* **52**, 249–259 (Cited on pages 109 and 191.)

Shapiro, A. (1986), Asymptotic theory of overparameterized structural models. *Journal of the American Statistical Association* **81**, 142–149 (Cited on page 42.)

Silverman, B. W. (1986), *Density estimation for statistics and data analysis*. Chapman and Hall (Cited on page 136.)

Silvey, S. D. (1970), *Statistical inference*. Penguin Books (Cited on pages 36, 63, and 77.)

Sturtz, S., U. Ligges and A. Gelman (2005), R2WinBUGS: A package for running WinBUGS from R. *Journal of Statistical Software* **12**(3), 1–16 (Cited on pages 83, 126, 175, and 231.)

Swartz, T. B., Y. Haitovsky, A. Vexler and T. Yang (2004), Bayesian identifiability and misclassification in multinomial data. *The Canadian Journal of Statistics* **32**, 285–302 (Cited on pages 125, 127, and 153.)

Thowsen, A. (1978), Identifiability of dynamic systems. *International Journal of Systems Science* **9**, 813–825 (Cited on pages 42, 164, and 165.)

Tonsing, C., J. Timmer and C. Kreutz (2018), Profile likelihood-based analyses of infectious disease models. *Statistical Methods in Medical Research* **27**, 1979–1998 (Cited on page 83.)

Vajda, S., K. R. Godfrey and H. Rabitz (1989), Similarity transformation approach to identifiability analysis of nonlinear compartmental models. *Mathematical Biosciences* **93**(2), 217–248 (Cited on page 162.)

Van Wieringen, W. N. (2005), On Identifiability of Certain Latent Class Models. *Probability Letters* **75**, 211–218 (Cited on page 64.)

Viallefont, A., J.-D. Lebreton and A.-M. Reboulet (1998), Parameter identifiability and model selection in capture-recapture models: a numerical approach. *Biometrical Journal* **40**, 313–325 (Cited on pages 29, 44, 71, 72, and 73.)

Visser, I., M. E. J. Raijmakers and P. C. M. Molenaar (2000), Confidence intervals for hidden Markov model parameters. *British Journal of Mathematical and Statistical Psychology* **53**, 317–327 (Cited on page 186.)

Wald, A. (1950), Note on the Identification of Economic Relations. In T. C. Koopmans (ed.), *Statistical Inference in Dynamic Economic Models*, pp. 238–244, John Wiley (Cited on page 42.)

Walter, E. (1982), *Identifiability of state space models*. Springer (Cited on pages 36, 38, 155, and 161.)

Walter, E. (1987), *Identifiability of parameteric models.* Pergamon Press (Cited on page 155.)

Walter, E. and Y. Lecourtier (1981), Unidentifiable compartmental models: what to do? *Mathematical Biosciences* **56**, 1–25 (Cited on page 161.)

Walter, E. and Y. Lecourtier (1982), Global approaches to identifiability testing for linear and non-linear state space models. *Mathematics and Computers in Simulation* **24**, 472–482 (Cited on pages 38 and 166.)

Walter, E. and L. Pronzato (1996), On the identifiability and distinguishability of nonlinear parametric models. *Mathematics and Computers in Simulation* **42**, 125–126 (Cited on pages 158 and 166.)

Walter, G. G. and M. Contreras (1999), *Compartmental modelling with networks.* Springer· (Cited on pages 155 and 166.)

Williams, B. K., J. D. Nichols and M. J. Conroy (2001), *Analysis and mangement of animal populations.* Academic Press (Cited on page 65.)

Xia, X. and C. H. Moog (2003), Identifiability of nonlinear systems with applications to hiv/aids models. *IEEE Transactions on Automatic Control* **48**, 330–336 (Cited on page 166.)

Youngflesh, C. (2018), *MCMCvis: tools to visualize, manipulate, and summarize MCMC output.* https://CRAN.R-project.org/package=MCMCvis, R Package Version 0.9.4 (Cited on page 136.)

Yu, C. (2015), *The use of mixture models in capture-recapture.* Ph.D. thesis, University of Kent (Cited on pages 215 and 216.)

Zhou, M., R. S. McCrea, E. Matechou, D. J. Cole and R. A. Griffiths (2019), Removal models accounting for temporary emigration. *Biometrics* **75**, 24–35 (Cited on pages 68, 94, 211, 212, 216, and 217.)

Zucchini, W., I. L. MacDonald and R. Langrock (2016), *Hidden Markov models for time series.* CRC Press (Cited on pages 185 and 186.)

Index

abundance state-space model, 170, 175, 176, 178–180
AIC, 31, 73, 88
at least locally identifiability, 37

Bayesian identifiability, 127
    definition, 127
    informative priors, 218
    symbolic method, 146
BIC, 31
binary distribution, 2, 5
binary latent class model, 23, 24, 29, 39, 45, 51
binary logistic regression, 21
boundary estimates, 28

capture-recapture, 109, 110, 191, 193
    multi-state model, 196–198, 213
    trap dependence, 60
capture-recapture-recovery, 194–196
compartmental models, 9, 10, 19–21, 40, 63, 85, 155
    exhaustive modelling, 161, 162
    linear compartmental models, 9, 92, 110–112, 156, 157, 159, 160, 162, 164
    non-linear compartmental models, 156, 158, 165
    similarity of transformation, 161, 162
    Taylor series expansion, 163–165
    transfer function, 158–160
conditionally full rank, 36, 56–59
    definition, 36
constraints, 212–215, 218

continuous state-space model, 155
    SIR model, 185
Cormack Jolly Seber (CJS) model, 109, 110, 116, 191, 193
covariates, 101–105, 218

data cloning, 79–84, 143–147, 176, 230
deficiency, 43–46, 50, 51, 58, 59, 88, 103, 110, 117, 191, 196, 199, 203, 207
    definition, 43
derivative matrix, 12, 42, 43, 88
differential equation model, 4
discrete state-space models, 169
    abundance model, 170, 175, 176, 178–180
    linear discrete state-space model, 169–171
    multi-site abundance, 171, 182, 183
    non-linear discrete state-space model, 170, 173
    SIR model, 171, 172, 184
    stochastic Ricker stock recruitment model, 173, 174, 180
    stochastic volatility, 172, 173, 181, 182

essentially full rank, 36, 56, 57
    definition, 36
estimable parameter combinations, 48–51, 88, 103, 160, 166
    null space, 50
    partial differential equations, 50
    subset profiling, 117–122

exhaustive modelling approach, 161,
        162
exhaustive summary, 38, 41, 48, 87,
        105, 107, 146, 148–153,
        177, 186, 193, 195–197,
        200, 202, 207–209, 213
    definition, 38
    example, 38, 40, 41
    expansion, 179–182
    identifiability theorem, 42
    limited by number of terms,
        190
    log-likelihood, 46–48
    parameter redundancy
        theorem, 42
    reduced form, 107–112, 115,
        116
    state-space models, 179
    z-transform, 178
expansion exhaustive summary,
        179–182
experimental design, 210–212, 217
exponential family models, 44, 62
extended symbolic method, 105,
        107–112, 115, 116
extension theorem, 52–55, 115, 116,
        122, 123, 150, 153,
        164–166, 180, 190, 194, 207
    trivial application of, 52–54
extrinsic parameter redundancy, 37,
        38, 41, 47, 48, 201
    definition, 37

Fisher information matrix, 28, 29,
        43, 44, 46, 51
fisheries model, 112
flat ridge, 14, 83
    posterior, 128–130
full rank, 36
    conditionally full rank, 36, 56,
        57
    essentially full rank, 36, 56, 57
    definition, 36

gamma distribution, 2
general results, 51, 89

global identifiability, 8, 9, 12, 14, 15,
        21, 36, 42, 61, 63, 108, 187
    exponential family models, 62

Hessian matrix, 29, 43, 44, 46, 72,
        74
    numerical approximation, 72,
        227, 228
Hessian method, 71–77, 174, 227,
        228, 230
hidden Markov models, 185–187
hybrid symbolic-numerical method,
        12, 90–92, 117, 182, 183,
        200, 201, 216
    general results, 122, 123

identifiability
    Bayesian identifiability, 127
    definition, 36
    global identifiability, 8, 9, 12,
        14, 15, 21, 36, 42, 61, 63,
        187
    local identifiability, 8–10, 12,
        14, 15, 18, 36, 42, 61, 63,
        161, 163, 165, 187
    non-identifiability, 1, 8, 10–12,
        160, 162, 164, 166
    posterior identifiability, 127
information criterion, 73
informative priors, 218
integrated population models,
        201–210, 212, 219
intrinsic parameter redundancy, 37,
        38, 41, 46–48
    definition, 37

Jacobian matrix, 43

Kalman filter, 174

label switching, 18, 187
latent class model, 23, 24, 29, 30,
        39, 45, 51, 104, 105
least squares, 19–21, 85
likelihood, 3
    subset profiling, 117–122

contour plot, 14, 15
flat ridge in surface, 14, 83
profile log-likelihood, 14–18,
    83–86, 174, 175
three dimensional plot, 14, 15
linear compartmental model, 9, 40,
    92, 110–112, 156, 157, 159,
    160, 162, 164
linear discrete state-space models,
    169–171
linear regression model, 6, 8, 62, 76,
    77
local identifiability, 8–10, 12, 14, 15,
    18–21, 36, 42, 61, 63, 85,
    161, 163, 165, 187
at least, 37
locally identifiable
    reparameterisation, 48, 49,
        63, 160, 166
    null space, 50
    partial differential equations,
        50
log-likelihood, 3
logistic regression, 132–134, 142,
    146, 147, 150, 151

Maple code, 43
deficiency, 221
derivative matrix, 221
estimable parameter
    combinations, 222
hybrid symbolic-numerical
    method, 224
matrices and vectors, 221
occupancy model, 221, 223, 224
PLUR decomposition, 224
rank, 221
mark-recovery, 6, 7, 16, 25, 26, 30,
    32, 33, 40, 41, 47, 48,
    53–55, 58, 59, 68, 69,
    74–76, 84–89, 95–98, 102,
    103, 105, 107, 120–123,
    129–131, 138–141,
    151–153, 189–192
mathematical models, 1, 4

maximum likelihood, 3
MCMC, 126, 128
memory models, 199–201, 213–215
minimal parameter set, 48, 49
mixture model, 18
model selection, 31
    AIC, 31, 91
    BIC, 31
    score test, 32, 33
multi-site abundance state-space
    model, 171, 182, 183
multi-state capture-recapture,
    196–198, 213
Multiple-input Multiple-output
    (MIMO) models, 10, 50

N-mixture model, 26–28
near-redundancy, 28, 67, 68, 93–95,
    128, 151
    definition, 37
    Hessian method, 94, 95
nested models, 58, 59
non-identifiable, 1, 8, 10–12, 14, 15,
    160, 162, 164, 166
non-linear compartmental model,
    156, 158, 165
non-linear discrete state-space
    models, 170, 173
normal distribution, 2, 3
numerical methods, 71

occupancy model, 5, 16, 38, 44, 50,
    73, 74, 78, 79, 81, 82, 92,
    129, 130, 137, 138,
    143–145, 148, 149, 211,
    228, 230

parameter redundancy, 1, 4–7, 11,
    13–16, 36, 42
    data cloning, 79–84, 230
    deficiency, 43
    definition, 36
    full rank, 36
    general results, 191, 192
    Hessian method, 71–77, 174,
        227, 228, 230

hybrid symbolic-numerical
method, 90–92
likelihood, 13
multiple maximum likelihood
estimates, 22, 24–26
near-redundancy, 28
nested models, 58, 59
numerical methods, 71
profile log-likelihood, 83–86,
174, 175
simulation, 77–79
standard errors, 23, 28–30, 44
steps to determine parameter
redundancy, 86
subset profiling, 117–122
symbolic method, 86–89
theorem, 43
parameterisation, 2, 3
perfect separation, 21
PLUR decomposition, 57–61, 151,
224
poor mixing, 128
population growth model, 4, 5, 82–84
posterior distribution, 125
posterior identifiability
definition, 127
symbolic method, 146, 148–153
practical identifiability, 67–69,
96–98
definition, 38
prediction, 183, 184
prior distribution, 126, 128, 131,
133, 134
prior posterior overlap, 135–142,
175, 209, 210
probability distribution, 1–3
profile log-likelihood, 14–18, 83–86,
174, 175
subset profiling, 117–122

random effects, 167, 218
reduced echelon form, 57
reduced form exhaustive summary,
107–112, 115, 116
definition, 107

regression, 6, 8, 62
removal model, 211
reparameterisation, 3, 7, 17, 107
reparameterisation theorem,
108–112, 115, 116, 182
robust design, 211

score test, 32
similarity of transformation
approach, 161, 162
simulation, 77–79
SIR state-space model, 171, 172,
184, 185
sloppiness, 67, 68
definition, 37
standard errors, 23, 28–30
state-space models, 179
statistical models, 1
stochastic Ricker stock recruitment
state-space model, 173,
174, 180
stochastic volatility state-space
model, 172, 173, 181, 182
symbolic method, 86–89, 177–182,
190
Bayesian identifiability, 146
extended, 105, 107–112, 115,
116
Fisher information matrix, 44
posterior identifiability, 146,
148–153
theorem, 43

Taylor series expansion approach,
163–165
transfer function approach, 40,
158–160
Turing factorisation, 57–59

weak identifiability, 128, 135–142,
151, 175, 209, 210
definition, 128

z-transform exhaustive summary,
178

Printed in the United States
by Baker & Taylor Publisher Services

Printed in the United States
by Baker & Taylor Publisher Services